PATTERN RECOGNITION IN BIOLOGY

PATTERN RECOGNITION IN BIOLOGY

MARSHA S. CORRIGAN
EDITOR

Nova Science Publishers, Inc.
New York

For permission to use material from this book please contact us:
Telephone 631-231-7269; Fax 631-231-8175
Web Site: http://www.novapublishers.com

LIBRARY OF CONGRESS CATALOGING-IN-PUBLICATION DATA

Pattern recognition in biology / Marsha S. Corrigan (editor).
p. ; cm.
Includes bibliographical references and index.
ISBN-13: 978-1-60021-716-6 (hardcover)
ISBN-10: 1-60021-716-8 (hardcover)
1. Optical pattern recognition. I. Corrigan, Marsha S.
[DNLM: 1. Pattern Recognition, Physiological. 2. Pattern Recognition, Visual.]
TA1650.P39 2007
570.285'64--dc22

2007017812

Published by Nova Science Publishers, Inc. ✣ New York

Contents

PREFACE

Pattern recognition is the research area that studies the operation and design of systems that recognize patterns in data. It encloses subdisciplines like discriminant analysis, feature extraction, error estimation, cluster analysis (together sometimes called statistical pattern recognition), grammatical inference and parsing (sometimes called syntactical pattern recognition). Important application areas are image analysis, character recognition, speech analysis, man and machine diagnostics, person identification and industrial inspection. This new book presents leading-edge research from around the world.

Chapter 1 - Planning behaviours adapted to rapid changes occurring in the environment requires a continuous integration of various incoming information. Thus, perceiving the world through patterns ensures the rapid deployment of efficient behaviours by reducing the computational resources to be invested in action. Faces constitute a very special class of patterns, playing a crucial role in the regulation of social behaviours and interactions. The specificity of face-patterns is confirmed by the remarkable ease with which human beings recognize and categorize their conspecifics' faces. For many years, this particular expertise has been thought to be rooted in the specific types of processing applied to faces. Moreover, both neuroimaging and neuropsychological studies provided evidence for specialization of some brain structures for face recognition. Recent works however contributed to question that domain-specificity. Indeed, it has been found that some objects may share with faces similar functional properties, in particular those for which the authors developed expertise in distinguishing them at the subordinate – i.e. individual – level. The assumption that any class of object can, under some conditions, become as special as faces strongly argues for the existence of a higher principle that would organize, beyond their traditional division, the perception of both facial and non-facial patterns. Such an assumption could be helpful to understand how faces may have progressively become, in our evolutionary past, a special class of object. It could also have important implications for identifying mechanisms at stake in patients suffering from facial and non-facial pattern recognition disorders as it may provide, in future, relevant tools to account for their putative interactions.

Chapter 2 - Detecting colliding objects in complex dynamic scenes in real time is a difficult task for conventional computer vision techniques, especially with a limited computing source. However, visual neural processing mechanisms revealed in animals such as insects have provided very simple and effective solutions for recognizing colliding objects in complex dynamic scenes. In this chapter, two different bio-inspired collision recognition neural networks are described.

One is inspired by a lobula giant movement detector (LGMD) in locusts. The LGMD inspired neural network (the LGMD hereafter) responds to objects on a direct collision course strongly with high levels of excitation and spikes. It either responds less vigorously to approaching objects that are off a direct collision course. In driving situations it can be tuned to provide a warning signal when collisions are imminent. The LGMD has also been modified to provide guidance to a small mobile robot to enable it to cope with complex situations under challenging light conditions.

The other collision recognition network is made up of directional selective neural networks (DSNNs) inspired by directional selective neurons in insects' visual pathway. DSNNs respond to translating visual motion and characteristically respond with excitation to a particular direction of motion. Studies, using genetic algorithms to tune and select the best networks, show that these DSNNs with their different preferred directions can be combined so that they signal collision. The combined DSNNs have demonstrated their reliability in tests in two different visual environments, i.e., a driving environment where the scenes are from a camera mounted behind the windscreen of a driving car and a robotic laboratory environment where the scenes are from the eye of a mini mobile robot playing with a black ball on a desk. In both of the environments, the best adapted DSNNs are good at coping with each environment.

Benefiting from the life-like spatial-temporal computation structure, the above bio-inspired collision recognition systems could be realized as analogue VLSI chips for cheap and massive production.

Chapter 3- Automated human face recognition is required in numerous applications. While considerable progress has been made in color/two dimensional (2D) face recognition, three dimensional (3D) face recognition technology is much less developed. 3D face recognition approaches based on the appearance of range images and geometric properties of the facial surface have been proposed. Methods that combine 2D and 3D modalities also exist. These innovations have advanced the field and have created novel areas of investigation. The purpose of this chapter is to provide a summary and critical analysis of the progress in 3D and 2D+3D face recognition. The chapter also identifies open problems and directions for future work in the area.

Chapter 4 - The traditional closest point criterion (CPC) has been widely used for 3D image registration, object recognition, and shape matching. In this chapter, the authors employ psychological studies to test the hypothesis whether human beings often either consciously or subconsciously apply this criterion to establish correspondences between overlapping 3D free form shapes captured from different nearby viewpoints. To this end, the authors design four sets of data with different tasks and invite a class of year 2 undergraduates at a university and two classes of year 7 and year 9 children at a comprehensive school respectively for a test.

After experiments, the answer books were collected and carefully compared against the standard answers determined at the phase of data design. The comments on why participants perform the given tasks like that were collected and summarized. Both correlation analyses and significance tests were performed between the average marks on each question and the participants in different age groups, showing that age has no significant impact on the average performance of participants.

What the participants drew and wrote on the answer papers and commented on why they performed the given tasks like that show clearly that no matter whether the participants in the different age groups realized or not, (1) the CPC has been widely used to perform the 3D free form shape matching tasks in one way or another, (2) the CPC is more likely to be inconsistently used especially by younger participants, and (3) the CPC is often used by younger participants in its brute force form as one-to-one mapping, many-to-one mapping and one-to-many mapping. As a side effect of these studies, a number of useful criteria such as one-to-one mapping, relative motion and distance, and overlapping area maximization have been revealed. The psychological studies performed thus indeed deepen our understanding of the traditional CPC for 3D free form shape matching and are fruitful, as expected. The findings from the psychological studies described in this chapter will be useful for the development of novel automatic 3D free form shape matching algorithms.

Chapter 5 - In this chapter the authors explore the nature of the information processing differences captured in studies of procedural learning and other types of implicit learning conducted with individuals with obsessive-compulsive disorder (OCD). The authors propose that findings from this line of research might function to help integrate several seemingly discrepant theories that have been proposed to explain the development and maintenance of obsessional disorders. The authors present an overview of this common psychiatric disorder and review cognitive, metacognitive, and neurobiological models of OCD. Following an extended review of procedural learning studies in OCD, they suggest that the implicit learning differences that characterize OCD are importantly related to a tendency to be hyperaware of thought processes. This propensity to focus on thought processes may be a possible cognitive risk factor for OCD. Exploration of the nature of this cognitive vulnerability and its relationship to procedural learning might help bridge theories and empirical findings, and help advance the treatment of this common psychiatric disorder.

Chapter 6 - This chapter presents new results on modeling 24 hour (circadian) human heart rate data collected with the LifeShirt system using a variety of linear regression and neural network models. Such modeling is important in biopsychology, chronobiology, and chronomedicine where signals collected continuously from human subjects for one or several days need to be interpreted. Ambulatory heart rate is influenced by a variety of factors, including physical activity, posture, and respiration, and our models try to predict heart rate based on these factors. The analyses described in the chapter indicate that neural and especially neuro-fuzzy techniques provide better results in the modelling of human heart rate at the circadian scale than conventional linear regression. The advantages of the neuro-fuzzy approaches consist in their computational efficiency, better interpretability, and the possibility of incorporation of prior knowledge for easier model construction.

Chapter 7 - Although it is difficult to recognize visual patterns precisely detected in the peripheral visual field, obtained information might serve an important role for an efficient visual search. Retinal signal triggers a rotation of eyeballs toward a blurred visual image in the peripheral visual field. At that time, features of a detected pattern might influence the direction of eyeball rotation. In this chapter, the authors focus on the recognition of geometric figures in the peripheral visual field. The results of behavioral studies are reviewed in the first part of this chapter. Factors affecting visual discrimination are discussed, e.g., different retinal locations, shape of the pattern, deterioration of visual acuity at the fovea, and developmental effects on pattern recognition in the peripheral visual field. In the latter part of this chapter, the neurophysiological aspects of geometric figure discrimination are discussed. Geometric figure presentation in the peripheral visual field provokes electrical brain responses. The later component of response seems to be related to the perception and recognition of visual objects in the peripheral visual field. Furthermore, neurophysiological dynamics of pattern recognition in the peripheral visual field are discussed.

Chapter 8 - For many decades, one of the persisting paradigms of immunology stated that cells of the adaptive immune system recognize specific antigens with the help of clonally distributed antigen receptors, while cells of the innate immune system simply recognize "something foreign." In the late 1990's, however, this paradigm started to change. It became apparent that various cells, including cells of the immune system, express a family of germline-encoded receptors homologous to Drosophila protein Toll (Toll-like receptors, TLRs). These receptors recognize particular pathogen-associated molecular patterns (PAMPs) as well as the so-called "danger-associated molecular patterns" (DAMPs). The interaction between PAMPs and DAMPs on the one hand and TLRs on the other leads to induction of the innate and the specific (adaptive) immune responses. In this review, the authors discuss data on the tissue distribution of different TLRs, and on the ligands with which they interact. The authors analyze events that follow the interaction of these ligands with TLRs expressed on immune cells (B and T lymphocytes, macrophages, dendritic cells) and on cells outside of the immune system. The authors briefly discuss how abnormal TLR signaling can lead to disease. Finally, they review some recent studies pointing to the role of TLRs in such widespread and dangerous human disease as atherosclerosis.

In: Pattern Recognition in Biology
Editor: Marsha S. Corrigan, pp. 1-19
ISBN: 978-1-60021-716-6

Chapter 1

VISUAL PATTERN RECOGNITION: WHAT MAKES FACES SO SPECIAL?

Valérian Chambon[1,2*], Mathilde Vernet[1,2], Flavie Martin[3], Jean-Yves Baudouin[4], Guy Tiberghien[3] and Nicolas Franck[1,5]

[1] Centre de Neuroscience Cognitive, UMR-5229 CNRS,
Institut des Sciences Cognitives, Bron, France
[2] Université Lumière Lyon 2, France
[3] L2C2, UMR-5230 CNRS, Institut des Sciences Cognitives, Bron, France
[4] SPMS, Université de Bourgogne, Dijon, France
[5] Centre Hospitalier le Vinatier and Université Claude Bernard Lyon 1, France

ABSTRACT

Planning behaviours adapted to rapid changes occurring in the environment requires a continuous integration of various incoming information. Thus, perceiving the world through patterns ensures the rapid deployment of efficient behaviours by reducing the computational resources to be invested in action. Faces constitute a very special class of patterns, playing a crucial role in the regulation of social behaviours and interactions. The specificity of face-patterns is confirmed by the remarkable ease with which human beings recognize and categorize their conspecifics' faces. For many years, this particular expertise has been thought to be rooted in the specific types of processing applied to faces. Moreover, both neuroimaging and neuropsychological studies provided evidence for specialization of some brain structures for face recognition. Recent works however contributed to question that domain-specificity. Indeed, it has been found that some objects may share with faces similar functional properties, in particular those for which we developed expertise in distinguishing them at the subordinate – i.e. individual – level. The assumption that any class of object can, under some conditions, become as special as faces strongly argues for the existence of a higher principle that would organize, beyond their traditional division, the perception of both facial and non-facial patterns. Such an assumption could be helpful to understand how faces may have progressively become, in our evolutionary past, a special class of object. It could also have important implications for identifying mechanisms at stake in patients suffering from facial

[*] E-mail address: valerian.chambon@isc.cnrs.fr, Fax: +33 437 911 210. Corresponding author: Valérian Chambon, Institut des Sciences Cognitives, 67, bd Pinel, 69 675 BRON Cedex, France

and non-facial pattern recognition disorders as it may provide, in future, relevant tools to account for their putative interactions.

Key words: facial pattern recognition, domain-specificity, expertise hypothesis

INTRODUCTION

Planning behaviours adapted to rapid changes occurring in the environment requires a continuous integration of various incoming information. Thus, perceiving the world through patterns ensures the rapid deployment of efficient behaviours by reducing the computational resources to be invested in action. Faces constitute a class of patterns that is special on two accounts: i) first, they play a crucial role in the regulation of our social behaviours and interactions, ii) second, we show a remarkable ease to recognize and categorize our conspecifics' faces. The huge amount of data collected these last two decades in experimental psychology and cognitive neuroscience, indicates that this particular expertise is rooted in the specific types of processing (i.e. holistic, first- and second-order configural processing) applied to faces. Indeed, converging evidence revealed that facial pattern recognition is severely jeopardized when such mechanisms are disturbed, either when they are selectively impaired following a neurological (e.g. prosopagnosia) or psychiatric (e.g. schizophrenia) disorder or, in healthy adults when using specific experimental settings (e.g. whole-part effect, composite faces, inversion paradigm). Moreover, facial configuration processing has long been thought to rely on specific cerebral networks that seem not to be involved in non-facial pattern recognition.

Recent works, however, contributed to call into question that apparent domain-specificity. Indeed, previous studies on face recognition prove to be insufficient to account for troubling effects that occur in some face recognition disorders or during development. The reason-why could be rooted in some methodological confusion that have led in wrongly attributing to faces what might be the by-product of an *expertise*-related operation. Under some conditions, indeed, it has been shown the "unique" functional properties of face recognition can be shared by the recognition of other object classes, such as those with which we have privileged contact or experience. These data still require further investigations, but they may prove to be of great importance however. By suggesting that the organization of the visual recognition system is subordinated to a higher 'expertise' principle, they encourage to go beyond the traditional distinction between face and object realms.

Some of these points are reviewed in the present chapter, based on literature data and preliminary results from our research group. In total, we will see that understanding what makes faces so special patterns cannot avoid a detailed investigation on how the brain structures deal with objects, especially these ones that share with faces similar functional properties. Such an investigation may be especially valuable on several accounts, notably to understand how faces, in our evolutionary past, may have become a special class of object. It could also have important implications for identifying mechanisms at stake in patients suffering from facial and non-facial pattern recognition disorders as it may provide, in future, relevant tools to account for their occasional interactions (e.g. prosopagnosia and other visual agnosia).

1. What Makes a Face a Face?

A). Putting Faces in Context: Social Interaction and Expertise

Faces constitute a well peculiar category of objects: both marker of personal identity and social status (e.g. age), the face is also a way of pragmatic communication (e.g. facial emotion), the content of which often exceeds what words can tell. As a consequence, it is long admitted that the interpretation of facially conveyed information is vitally important for the regulation of our complex social behaviours, as shown by a poor social functioning in pathologies with selective or diffuse facial recognition disorders, such as in prosopagnosia (Barton et al., 2003), in schizophrenia (Walker et al., 1984), or in autistic spectrum disorders (Barton et al., 2004).

As a consequence of the role that facial information plays in our social interactions, we show a remarkable ease in recognizing the identity of our fellow humans, inferring an emotional state or categorizing sex from a face. When compared to non-facial patterns, this peculiar expertise is translated into greater accuracy and shorter processing times, whatever the type of information that is processed (identity, emotion, gender, or age). Moreover, these performances are robust to various distortions, since they dramatically resist to the presentation of degraded facial stimuli, seen from different viewpoints, at distance, in poor lightening or after 10 year of aging. At last, it is noteworthy that this particular expertise occurs very early: young infants, even newborns, show a response selective to human faces and pseudo-faces (*vs.* non-facial patterns) and are specifically sensitive to face orientation (Slater et al., 2000; Gallay et al., 2006; Turati et al., 2004). This last observation – otherwise consistent with what the chomskyan framework argues about the human faculty of language– is a matter of importance: if young infants show a particular sensitivity to facial stimuli, it seems hard therefore to fully attribute the cause of our expertise to our faces' everyday life experience. As we will see, however, such observation does not preclude to be occurred, during the development, a gradual maturation of innate mechanisms related to that expertise.

B). Whole-Part Effect, Composite Faces, and Inversion Paradigm: the Role of Configural Information in Face Processing

In the field of face recognition, the question on the underlying mechanisms proved to be central for understanding what make faces so special. Thus, it has long been suggested that our expertise for faces is rooted in the specific type of processing to be applied. This processing, termed "configural" or "relational" processing, was first used by Carey and Diamond (1977) to account for the strong relationships between the different facial features (e.g., the relative shape and positioning of the mouth in relation to the shape and positioning of the nose, eyes, etc.). Historically, first-order and second-order relational information has been described (Diamond & Carey, 1986). The former refers to simple, canonical arrangements of facial components, namely, two horizontally positioned eyes under a forehead and above the nose and the mouth, etc. This type of information is sufficient to distinguish a face from something else (a shoe, or a house), but it is not enough to distinguish one face from another. The distinction between facial models requires second-order relations that describe the relative distance of facial internal features between each other. The spacing of features is highly variable among individuals and has then been regarded, not only fundamental for distinguishing between facial exemplars, but also as the base of our expertise (Diamond & Carey, 1986; for a review, see Maurer et al., 2002). Finally, a third type of facial information has been later identified, referring to as "holistic processing" – a processing by which the features are fusing together into a whole, or gestalt (Tanaka & Farah, 1993).

The relational or configural processing of face is usually regarded as more crucial for faces perceptual analysis than featural processing, which is on the physical characteristics of the facial components, taken independently of each other (nose shape, eyes color…). Indeed, if a face can be perceived through the physical characteristics of its individual components, what makes that a face is a face is primarily a certain arrangement, or "configuration" of its features between them. Numerous evidence support the observation that, in the case of faces, the overall configuration prevails over the part-based analysis for various facial information, such as identity, gender (Baudouin & Humphreys, 2006a) or emotion (Calder et al., 2000). For example, it is an accepted fact that the perception of a facial feature (such as the eyes, the mouth, or the nose) is strongly influenced by the "status" of all other features. Such effect attest that facial features are not processed only in sequential way, but are also processed in parallel, in such a way that the face is perceived as an integrated whole, i.e. a system of interlinked features rather than a collection of isolated ones. In this way, it has been many times replicated that the perceptual analysis of a particular facial component is facilitated when it is presented in the context of a whole face (*vs.* presented in isolation) (Tanaka & Farah, 1993). For example, it is easier to recognize the identity of a feature (Peter's nose) presented in a facial context (e.g. Peter's face with Peter's nose versus Bill's nose) rather than as an isolated feature (Peter's nose versus Bill's nose). If the general configuration of a face appears to strongly affect the perception of its features, such a benefit is not found, however, with houses or scrambled faces, suggesting that this 'whole-part recognition effect' is specific to facial patterns. Regarding the processing specifically involved in facial patterns, the whole-part effect constitutes a strong behavioral marker for a specific holistic processing of faces. Such a processing appears to be irrepressible, since as soon as an adult detects the first-order relations of a face, he automatically tends to process the stimulus as a gestalt, making it harder to process individual features. It is why the short presentation of a face impedes the feature-by-feature comparison, the face being at once processed as a pattern of strongly integrated features (Hole, 1994).

A second convincing illustration of this phenomenon comes from the paradigm of composite faces. In a famous experiment, Young et al. (1987) designed an experiment consisting in manipulating artificial faces where the top half belonged to one celebrity and the bottom half to another celebrity ("composite" faces). When the two segments of the composite face were misaligned (non-composite or control condition that disrupts holistic processing), it was quite easy to recognize the identity of the top or bottom part. But when the two halves were properly aligned, they fused to create a "new" identity: the subjects were captivated by the global configuration and tended to perceive an unknown face. Because they processed the face as a whole, they had trouble recognizing the identities of the two different parts of the composite face.

In a paradigm close to that of Young and collaborators, Calder et al. (2000) showed that such a phenomenon also occurred as the subject had to process, not the identity anymore, but the facial emotions. In this case, the subjects' task consisted in judging the expression of a composite face (bottom and top halves aligned) or a non-composite face (two halves misaligned). In both conditions (composite and non-composite), the two halves did not express the same emotion. The authors showed that alignment impaired the identification of the expression on the top or bottom half: the composite face generated a new facial emotional configuration that interfered with the participant's ability to assess each separate emotion on one half or the other. However, when the facial configuration was broken up, the subjects immediately recognized the emotion expressed on the half they were asked to judge, and the emotion expressed by the other half did not interfere. This composite effect that occurs for both facial identity and emotion proves to be highly robust, since it has also been observed in faces after negation, an operation of suppressing high frequencies that are known to carry face's second-order relations (Hayes et al., 1986).

Now, it is possible to experimentally disrupt the benefit of whole-part effect and to nullify the composite effect by *inverting* the presented faces (i.e., rotating them by 180°). In upside-down composite faces, the emotion expressed by the bottom and top halves did not interfere anymore, even when the parts were aligned. In this case, the new emotional configuration is not detected by subjects who are then able to selectively process the emotion of the targeted part, without interfering with the other emotional part. The loss of composite effect in upside-down faces can be explained by the fact that the inversion hinders, if not completely disrupts, the relational processing of faces and prevents glueing together the bottom and top halves of composite faces (Bartlett & Searcy, 1993; Searcy & Bartlett, 1996).

In upside-down faces, the disruption of relational information in general, and more specifically of the second-order relations, has often been replicated since Yin's original study (1969). In an upright face, for example, the alteration of second-order relations information is readily detected (modified eye position or orientation in relation to the canonical position or the orientation of other features), whereas it becomes virtually invisible with inverted faces (i.e. "thatcherized faces"; Thompson, 1980). Nevertheless, a striking fact is that the inversion procedure, if seriously disrupts our ability to perceive the misalignment of features - and hence our ability to process the relational properties of the face - spares first-order relations and componential ('featural') processing. For example, in grey-scale photographs of inverted faces, adults have difficulty recognizing the identity of the face based on the spacing of features but continue to detect first-order relations, that is, to perceive the face as a face (Freire et al., 2000). Likewise, modifying a particular feature of an upside-down face (such as eye color) is readily detected by subjects (see Leder & Bruce, 1998). Preservation of featural processing in inverted faces suggests that inversion, by disrupting facial configuration processing, would force a feature-by-feature perceptual analysis like that generally performed for objects. Data collected via fMRI seem to confirm this hypothesis, since various studies have shown that the presentation of inverted faces increases the contribution of cortical areas involved in normal-object perception (Aguirre et al., 1999; Haxby et al., 1999).

C). Distinguishing Faces from Objects: Some Methodological Considerations

Because the three types of face's relational information behave differently after inversion, such a procedure has long constituted strong evidence in favour of their separability. As a consequence, the inversion paradigm has also been regarded as an objective way to demonstrate the face-specificity of configural processing. Indeed, numerous studies showed that the inversion effect is disproportionate for facial stimuli, suggesting that configural information is more crucial for face processing than for the processing of other objects. For instance, inverting a non-facial, mono-oriented, complex stimulus (like a picture of a house or a shoe) does not affect recognition as much as the inversion of a face (Farah et al., 1995a; Valentine, 1988). So can the inversion effect be definitively considered as a "diagnostic feature" of the unique nature of face recognition? Controversy still remains however. For example, whether inverting faces impairs second-order rather than holistic information processing is still to be determined (for a review, see Maurer et al., 2002). Likewise, it has been suggested that inversion could also perturbed the extraction of featural information, particularly in facial emotional tasks (Baudouin & Humphreys, 2006b). Moreover, under some conditions, inversion could even interfere with the recognition of other mono-oriented stimuli (Diamond & Carey, 1986). In fact, to show that inversion is a good predictor of the face's "specialness", there still needs to demonstrate, as underlined Maurer and collaborators, that "the inversion effect behaves differently [at least] for two different types of face processing or for faces compared to another class of objects that, like faces, are mono-oriented, homogeneous, and normally identified at the subordinate level" (Maurer et al., 2002). We will see that some recent studies conducted with psychiatric and neurological patients, or using original experimental settings, showed that it could be not always the case.

2. Should We Revise Our Methods? Lessons from Development and Face Recognition Disorders

As often argue by neuropsychologists, an efficient strategy to study mechanisms underlying a particular ability, or function, consists in examining the profile of performance in populations where this ability is still not fully mature, or is impaired. Regarding the separability of facial processing mechanisms and their role in the processing of face patterns, studies conducted in young infants and in some psychiatric and neurological patients may contribute clarifying many unresolved questions:

1) First, how do evolve these mechanisms during development? Is their acquisition hierarchized, following an unvarying gradient of specialization? For example, are holistic and second-order relations processing following and depending upon detection of first-order facial relations?

2) Second, if such mechanisms are crucial for our facial expertise, is their acquisition correlated to an improvement in face-processing skills with age? Likewise, one should be able to reveal that such mechanisms are impaired in patients with face recognition disorders.

3) Finally, it has been shown that objects recognition may depend on analytic, componential (*vs.* configural) processing. Its role in facial expertise, however, remains unclear. This point deserves careful thought, since here is at stake the assumption of a putative continuity between objects and faces processing mechanisms. Thus, how may be componential type of analysis related to configural processing? Is it conceivable that both the mechanisms and cerebral circuitry primarily dedicated to object recognition have progressively specialized for facial patterns processing during the evolution of species?

A). Developmental Trajectories in Acquiring Facial Expertise

The ability to perceive faces configurationally is gradually acquired during development (Tanaka & Farah, 1993; Leder & Bruce, 1998) and, although in young infancy there is evidence for featural and relational processing, they later mature at different rates however.

Holistic processing seems to occur very early during the development since composite effect was observed in 4-year-old children of similar amplitude to older children (De Heering et al., 2007; for a similar observation from 6- to 10-year-old children, see also Carey & Diamond, 1994) as well as a whole-part advantage effect in 4- and 6-year-old children (Pellicano & Rhodes, 2003; Tanaka et al., 1998). Cashon and Cohen reported that 4-month-old infants look longer at a composite face made up of the internal features of a familiar face and the external features of another familiar face than at each whole familiar face (Cohen & Cashon, 2001; Cashon & Cohen, 2003). This observation was extended to 8-month-old infants by switching single features rather than all internal features; infants reacted when only the eyes, nose, or mouth of a familiar face were put into another familiar face (Schwarzer & Zauner, 2003). In a paradigm with emotional composite faces, Durand et al. (2007) also showed that children as young as 6 years processed facial emotion in a holistic way, as they usually process facial identity (Diamond & Carey, 1994) and as adults themselves process facial emotion (Durand et al., 2007; Calder et al., 2000).

Despite early signs of both featural and holistic processing of face, the ability to process second-order configural information in a fine-tuned way however appears to gradually develop in many studies (Freire & Lee, 2001; Mondloch et al., 2002; Mondloch et al., 2004; Mondloch{ XE "Mondloch, C. J." } et al., 2003; Schwarzer, 2000). This later development may explain why children, compared to adults, show poorer general performances in facial tasks, since second-order relations are often considered as more crucial than other types of relational information in acquiring facial expertise. At which age configural processing fully mature is however still under debate and results prove to be heterogeneous, notably when tested inversion effect. Thus, Carey and Diamond (1977) suggested that children do not use configural information until they are 10 years old; children younger than 10 would tend to base their judgments on local information rather than on relational ones. Numerous studies however challenged Carey and Diamond's (1977) observation. Flin (1985), for example, found that inversion altered children's performance as early as 7 years in the same way as older children and adults. In the study by Brace et al. (2001), children as young as 5 years showed an inversion effect, but not 2- to 4-year olds. In general, the ability to process faces configurationally is said not to mature before the age of 6 but Sangrigoli and de Schonen (2004) recently showed that even 3-year-olds could be affected by rotating faces upside-down.

In total, the acquisition of second-order relations show to be important for faces processing, but however not crucial since young infants, based on holistic processing, exhibit general performances above chance threshold in a broad range of facial tests. Children set up featural and holistic processing earlier than relational processing based on second-order relations but heterogeneous results encourage questioning the specificity of the inversion paradigm. Indeed, some authors have hypothesized that rotating faces by 180° could also affect in children featural processing and, therefore, would not be such a pure measure of relational processing than it seems to (Durand et al., 2007). Besides, if the ability to process faces configuration gradually develops with age, very few is known on how the child's environment interacts with the development of this ability. Two studies with adolescents treated for bilateral congenital cataracts indicate that early visual deprivation spares the development of featural processing but permanently impairs the later development of sensitivity to second-order relations (Le Grand et al., 2001; Geldart et al., 2002). The data are a matter of importance since they suggest that the acquisition of second-order relations could be sensitive to experiencing such relations. Indeed, facial expertise is impaired as this experience is jeopardized, what claims for a decisive role of empirically experiencing facial patterns in acquiring what makes them so special.

B). The Case for Schizophrenia: Configural Processing, Visual Scanpath, and Compensatory Strategies

Schizophrenia deficits in facial recognition and discrimination have been studied extensively, and most investigators have pointed out that patients with schizophrenia perform less well than non-patients and psychiatric controls in numerous facial paradigms, including facial identity, emotion and age recognition tests (Feinberg et al, 1986; Berndl et al., 1986; Archer et al., 1992; Kerr & Neale, 1993; Martin et al., 2005). Moreover, these disorders are face-specific and appear independently of a more fundamental problem in complex object processing (Williams et al., 1999). The extent of the schizophrenic deficit in facial recognition suggests the alteration of a processing mechanism common to all kinds of facial information, and the configural information extraction process has then been regarded as a probable candidate. Nevertheless, only a few studies directly tested the hypothesis. Among them, Schwartz et al. (2002) reported results that indicate an inversion effect for faces in schizophrenic patients which tended to be stronger than that observed for houses. In a paradigm with composite faces, they also reported an interference of the counterpart when the schizophrenics had to pay attention to half of the face. In 2006, we conducted in our group a similar experiment also using the inversion paradigm (Chambon et al., 2006). 26 schizophrenic patients and healthy controls volunteered in the study, that consisted in identify the emotion expressed on a set of 48 upright and 48 upside-down faces. As the nature of what the inversion effect reveals is still thought to be controversial, we also examined which decision criterion patients used in both upright and upside-down conditions (signal decision theory, see Grier, 1971). We computed this criterion in order to determine whether participants, as it could be expected, did shift from a configurational to a featural perceptual strategy after inverting faces.

As in Schwartz's study, patients with schizophrenia were found to be globally impaired in recognizing facial emotions in both conditions. We also found an inversion effect that was as strong for schizophrenic patients as for controls. Nevertheless, we observed some particularities in the schizophrenia group that did not fit with the conclusion to a preserved configural processing. Notably, whereas controls shift their decision bias when emotion where displayed upside-down, patients did not; they adopted a similar pattern of decision bias for upright and upside-down faces. The absence of decision shifting we found in the upside-down condition attests a twofold phenomenon: first, patients adopt the same perceptual strategy in both upright and inverted faces; second, since their pattern of decision bias for upright faces is similar to the patterns controls adopted of inverted faces - i.e., when the processing of configural information was disturbed -, this strategy may be considered as based on local rather than relational information.

Relying on local parts of the face could be regarded in schizophrenic patients as a useful strategy to compensate for an earlier problem in the configural processing of faces. However, nothing indicates that patients are not sensitive to configural information. Indeed, the persistence of an inversion effect indicates that they do in fact process configural information, but in an incorrect way. Such configural information of poor quality may influence the processing of local parts of faces, leading to an overall failure in integrating critical features of the face into a whole.

This impairment in the implementation of strategies for relational properties has been otherwise confirmed in a recent study by Baudouin, Vernet and Franck that did not use the inversion paradigm (Baudouin et al., submitted). In this study, second-order relations were directly manipulated by modifying, with different intensities, the distance between eyes of similar side-by-side faces. The purpose of the study was to compute the minimal distance participants were able to discriminate. The results showed that schizophrenic patients needed a distance that was two times more important than controls, suggesting without using the traditional inversion paradigm that schizophrenic patients do fail in correctly processing second-order relations of the face.

Such an abnormal configural processing has been hypothesised to rely on an inappropriate exploration of faces. In healthy subjects, the visual scanning of face follows a regular pathway: participants focus on main features, making shift between the different facial components (e.g., eyes, nose and mouth; Walker-Smith { XE "Walker-Smith, G. J." }et al.{ XE "Findlay, J. M." }, 1977). Schizophrenic patients, however, exhibit a restricted visual scanpath, with fewer foveal fixations of longer duration and shorter distances between fixations (Loughland et al., 2002; Streit et al., 1997; Williams et al., 1999). This abnormal pattern of visual exploration has clear incidence on the way patients extract information from face and could therefore reflect patients' over-reliance on sequential visual search strategies. Based on a hieratic processing of local parts of the face, such strategies may be used to the detriment of the processing of the relational properties, the ones that imply making saccades from one feature to another.

C). The Case for Prosopagnosia: Recognition Subordinate Level, Feature-Based Strategies, and Local Expertise Effect

Interestingly, a similar impairment in configural processing, associated with an over-reliance on local parts of the face, was also suggested in prosopagnosic patients. Syndrome of prosopagnosia is a disability in recognising familiar people from their faces (Bodamer, 1947) that follows lesions in the ventral temporal lobe (Meadows et al., 1974; Farah et al., 1995b). It is noteworthy that prosopagnosia is not a deficit in identifying faces at the category level (e.g. in detecting that a face is a face, and not a chair or a shoe) but in distinguishing between facial exemplars, that is, in recognizing faces at the *individual* level (Peter *vs.* Bill *vs.* ...) (Schweich & Bruyer, 1993). Since the capacity to categorize faces at the individual level – also called subordinate level – depends on second-order relations, that processing seems to be a good candidate to account for the prosopagnosia deficit in facial recognition[1]. The results remain unclear however : some studies showed that prosopagnosic patients were severely impaired in the recognition of upright faces but significantly better at inverted faces (e.g., Saumier et al., 2001; De Gelder & Rouw, 2000), whereas Baudouin and Humphreys, in a recent case-study

[1] However, there is still some confusion on whether the ability to discriminate faces at the subordinate level depends on second-order relations rather than on the holistic processing of the face. As suggested by Boutsen and Humphreys (2002), an impaired holistic processing with spared second-order relations may characterise the prosopagnosia profile; on the other hand, patients with schizophrenia would show the opposite pattern of deficit. Such dissociation may be helpful to account for differences that arise in facial expertise between both diseases.

with emotional composite faces, found a prosopagnosic patient (H.J.A) to be strongly affected by face inversion (Baudouin & Humphreys, 2006b), despite previous evidence for poor signs of configural processing (Boutsen & Humphreys, 2002). As already mentioned, featural processing may also play a part in face recognition (Cabeza & Kato, 2000) and inversion may disrupt such a feature-based processing (see Baudouin & Humphreys, 2006b). According to the authors, the same phenomenon would occur in H.J.A. patient, which would have developed a compensatory analysis strategy based on local information of the face, then disrupted by the inversion procedure. De facto, such a sensitivity of local information to inversion could explain in patients that strongly rely on componential strategies, why inverting faces affect them as much as healthy subjects. Besides, it must be especially true as the task required to extract facial emotions from local part of the face since some local facial features, as mouth for example, are highly sensitive to their familiar orientation. As these features are critical for recognizing facial emotion, then, inverting them might lead to poor emotional judgments.

These results deserve some further remarks. The literature abounds in studies demonstrating the syndrome of prosopagnosia as a pure face-related deficit. In support of this assumption, prosopagnosic patients usually show a spared ability in recognizing people from non-facial cues, such as their voice or their gait (for reviews, see De Renzi, 1997 and Young, 1992). Most of these evidence relied on the inversion paradigm, so far regarded as an objective measure for the role of relational information in patients with face recognition disorders. However, as shown by heterogeneous results obtained with that paradigm in developmental, psychiatric and neurological populations as well as its apparent sensitivity to local face-nonspecific information, these results need to be considered cautiously. Actually, to which extent may prosopagnosia be regarded as a pure deficit in face recognition? Prosopagnosic patients frequently show to be able to process other kinds of facial information, such as gender or age (Tranel et al., 1998), whilst the matching of unfamiliar faces can be performed accurately (Young & Bruce, 1991; for a review, see Nachson, 1995). Besides, they also show to display residual configural processing in studies that did not specifically use the inversion paradigm or that focused on other types of analysis (Boutsen & Humphreys, 2002; Duchaine, 2000; for a similar report in patients with schizophrenia, see also Chambon et al., 2006). At last, there is controversy over whether a pure prosopagnosia exists that has no effect on other aspects of visual object perception (Gauthier et al., 2000a); for example, prosopagnosia is sometimes accompanied by other selective agnosias, such as the inability to recognize and discriminate between different coins or between different cars[2] (Farah, 1996; Moscovitch et al., 1997). In total, these statements call for two comments:

1) first, if the inversion effect is not barely specific to relational information in general, and more particularly to second-order relations – considered as markers of facial processing -, then such an effect would not be "diagnostic" of the unique nature of face recognition; besides, it has been shown that, under some conditions, some non-facial patterns may also exhibit an inversion effect (Diamond & Carey, 1986). By extension, using the inversion paradigm in face recognition disorders could be inappropriate to reveal which faulty mechanisms are involved.

[2] Alternately, it has been suggested that prosopagnosia may require investigating higher levels of associative processing (episodic and semantic processing) rather than just focusing at the perceptual level (see Tiberghien & Clerc, 1986).

2) second, configural information itself could not be an infallible marker of what makes faces unique patterns, but could just be the by-product of an *expertise*-related operation.

This last assumption, as striking as it may sound, is actually rooted in several data from various fields of research. In a recent study, Bukach, Bub, Gauthier, and Tarr (2006) reported the case of a prosopagnosic patient showing a residual configuration processing over a local region of the face. An explanation for this 'local expertise effect' may be that relying continuously on facial local features to make perceptual decision can lead to develop a particular ability for processing these features. Such an expertise would then be related to the progressive reengagement of configural processing. This assumption, if justified, suggests that would benefit from configural processing, not exclusively the face category, but all classes of patterns for which we developed some kind of training. Thus, rather to be unique to faces, configural processing would be the rule for other object categories that, like faces, would be of a great exposure, constitute a homogenous class and for which we developed expertise in distinguishing members at the individual level (Bill *vs.* Peter, which requires more subtle discrimination than distinguishing a face from a shoe). As we will see, the best evidence for that strong assumption - configural processing of non-face objects - comes from some works conducted in the Isabel Gauthier's research group.

3. What it Means to be "Special"? The Specificity of Faces in Question

A). Greebles Strike back: the Expertise Hypothesis

In an original study, Gauthier and her collaborators trained naive subjects to discriminate between artificial patterns they called "greebles" (Gauthier & Tarr, 1997; Gauthier et al., 1998). Greebles are three-dimensionnal virtual objects, cylinder-like with protrusions of different size and shape. All individuals of the greebles' family share the same first-order relations, namely, the same basic arrangement of the protrusions, but the exact shape of their appendages vary between exemplars so as to define genders, families and individuals. Thus, greebles constitute a homogeneous class of objects the distinction among which requires, as in faces, the ability to perceive them at the subordinate level.

Gauthier and collaborators showed that, after extensive training, the advantage that trained observers over naive observers have in accuracy, reaction time, and sensitivity to changes in the configurations of the greebles, is disrupted by inversion. Interestingly, the trained observers displayed the same inversion effect for greebles that is "diagnostic" of the unique nature of face recognition. Using the technique of event-related potentials, Rossion and coworkers found in addition that a negative wave, called N170 (usually regarded as a neural correlate of detecting a face), occurred when upright greebles were presented and delayed for inverted greebles, suggesting that inversion, as in faces, nullified benefit from configural processing of greebles in trained subjects (Rossion et al, 2002).

Taken together, these results suggest that subjects do shift from feature-based to more configural processing as they become greebles experts. Such data are of crucial importance for two reasons : 1) first, they show that whereas experts use configural strategy for processing their category of expertise, novices may use a featural strategy; that is, novices may not access the subordinate level, whereas experts did so automatically, 2) second, they suggest that some of the properties of face recognition (such as configural processing) can be acquired by training, and thus, that recognition of any class of object can become special if enough time is spent practicing.

B). How Are Objects and Faces Represented in the Brain? Looking for a Superordinate Principle

These results naturally address the question of how is translated this "expertise effect" into the organization and functioning of the visual pattern recognition system in general. According to modular theories, the brain is organized into subcomponents, or "modules", each dedicated to processing and representing a particular type of information (see Fodor, 1983). One of the main characteristics of a module is to be informationally encapsulated insofar as it has minimal interactions with other modules and types of information. Thus, it is long admitted that ventral temporal cortex contained a limited number of modules specialized for the recognition of special categories such as places (*parahippocampal place area*, PPA, that responds selectively to spatial layout; Epstein & Kanwisher, 1998), or body parts (*extrastriate body part area*, EBA; Downing et al., 2001), whereas faces would elicit maximal response in a small region near the occipito-temporal junction, called FFA (*fusiform face area*; Kanwisher et al. 1997; Tong et al., 2000). From this highly compartmented organization can be logically deduced that object categories are clustered in ventral temporal cortex according to specificity of their visual attributes. More specifically, evidence for specialization in the fusiform gyrus for face perception strongly supported the assumption that the brain areas responsible for face and object processing can be dissociated (Haxby et al. 1994; Puce et al., 1995, 1997).

However, in a study using fMRI, Gauthier and collaborators showed that, as individuals acquire expertise with greebles, BOLD activation in the right FFA changes from being relatively indifferent to greebles to demonstrating a pattern of activity similar to that seen for upright faces (Gauthier et al., 1999). According to the authors, such results would demonstrate that specialization of the fusiform gyrus in humans may be the result, not of an innate bias in favour of the face category, but rather of our extensive experience with faces, and consequently, of the type of processing we applied to. Thus, configural processing would not be, strictly speaking, a type of coding exclusive to faces, but rather a by-product of expertise susceptible to apply to other class of non-facial patterns for the recognition of which we developed some training. In other words, FFA would not constitute a region for face recognition *per se*, but *"for subordinate identification of any object category that is automated by expertise"* (Gauthier et al., 1999).

Such an assumption has strong implications for the understanding of both the organization and functioning of the pattern recognition visual system. In particular, category-related visual attributes prove to be not the only factor that determines specialization of visual ventral system. Object representations are clustered according to the type of processing that is required, and the nature of that processing (featural *vs*. configural) may be modulated by experience. Moreover, we saw that faces are usually recognized at the individual levels whereas objects are more generally recognize at the category (generic) level. Therefore, both level of category *and* expertise may account for a large part of the activation differences between faces and objects.

This aim may help to understand why region dedicated to faces processing can sometimes be recruited by non-facial patterns as subjects developed expertise to recognize them at the subordinate level (such as in cars or birds' experts, see Gauthier et al., 2000b). Conversely, it explains why brain structures usually dedicated to object perception are sometimes recruited for face recognition as is apparent a loss of facial expertise, such as it has already been reported in autistic patients (Schultz et al., 2000). In such cases, lesions or at least functional impairments in FFA may lead at recruiting regions specialized in feature-based strategies and may account for the over-reliance to local parts of faces that occurs, as previously shown, in a number of pathologies.

More specifically, these data prove to be especially consistent with some specificity of the prosopagnosia deficit, such as an impairment for categorizing faces at the individual level (depending on second-order relations) rather than at the superordinate level (depending on first-order relations), or such as a failure to access the expertise, subordinate level to process faces, whereas a spared ability to discriminate them from other non-facial patterns at the species level (a human face *vs.* a pelican). At last, such a distinction between expertise-related categorisation levels could be helpful to account for putative interactions between prospagnosia and visual agnosia, as it sometimes occurs in prosopagnosic patients that show to be no longer able to identify car makes, or birds; that is, to discriminate among exemplars of a class composed of visually similar and semantically homogeneous objects (Lhermitte et al., 1972; see also Gauthier et al., 2000b).

C). Faces *vs.* Objects Recognition: Bridging the Gap?

The search for a higher principle underlying both face and object recognition has been sometimes viewed as the Holy Grail in the field of pattern recognition. The nature of such a principle is still under debate and, due to its strong implications on how may be organized the visual recognition system, still pits antagonist conceptions of mind against each other (for a review of the dispute opposing modular *vs.* distributed theories, see Cohen & Tong, 2001). By putting together faces and objects along a same continuum from featural to configural processing, the "expertise hypothesis" from Gauthier and collaborators however tells a story that seems particularly appropriate to account for similar effects in faces and objects processing, as well as for various phenomena occuring in the pathology or during development. Moreover, it appears to be consistent with what we know about both the functioning and organization of the ventral temporal cortex. Such hypothesis still requires further investigations however, and other competing, but not necessary contradictory, underlying principles are conceivable. As such, Malach and collaborators proposed that different category-related resolution needs could be an important factor in organizing object representations along the visual recognition brain system: for example, faces, as letters, appear to be associated with more central visual-field bias than other object categories, such as buildings (Malach et al., 2002). Thus, the eccentricity biases associated with object categories would determine the location of related areas along the visual temporal cortex. But, as suggested the authors themselves, the association between some object categories and specific eccentricity biases does not exclude, nevertheless, that this association might be guided by visual experience and the visual expertise of the subjects.

A challenging, though crucial, issue finally concerns the developmental implementation of the expertise principle. As we already mentioned, the early sensitivity of children to faces *vs.* objects suggests there exists an innate bias in favour to the face category. Such a predisposition would preexist to experiencing faces at the subordinate level. Likewise, it has been found that the Fusiform Face Area, despite its recruitment for expert recognition of non-facial patterns, exhibits a greater response to faces (Gauthier et al., 1999; for a critique of the expertise hypothesis, see also McKone et al., 2007). How does the Gauthier et al.'s model – that is based on the assumption of an empirical, experience-related expertise – integrate the existence of such an innate sensitivity to facial patterns? As already suggested, this putative innate bias for faces recognition does not necessarily prevent a later maturation of this ability. Indeed, such a predisposition may be necessary, but not sufficient to develop some kind of face expertise: as shown by adolescents with bilateral congenital cataracts, when the possibility to experience second-order relations is disrupted, the ability to discriminate between faces is severely jeopardized.

Determining to what extent is innate or acquired our expertise for face could require, more fundamentally, shifting from an ontogenetic to a phylogenetic point of view. In the first *Homo sapiens* lineage's members, nothing prevents us thinking that featural processing was the rule and that those who developed a more efficient configural coding for faces may be favoured, due to the selective advantage rapidly processing such evolutionary crucial stimuli. Then, genotypic variations related to a bias toward faces configural processing may have spread through species and be transmitted to descendants. Finally, considering configural processing of face as a 'phylogenetically acquired strategy' may account for the pervasiveness of such innate bias toward faces in human people, even in newborns, as it may explain why a differential activation in FFA for face processing persists over expert recognition of non-facial patterns.

REFERENCES

Aguirre, G.K., Singh, R., D'Esposito, M. (1999). Stimulus inversion and the responses of face and object-sensitive cortical areas. *NeuroReport,* **10**, 189–194.

Archer{ XE "Archer, J." }, J., Hay{ XE "Hay, D. C." }, D. C., & Young{ XE "Young, A. W." }, A. W. (1992). Face processing in psychiatric condition. *British Journal of Clinical Psychology,* **31**, 45-61.

Bartlett, J.C., & Searcy, J. (1993). Inversion and configuration of faces. *Cognitive Psychology,* **25**, 281–316.

Barton, J.J., Cherkasova, M.V., Press, D., Intriligator, J., & O'Connor, M. (2003). Developmental prosopagnosia: a study of three patients. *Brain and Cognition,* **51**, 12–30.

Barton, J.J., Cherkasova, M.V., Hefter, R., Cox, T.A., O'Connor, M., & Manoach, D.S. (2004). Are patients with social developmental disorders prosopagnosic? Perceptual heterogeneity in the Asperger and socio-emotional processing disorders. *Brain,* **127**, 1706-16.

Baudouin{ XE "Baudouin, J.-Y." }, J.-Y., & Humphreys{ XE "Humphreys, G. W." }, G. W. (2006a). Configural information in gender categorisation. *Perception,* **35**, 531-540.

Baudouin, J.-Y., & Humphreys, G. W. (2006b). Compensatory strategies in processing facial emotions: Evidence from prosopagnosia. *Neuropsychologia,* **44**, 1361-1369.

Berndl, K., von Cranach, M., & Grusser, O.J. (1986). Impairment of perception and recognition of faces, mimic expression and gestures in schizophrenic patients. *European Archiv of Psychiatry and Neurological Sciences,* **235**, 282-291.

Bodamer, J. (1947). Die Prosopagnosie. *Archiv für Psychiatrie une Nervenkrank,* **179**, 6-54.

Boutsen, L., & Humphreys, G.W. (2002). Face context interferes with local part processing in a prosopagnosic patient. *Neuropsychologia,* **40**, 2305–2313.

Brace, N. A., Hole, G. J., Kemp, R. I., Pike, G. E., Van Duuren, M., & Norgate, L. (2001). Developmental changes in the effect of inversion: Using a picture book to investigate face recognition. *Perception,* **30**, 85–94.

Bukach, C.M., Bub, D.N., Gauthier, I., & Tarr, M.J. (2006). Perceptual expertise effects are not all or none: Spatially limited perceptual expertise for faces in a case of prosopagnosia, *Journal of Cognitive Neuroscience,* **18**, 48–63.

Cabeza, R. & Kato, T. (2000). Features are also important: Contributions of featural and configural processing to face recognition. *Psychological Science,* **11**, 429–433.

Carey, S., & Diamond, R. (1977). From piecemeal to configurational representation of faces. *Science,* **195**, 312–314.

Carey{ XE "Carey, S." }, S., & Diamond{ XE "Diamond, R." }, R. (1994). Are faces perceived as configurations more by adults than by children? *Visual Cognition,* **1**, 253-274.

Calder, A.J., Young, A.W., Keane, J., & Dean, M. (2000). Configural information in facial expression perception. *Journal of Experimental Psychology: Human Perception and Performance,* **26**, 527-551.

Calder, A. J., & Jansen, J. (2005). Configural coding of facial expressions: The impact of inversion and photographic negative. *Visual Cognition,* **12**, 495-518.

Cashon, C. H., & Cohen, L.B. (2003). The construction, deconstruction, and reconstruction of infant face perception. In A. Slater, & O. Pascalis (Eds.). *The development of face processing in infancy and early childhood: Current Perspectives.* New York: NOVA Science.

Chambon{ XE "Chambon, V." }, V., Baudouin{ XE "Baudouin, J.-Y." }, J.-Y., & Franck{ XE "Franck, N." }, N. (2006). The role of configural information in facial emotion recognition in schizophrenia. *Neuropsychologia,* **44**, 2437-2444

Cohen{ XE "Cohen, L. B." }, L. B., & Cashon, C. H. (2001). Do 7-month-old infants process independent features or facial configurations? *Infant and Child Development,* **10**, 83-93.

Cohen, J.D. & Tong, F. (2001). The Face of Controversy. *Science,* **293**, 2405-2407.

De Gelder, B., & Rouw, R. (2000). Paradoxical configuration effects for faces and objects in prosopagnosia. *Neuropsychologia,* **38**, 1271–1279.

De Heering, A., Houthuys, S., & Rossion, B. (2007). Holistic face processing is mature at 4 years of age: Evidence from the composite face effect. *Journal of Experimental Child Psychology,* **96**, 57-70.

De Renzi, E. (1997). Prosopagnosia. In T.E. Feinberg & M.J. Farah (Eds). *Behavioural neurology and neuropsychology* (pp 245-255). New York: McGraw-Hill.

Diamond, R., & Carey, S. (1986). Why faces are and are not special: An effect of expertise. *Journal of Experimental Psychology: General,* **115**, 107–117.

Downing, P., Jiang, Y., Shuman, M., & Kanwisher, N. (2001). A Cortical Area Selective for Visual Processing of the Human Body. *Science,* **293**, 2470-2473.

Duchaine, B.C. (2000). Developmental prosopagnosia with normal configural processing. *Neuroreport,* **11**, 79-83.

Durand, K., Gallay, M., Seigneuric, A., Robichon, F., & Baudouin, J.-Y. (2007). The Development of Facial Emotion Recognition: The Role of Configural Information. *Journal of Experimental Child Psychology,* in press.

Epstein, R., & Kanwisher, N. (1998). A Cortical Representation of the Local Visual Environment. *Nature*, **392**, 598-601.

Farah, M.J., Tanaka, J.W., & Drain, H.M. (1995a). What causes the inversion effect? *Journal of Experimental Psychology: Human Perception and Performance,* **21**, 628–634.

Farah, M.J., Levinson, K.L., & Klein, K.L. (1995b). Face perception and within-category discrimination in prosopagnosia. *Neuropsychologia,* **33**, 661-74.

Farah, M.J. (1996). Is face recognition 'special'? Evidence from neuropsychology. Behavioural Brain Research, **76**, 181–189.

Feinberg, T.E., Rifkin, A., Schaffer, C., & Walker, E. (1986). Facial discrimination and emotional recognition in schizophrenia and affective disorders. *Archives of General Psychiatry,* **43**, 276-279.

Flin{ XE "Flin, R. H." }, R. H. (1985). Development of face recognition: An encoding switch? *British Journal of Psychology,* **76**, 123-134.

Fodor, J. (1983). *The Modularity of Mind: An Essay on Faculty Psychology*. Cambridge, MA: MIT Press.

Freire, A., Lee, K., & Symons, L. (2000). The face-inversion effect as a deficit in the encoding of configural information: direct evidence. *Perception,* **29**, 159–170

Freire{ XE "Freire, A." }, A., & Lee{ XE "Lee, K." }, K. (2001). Face recognition in 4- to 7-year-olds: Processing of configural, featural, and paraphernalia information. *Journal of Experimental Child Psychology,* **80**, 347-371.

Gallay, M., Baudouin, J.-Y., Durand, K., Lemoine, C., & Lécuyer, R. (2006). Qualitative differences in the exploration of upright and upside-down faces in four-month-old infants: An eye-movement study. *Child Development,* **77**, 984-996.

Gauthier, I., & Tarr, M.J. (1997). Becoming a "greeble" expert: exploring the face recognition mechanisms. *Vision Research,* **37**, 1673-1682.

Gauthier, I., Williams, P., Tarr, M.J., & Tanaka, J.W. (1998). Training "greeble" experts: a framework for studying expert object recognition processes. *Vision Research,* **38**, 2401-2428.

Gauthier, I., Tarr, M.J., Anderson, A.W., Skudlarski, P. & Gore, J.C. (1999) Activation of the middle fusiform 'face area' increases with expertise in recognizing novel objects. *Nature Neuroscience,* **2**, 568–573.

Gauthier, I., Tarr, M.J., Moylan, J., Skudlarski, P., Gore, J.C., & Anderson, A.W. (2000a). The "fusiform face area" is part of a network that processes faces at the individual level. *Journal of Cognitive Neuroscience,* **12**, 495-504.

Gauthier, I., Skudlarski, P., Gore, J.C., & Anderson, A.W. (2000b). Expertise for cars and birds recruits areas involved in face recognition. *Nature Neuroscience,* **3**, 191-197.

Geldart, S., Mondloch, C.J., Maurer, D., de Schonen, S., Lewis, T.L., & Brent, H.R. (2002). The effect of early visual deprivation on the development of face processing. *Developmental Science,* **5**, 490-501.

Grier, J.B. (1971). Nonparametric indexes for sensitivity and bias: Computing formulas. *Psychological Bulletin,* **75**, 424-429.

Haxby, J.V., Horwitz, B., Ungerleider, L.G., Maisog, J.M., Pietrini, P., & Grady, C.L. (1994). The functional organization of human extrastriate cortex: A PET-rCBF study of selective attention to faces and locations. *Journal of Neuroscience,* **14**, 6336-6353.

Haxby, J.V., Ungerleider, L.G., Clark, V.P., Schouten, J.L., Hoffman, E.A., & Martin, A. (1999). The effect of face inversion on activity in human neural systems for face and object perception. *Neuron,* **22**, 189–199.

Hayes, T., Morrone, M.C., & Burr, D.C. (1986). Recognition of positive and negative bandpass-filtered images. *Perception,* **15**, 595–602.

Hole, G. (1994). Configurational factors in the perception of unfamiliar faces. *Perception,* **23**, 65–74.

Kanwisher, N., McDermott, J., & Chun, M. (1997). The Fusiform Face Area: A Module in Human Extrastriate Cortex Specialized for the Perception of Faces. *Journal of Neuroscience*, **17**, 4302-4311.

Kerr, S.L., & Neale, J.M. (1993). Emotion perception in schizophrenia: specific deficit or further evidence of generalized poor performance? *Journal of Abnormal Psychology,* **102**, 312-318.

Leder, H., & Bruce, V. (1998). Local and relational aspects of face distinctiveness. *Quarterly Journal of Experimental Psychology,* **51**, 449-473.

Le Grand, R., Mondloch, C.J., Maurer, D., & Brent, H.P. (2001) Early visual experience and face processing. *Nature,* **410**, 890-899.

Lhermitte, J., Chain, F., Escourolle, R., Ducarne, B., & Pillon, B. (1972). Etudes anatomo-cliniques d'un cas de prosopagnosie. *Revue Neurologique,* **126**, 329-346.

Loughland, C.M., Williams, L.M., & Gordon, E. (2002). Visual scanpath to positive and negative facial emotions in an outpatient schizophrenia sample. *Schizophrenia Research,* **55**, 159-170.

Malach, R., Levy, I., & Hasson, U. (2002). The topography of high-order human object areas. *Trends in Cognitive Sciences,* **6**, 176-184.

Martin, F., Baudouin, J.-Y., Tiberghien, G., & Franck, N. (2005). Processing of faces and emotional expression in schizophrenia. *Psychiatry Research,* **124**, 43-53.

Maurer, D., Le Grand, R., & Mondloch, C.J. (2002). The many faces of configural processing. *Trends in Cognitive Sciences,* **6**, 255-60.

McKone, E., Kanwisher, N., & Duchaine, B.C. (2007). Can generic expertise explain special processing for faces? *Trends in Cognitive Sciences,* **11**, 8-15.

Meadows, J.C. (1974). The anatomical basis of prosopagnosia. *Journal of Neurology, Neurosurgery & Psychiatry,* **37**, 489-501.

Mondloch{ XE "Mondloch, C. J." }, C. J., Le Grand, R., & Maurer{XE "Maurer, D." }, D. (2002). Configural face processing develops more slowly than featural face processing. *Perception,* **31**, 553-566.

Mondloch, C. J., Geldart, S., Maurer, D., & Le Grand, R. (2003). Developmental changes in face processing skills. *Journal of Experimental Child Psychology,* **86**, 67-84.

Mondloch, C. J., Dobson, K. S., Parsons, J., & Maurer, D. (2004). Why 8-year-olds cannot tell the difference between Steve Martin and Paul Newman: Factors contributing to the slow development of sensitivity to the spacing of facial features. *Journal of Experimental Child Psychology,* **89**, 159-181.

Moscovitch, M.,Winocur, G., & Behrmann, M. (1997). What is special about face recognition? Nineteen experiments on a person with visual object agnosia and dyslexia but normal face recognition. *Journal of Cognitive Neuroscience,* **9**, 555-604.

Nachson, I. (1995). On the modularity of face recognition: The riddle of domain specificity. *Journal of Clinical and Experimental Neuropsychology* **17**, 256–275.

Pellicano, E., & Rhodes, G. (2003). Holistic processing of faces in preschool children and adults. *Psychological Science,* **14**, 618-622.

Puce, A., Allison, T., Gore, J.C., & McCarthy, G. (1995). Face Perception in Extrastraite Cortex Studied by Functional MRI. *Journal of Neurophysiology,* **74**, 1192-1199.

Puce, A., Allison, T., Spencer, S.S., Spencer, D.D., & McCarthy, G. (1997). A Comparison of Cortical Activation Evoked by Faces Measured by Intracranial Field Potentials and Functional MRI: Two Case Studies. *Human Brain Mapping,* **5**, 298-305.

Rossion, B., Gauthier, I. Goffaux, V., Tarr, M.J., & Crommelinck, M. (2002).Expertise training with novel objects leads to facelike electrophysiological responses. *Psychological Science,* **13**, 250-257.

Sangrigoli, S., & de Schonen, S. (2004). Effect of visual experience on face processing: A developmental study of inversion and non-native effects. *Developmental Science,* **7**, 74-87.

Saumier, D., Arguin, M. & Lassonde, M. (2001). Prosopagnosia: A case study involving problems in processing configural information. *Brain and Cognition,* **46**, 255–316.

Schultz, R.T., Gauthier, I., Klin, A., Fulbright, R.K., Anderson, A.W., Volkmar, F., Skudlarski, P., Lacadie, C., Cohen, D.J., & Gore, J.C. (2000). Abnormal ventral temporal cortical activity during face discrimination among individuals with autism and Asperger syndrome. *Archives of General Psychiatry,* **57**, 331-340.

Schwarzer, G. (2000). Development of face processing: The effect of face inversion. *Child Development,* **71**, 391-401.

Schwartz, B.L., Marvel, C.L., Drapalski, A., Rosse, R.B., & Deutsch, S.I. (2002). Configural processing in face recognition in schizophrenia. *Cognitive Neuropsychiatry,* **7**, 15–39.

Schwarzer, G., & Zauner, N. (2003). Face processing in 8-month-old infants: Evidence for configural and analytical processing. *Vision Research,* **43**, 2783-2793.

Schweich, M., & Bruyer, R. (1993). Heterogeneity in the cognitive manifestation of prosopagnosia: The study of a group of single cases. *Cognitive Neuropsychology,* **10**, 529-547.

Searcy, J.H., & Bartlett, J.C. (1996). Inversion and processing of component and spatial-relational information in faces. *Journal of Experimental Psychology: Human Perception & Performance,* **22**, 904–915.

Slater, A., Bremner, G., Johnson, S. P., Sherwood, P., Hayes, R., & Brown, E. (2000). Newborn infants' preference for attractive faces: The role of internal and external facial features. *Infancy,* **1**, 265-274.

Streit, M., Wolwer, W., & Gaebel, W. (1997). Facial-affect recognition and visual scanning behaviour in the course of schizophrenia. *Schizophrenia Research,* **24**, 311-317.

Tanaka, J.W., & Farah, M.J. (1993). Parts and wholes in face recognition. *Quarterly Journal of Experimental Psychology,* **46**, 225–245.

Tanaka{ XE "Tanaka, J. W." }, J. W., Kay, J. B., Grinnell, E., Stansfield, B., & Szechter, L. (1998). Face recognition in young children: When the whole is greater than the sum of its parts. *Visual Cognition,* **5**, 479-496.

Thompson, P. (1980). Margaret Thatcher: a new illusion. *Perception,* **9**, 483–484.

Tiberghien, G., & Clerc, I. (1986). The cognitive locus of prosopagnosia. In R. Bruyer (Ed.), *The neuropsychology of face perception and facial expression* (pp. 39-62). Hillsdale, NJ: Lawrence Erlbaum Associates.

Tranel, D., Damasio, A.R., & Damasio, H. (1988). Intact recognition of facial expression, gender and age in patients with impaired recognition of face identity. *Neurology,* **39**, 690-696.

Tong, F., Nakayama, N., Moscovitch, M., Weinrib, O., & Kanwisher, N. (2000). Response Properties of the Human Fusiform Face Area. *Cognitive Neuropsychology,* **17**, 257-279.

Turati, C., Sangrigoli, S., & de Schonen, S. (2004). Evidence of the inversion effect in 4-month-old infants. *Infancy,* **6**, 275-297.

Valentine, T. (1988). Upside-down faces: A review of the effect of inversion upon face recognition. *British Journal of Psychology,* **79**, 471–491.

Walker, E., McGuire, M., & Bettes, B. (1984). Recognition and identification of facial stimuli by schizophrenics and patients with affective disorders. *British Journal of Clinical Psychology,* **23**, 37-44.

Walker-Smith{ XE "Walker-Smith, G. J." }, G. J., Gale{ XE "Gale, A. G." }, A. G., & Findlay{ XE "Findlay, J. M." }, J. M. (1977). Eye movement strategies involved in face perception. *Perception,* **6**, 313-326.

Williams, L.M., Loughland, C.M., Gordon, E., & Davidson, D. (1999). Visual scanpaths in schizophrenia: Is there a deficit in face recognition? *Schizophrenia Research,* **40**, 189-199.

Yin, R.K. (1969). Looking at upside-down faces. *Journal of Experimental Psychology,* **81**, 141-145.

Young, A.W. (1992). Face recognition impairments. In V. Bruce, A. Cowey, A.W. Ellis & D.I. Perrett (Eds.). *Processing the facial image* (pp. 47–54). Oxford: Alden Press.

Young, A.W., & Bruce, V. (1991). Perceptual categories and the computation of Grandmother. *European Journal of Cognitive Psychology,* **3**, 5–49.

Young, A.W., Hellawell, D., & Hay, D.C. (1987). Configurational information in face perception. *Perception,* **16**, 747–759.

In: Pattern Recognition in Biology
Editor: Marsha S. Corrigan, pp. 21-61
ISBN: 978-1-60021-716-6

Chapter 2

COLLISION RECOGNITION INSPIRED BY VISUAL NEURAL SYSTEMS IN INSECTS

Shigang Yue[1*] and F. Claire Rind[2**]
[1]Brain Mapping Unit, Sir William Hardy Building, Downing Site, University of Cambridge, CB2 3EB UK
[2]Ridley Building, School of Biology and Psychology, University of Newcastle upon Tyne, NE1 7RU UK

ABSTRACT

Detecting colliding objects in complex dynamic scenes in real time is a difficult task for conventional computer vision techniques, especially with a limited computing source. However, visual neural processing mechanisms revealed in animals such as insects have provided very simple and effective solutions for recognizing colliding objects in complex dynamic scenes. In this chapter, two different bio-inspired collision recognition neural networks are described.

One is inspired by a lobula giant movement detector (LGMD) in locusts. The LGMD inspired neural network (the LGMD hereafter) responds to objects on a direct collision course strongly with high levels of excitation and spikes. It either responds less vigorously to approaching objects that are off a direct collision course. In driving situations it can be tuned to provide a warning signal when collisions are imminent. The LGMD has also been modified to provide guidance to a small mobile robot to enable it to cope with complex situations under challenging light conditions.

The other collision recognition network is made up of directional selective neural networks (DSNNs) inspired by directional selective neurons in insects' visual pathway. DSNNs respond to translating visual motion and characteristically respond with excitation to a particular direction of motion. Studies, using genetic algorithms to tune and select the best networks, show that these DSNNs with their different preferred directions can be combined so that they signal collision. The combined DSNNs have demonstrated their reliability in tests in two different visual environments, i.e., a driving environment where the scenes are from a camera mounted behind the windscreen of a driving car and a robotic laboratory environment where the scenes are from the eye of a mini mobile robot playing with a black ball on a desk.

---

[*] E-mail address: shigang.yue@ieee.org or sy262@cam.ac.uk
[**] E-mail address: claire.rind@ncl.ac.uk

In both of the environments, the best adapted DSNNs are good at coping with each environment.

Benefiting from the life-like spatial-temporal computation structure, the above bio-inspired collision recognition systems could be realized as analogue VLSI chips for cheap and massive production.

Keywords: collision, detection, visual neural networks, insects, LGMD, directional selective neurons, pattern recognition, evolution

1. Introduction

Detecting colliding objects in complex dynamic scenes in real time is a difficult task for conventional computer vision techniques, especially with current computing power. For autonomous mobile robots, the ability to avoid collision is a very important basic skill. Mobile robots have used several kinds of sensors, such as visual, ultrasound, infra-red, laser, and mini-radar, for object detection (for example, [Everett 1995, Adams 1998, Wichert 1999, and Manduchi et. al. 2005]). However, it is still very difficult for a robot to run autonomously without collision in complex, outdoor environments without human intervention. In another application field, to reduce or alleviate the impact of road collisions and the number of casualties in driving scenarios, the reliable technique for visual based collision detection is badly needed (for example, [Vahidi and Eskandarian 2003, Yue et. al 2006a]). Unfortunately, many traditional artificial vision systems have not yet been able to extract the relevant data from the wealth of information in visual scenes for rapid collision prediction [Indiveri and Douglas 2000, DeSouza and Kak 2002, Ruichek 2005].

On the other hand, nature has provided a rich source of inspiration for artificial visual systems. Many animals use their visual systems to successfully avoid collision in the real world. Insects in particular, with their rapid reactions to dynamic scenes use only a small amount of neural hardware, are very attractive as sources of inspiration (for example, [Huber et. al. 1999, Harrison and Koch 2000, Iida 2003, Web and Reeve 2003], reviewed by [Rind 2005]). In insects' visual pathways, identified specialized neurons have been known for several decades (for example, [O'Shea et. al 1974, Rind 1990a and 1990b]). The properties revealed can be used to produce unique computing efficient models for visual sensors for collision detection.

In the locust, an identified neuron, the lobula giant movement detector (LGMD), and its postsynaptic partner, the DCMD (descending contralateral movement detector), have been found to respond vigorously to looming stimuli [O'Shea et. al 1974, Rowell et. al 1977, Rind and Simmons 1992 and 1999, Schlottereer 1977, Simmons and Rind 1992]. The input circuitry of the LGMD neuron has been used as the basis for an artificial visual system for collision avoidance in robots and more recently in cars [Blanchard et. al 1999, 2000, Rind and Bramwell 1996, Rind 2002, Rind et. al. 2003, Santer et. al. 2004, Yue and Rind 2005, Yue et. al. 2006, Yue and Rind 2006a, 2006b, and 2006c]. The modified LGMD based neural network, as adapted for use as a VLSI circuit in a sensor for the car, also has enable a robot's visual system working at complex environments, under extreme illumination conditions [Yue et. al. 2006, Stafford et. al. 2005].

Several feature selective neurons may also be combined to provide a robust collision detecting visual system. Direction selective neurons, for example, have been found in animals for decades, for example, in insects such as the locust [Rind 1990a, 1990b], beetle and fly [Hassenstein and Reichardt 1956, Borst and Haag 2002], also in vertebrates such as the rabbit ([Barlow and Hill 1963, Barlow and Levick 1965, Stasheff and Masland 2002], reviewed by [Vaney and Taylor 2002]) and the cat (for example, [Priebe and Ferster 2005, Livingstone 2005]). Such directionally selective neurons could be used to signal looming (for example, [Horridge 1992]). When organised in an asymmetrical layered network, these direction selective neurons can produce a neural network specialized for collision detection [Yue and Rind 2006c]. By training and then testing in either a driving situation or in a robotic laboratory, the DSNN was shown to reliably detect collisions in dynamic scenes.

In this chapter, we will introduce the two neural systems, i.e., the LGMD based neural network and the directional selective based neural networks- all inspired by insects. The test scenarios are either from a robotic laboratory or from a real driving environment, or they were taken from both.

2. The LGMD for Collision Detection

The LGMD based neural network (LGMD hereafter) has been integrated into the control system of a small robot first by Blanchard, Verschure and Rind in 1999. Experiments conducted within an enclosed arena showed the performance of this biological inspired visual collision detection mechanism. Recently we have investigated the suitability of the LGMD network for collision detection in a car. The LGMD network would be required to operate in outdoor environments, in which objects and lighting conditions are complex and out of control.

2.1. Formulation of the System

The LGMD [Rind and Bramwell 1996, Blanchard and Rind 2000, Rind et. al 2004] was composed of four groups of cells - photoreceptor, excitatory, inhibitory and summing, and two single cells - feed-forward inhibition and LGMD [Rind and Bramwell 1996]. These groups of cells were also used as a basic for the modified neural network (Figure 1). To improve the robustness of the LGMD in situations where the background of a visual scene was complex, a new layer of grouping cells (G cells) was added to enhance the visual feature defining a colliding object and filter visual details irrelevant to the collision detection task. When integrated with a robot, a new cell feed-forward modulator (FFM), was introduced to modulate the response of the LGMD cell by varying its threshold to cope with extreme luminance conditions. The proposed neural network (shown in Figure 1) used in the chapter will be described in detail in the next part (please note that the G cells and FFM cell may not have exact counterparts in real locusts).

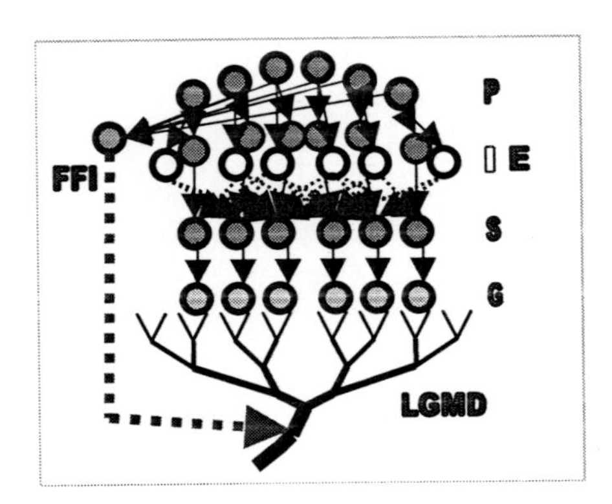

Figure 1. A schematic illustration of the LGMD based neural network for collision detection. There are five groups of cells and two single cells: photoreceptor cells (P); excitatory and inhibitory cells (E and I); summing cells (S); grouping cells (G); the LGMD cell and the feed forward inhibition cell (FFI). The input of the P cells is the luminance change. Lateral inhibition is indicated with dotted lines and has one frame delay. Excitation is indicated with black lines and has no delay. The FFI also has one frame delay. The input to FFI is luminance change from photoreceptor cells.

P Layer

The first layer of the neural network consisted of the photoreceptor *P* cells arranged in a matrix; the luminance L_f of each pixel in the input image was captured by each photoreceptor cell, the change of luminance P_f between frames of the image sequence was calculated and forms the output of this layer. The output of a cell in this layer was defined by equation:

$$P_f(x,y) = \sum_{i}^{n_p} p_i P_{f-i}(x,y) + (L_f(x,y) - L_{f-1}(x,y)) \tag{1}$$

where $P_f(x,y)$ was the change of luminance corresponding to pixel (x,y) at frame f, x and y were the indices of the matrix, L_f and L_{f-1} are the luminance, subscript f denoted the current frame and $f\text{-}1$ denoted the previous frame, n_p defined the maximum number of frames (or time steps) the luminance change could last, the persistence coefficient $p_i \in (0,1)$ and

$$p_i = (1 + e^{\mu i})^{-1} \tag{2}$$

where $\mu \in (+\infty, -\infty)$ and i indicated the previous i^{th} frame counted from the current frame f. Note that the LGMD neural network detected potential collision by responding to expansion of the image edges, a computational strategy rather than a strategy relying on object identification. If there is no difference between successive images, the *P* cells were not excited.

I E Layer

The output of the *P* cells formed the inputs to two separate cell types in the next layer. One type were the excitatory cells, through which excitation was passed directly to their retinotopic counterpart in the third layer, the *S* layer. The excitation $E(x, y)$ in an *E* cell had the same value as that in the corresponding *P* cell. The second cell type were lateral inhibition cells, which passed inhibition, after 1 image frame delay, to their retinotopic counterpart's neighbouring cells in the *S* layer. The gathered strength of inhibition of a cell in this layer was given by:

$$I_f(x,y) = \sum_i \sum_j P_{f-1}(x+i, y+j) w_I(i,j), \; (if\; i = j,\; j \neq 0) \tag{3}$$

where $I_f(x,y)$ was the inhibition corresponds to pixel (x,y) at current frame f, $w_I(i, j)$ was the local inhibition weight. Please note i and j were not allowed to be equal to zero simultaneously. This meant inhibition was only allowed to spread out to its neighbouring cells in the next layer rather than to its direct counterpart in the next layer.

S Layer

The excitatory flow from the *E* cells and inhibition from the *I* cells was summed by the S cells using the following equation:

$$S_f(x,y) = E_f(x,y) - I_f(x,y) W_I \tag{4}$$

where W_I was the inhibition weight.

G Layer

In the previous LGMD neural networks [Rind and Bramwell 1996, Blanchard and Rind 2000] *S* cells connected directly with the LGMD and the LGMD summed input from all the *S* cells. In the LGMD based neural network, the expanded edges were represented by clustered excitations. In order to enhance the ability to extract features of a colliding object against a complex background we have added a new layer, consisting of *G* cells, between the *S* cells and LGMD (Figure 1). This allowed clusters of excitation in the *S* cells to easily reach the *G* layer and therefore provided a greater input to the membrane potential of the LGMD neuron compared with the input of a single *S* cell (as illustrated in Figure 2). To implement the new mechanism, the excitation in an *S* cell that was passed to the *G* layer was multiplied by a passing coefficient *Ce*. The coefficient was determined by the cell's surrounding neighbours, i.e., defined by a convolution process

$$Ce_f(x,y) = \sum_{i=-1}^{1} \sum_{j=-1}^{1} S_f(x+i, y+j) w_e(i,j) \tag{5}$$

where $w_e(i, j)$ represented the influence of its neighbours. This operation could be simplified as a convolution mask [Davies 1997] and the passing coefficients computed in a matrix,

$$[w_e] = \frac{1}{9}\begin{bmatrix} 1 & 1 & 1 \\ 1 & 1 & 1 \\ 1 & 1 & 1 \end{bmatrix} \tag{6a}$$

$$[Ce]_f = [S]_f \otimes [w_e] \tag{6b}$$

where $[w_e]$ was the convolution mask, $[C_e]_f$ was the passing coefficient matrix, $\otimes$ denoted the convolution operation and $[S]_f$ was the excitation matrix in the S layer.

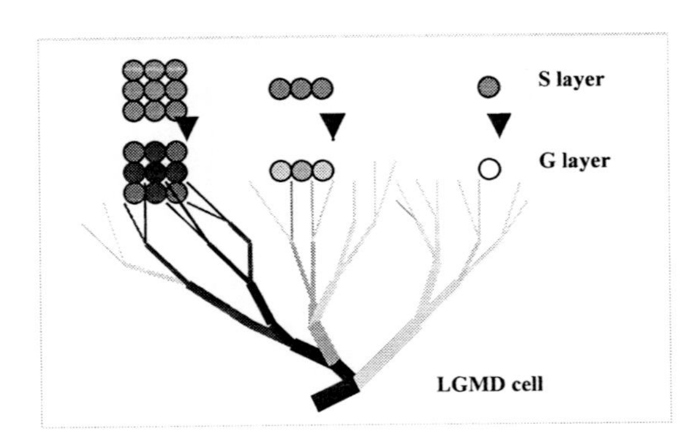

Figure 2. A schematic illustration of the grouped excitation processing mechanism. The cells(or pixels in an image) surrounded by strong excitations gain bigger passing coefficients and the isolated excitations get smaller passing coefficients and may be ruled out in the next layer G by threshold. Circles represent excitation in S and G layers. The strength of excitation in S layer, G layer and LGMD cell was indicated by grey levels where black represents the strongest excitation. S layer and G layer had one-to-one mapping.

When it reached the G layer, the excitation corresponding to cell (x, y) became,

$$G_f(x,y) = (S_f(x,y))Ce_f(x,y)\omega^{-1} \tag{7}$$

where ω was a scale and computed at every frame by

$$\omega = \Delta c + \max(abs[Ce]_f)C_w^{-1} \tag{8}$$

C_w was a constant, Δc was a small real number and $max(abs[Ce]_f)$ was the largest element in matrix $abs[Ce]_f$. As shown in Figure 2, the grouped excitations in the S layer (representing edges) became darker (stronger) when they reached the G layer whereas the isolated excitations became lighter (weaker).

From S layer to G layer, we set a threshold to filter decayed excitations,

$$\widetilde{G}_f(x,y) = \begin{cases} G_f(x,y) & if\ G_f(x,y)C_{de} \geq T_{de} \\ 0 & if\ G_f(x,y)C_{de} < T_{de} \end{cases} \tag{9}$$

where $C_{de}$was the decay coefficient and $C_{de} \in(0,1)$, T_{de} was the decay threshold. The presented grouped excitation (G) processing via equation (7) together with decay (D) processing via equation (9) not only enhanced the edges, but also filtered out excitation caused by background detail. The LGMD based neural network with grouped excitation and decay (GD) processing was used and compared in the later experiments.

LGMD Cell

The membrane potential of the LGMD cell K_f at frame f was summed after the G layer with a rectifying operation, which would turn negative values in the responses to positive before summing, as described by the following equation,

$$\mathrm{K}_f = \sum_x \sum_y abs(\widetilde{G}_f(x,y)) \tag{10}$$

The membrane potential of the *LGMD* cell K_f was then transformed into a sigmoid function:,

$$\kappa_f = (1 + e^{-\mathrm{K}_f n_{cell}^{-1}})^{-1} \tag{11}$$

where n_{cell} was the total number of the cells in G layer. Since K_f was greater than zero according to equation (10), the sigmoid membrane potential $\kappa_f \in (0.5{\sim}1)$.

FFM Cell

The collision detection system consisted of the LGMD based neural network and a CCD camera, which fed the input images to the network. The cameras of some robots have the ability to maintain a balanced image contrast (e.g., auto iris, K-team, Lausanne, Switzerland, http://www.k-team.com). Usually, a standard procedure (white point calibration) is used in the camera's hardware to normalize contrast within the image. This ability allowed the robot to see objects under both dark and bright conditions and could be very important for feeding accurate images to the LGMD based neural network when the robot was operating in either dark or bright environments. However, this mechanism had a major shortcoming, for example, objects may have little contrast against the background if very bright objects or light sources came into the view field. The network may respond to the colliding objects too late in this case or too early in the opposite scenario [Yue and Rind 2005]. To compensate for the drop of excitation in this situation, an adjustable threshold was needed. When integrated with a robot, threshold T_s could be used,

$$T_s = \alpha_{lt} T_{lt} + \alpha_{mp} T_{mp} \tag{12}$$

where T_{lt} was the adaptable part, T_{mp} was the constant part, α_{lt} and α_{mp} were the coefficients. $\alpha_{lt} > 0$ when integrated with the robot (the fed images are light compensated); otherwise, $\alpha_{lt} = 0$.

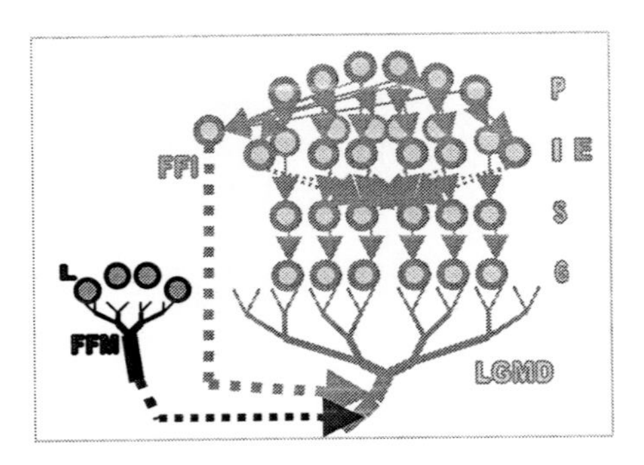

Figure 3. The feed forward modulator (FFM) cell was introduced to alter LGMD cell's threshold when the images fed to the neural network are captured by a camera with white point calibration. The input to FFM was luminance through luminance cells L.

The feed forward modulator (FFM) (Figure 3) was introduced to adapt the threshold in response to low contrast colliding objects in situations when most parts of the input image were dim,

$$T_{lt} = T_{lto} + \alpha_L \begin{cases} \Delta T_{lt} & if\ L_d > \Pi_{lt}^{u} \\ 0 & otherwise \\ -\Delta T_{lt} & if\ L_d < \Pi_{lt}^{l} \end{cases} \tag{13}$$

where T_{ito} was the initial value of the T_{it}, α_L was a coefficient, ΔT_{it} was a constant, Π_{it}^{u} and Π_{it}^{l} were the upper and lower boundary respectively, the L_d was gathered as,

$$L_d = (\sum_{i=1}^{n_r} [\max L]_f) + \sum_{i=1}^{n_c} [\max L^T]_f)\ )(n_c + n_r)^{-1} \tag{14}$$

where n_c and n_r were the number of columns and rows in the luminance matrix $[L]_f$, $[maxL]_f$ was the row of the largest elements in each column of $[L]_f$ and $[maxL^T]_f$ was the row of the largest elements in each column of $[L]_f^{\ T}$. With the above mechanism, the LGMD based neural network was more likely to respond to colliding objects as the threshold can be lowered.

Spiking Mechanism

The collision alarm was finally decided by a spiking mechanism. If the membrane potential κ_f exceeded the threshold T_s, a spike was produced,

$$S_f^{spike} = \begin{cases} 1 & if\ \kappa_f >= T_s \\ 0 & otherwise \end{cases} \tag{15}$$

where 1 represented a spike, 0 meant no spike. A collision was detected if there are n_{sp} spikes in n_{ts} time steps (n_{sp}.<= n_{ts}) [Rind et. al. 2004], i.e.,

$$C_{final} = \begin{cases} TRUE & if \sum_{f-n_{ts}}^{f} S_f^{spike} >= n_{sp} \\ FALSE & otherwise \end{cases} \quad (16)$$

where the value of C_{final} turns out to be TRUE when collision was detected. The robot's avoidance behaviour was initiated once collision was detected. However, spikes could be suppressed by the FFI cell when the robot was turning.

The Feed Forward Inhibition (FFI) Cell

If it was not suppressed during turning, the network may produce spikes and even false collision alerts because of the sudden change in the visual scene. The feed forward inhibition and lateral inhibition worked together to cope with such whole field movement [Santer et. al. 2004]. The FFI excitation at current frame was gathered from the photoreceptor cells with one frame delay,

$$F_f = \sum_{j}^{n_a} \alpha_{f-j}^F F_{f-j} + \sum_{x=1}^{n_r} \sum_{y}^{n_c} abs(P_{f-1}(x,y)) n_{cell}^{-1} \quad (17)$$

where $\alpha^F_{f\text{-}j}$ was the persistence coefficient for FFI and $\alpha^F_{f\text{-}j} \in (0,1)$, n_a defined how many time steps the persistence could last.

Once F_f exceeded its threshold T_{FFI}, spikes in the LGMD were inhibited immediately. The threshold T_{FFI} was also adaptable,

$$T_{FFI} = T_{FO} + \alpha_{ffi} T_{FFI_{f-1}} \quad (18)$$

where T_{FO} was the initial value of the T_{FFI} , the adaptable threshold was decided by the previous T_{FFI} and α_{ffi} was a coefficient.

As described in the above subsections, the LGMD detects collision based on pixel level comparisons; computationally expensive methods, such as object recognition or scene analysis, were not used. Because of this, the collision detection system was able to work in real time and was independent of object classification.

2.2. Test of the Collision Detection System

Two kinds of experiments were carried out to test the feasibility and robustness of the above collision detection system. One was an off-line test which was designed to test the effects of the additional excitation processing in the G layer using complex backgrounds from prerecorded video clips. Then, the neural network was integrated with a Khepera robot (K-Team, Switzerland, http://www.k-team.com) and was tested in real time experiments.

To obtain images, a K2D video turret (K-Team, Switzerland) with a CCD was mounted on top of the Khepera robot. The main properties of the camera are detailed in K2D video turret user manual, K-Team, Switzerland. We used the CCD camera, to sample images in real time. The camera was working at 25 frames per second in the experiment.

Parameters of the LGMD based collision detection system were set before the experiments. The input video images were 130 (in horizontal) by 100 (in vertical) pixels; images were grey scale ranging from 0 to 255 (parameter without unit, similar parameters hereafter will not be restated). The lateral inhibition spreads to its neighbour 1 layer away and with one frame delay. The local inhibition weight was set as shown in Figure 4. Other parameters were listed in Table 1. These parameters were tuned manually based on the early pilot experiments and were not changed unless stated.

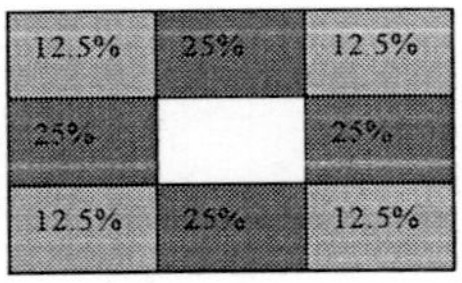

12.5%	25%	12.5%
25%		25%
12.5%	25%	12.5%

Figure 4. The local weights of inhibition that spread from the centre cell (in the IE layer) to neighbouring cells (in the S layer). The number in each cell represented the percentage of value it gained from the central pixel.

Table 1. The parameters of the LGMD based neural network

name	value	name	value	name	value
p	0	α_L	1	n_{sp}	4
W_I	0.3	ΔT_{lt}	0.03	n_{ts}	4
C_{de}	0.5	Π_{lt}^{u}	230	α_{f-1}^{F}	0
T_{de}	15	Π_{lt}^{l}	180	α_{ffi}	0.02
n_{cell}	13,000	T_{lto}	0	T_{FO}	10
α_{mp}	1	n_r	100	C_w	4
T_{mp}	0.86	n_c	130	Δc	0.01
n_p	1				

The LGMD based collision detection system was written in Matlab (the MathWorks, Inc., USA). The computer used in the experiments was a PC (Dell Precision 450) with one 2.40GHz CPU and Windows XP operating system. The communication between the computer and the robot was via serial port with Baud rate at 9600 bits/s. A USB frame grabber (Hauppauge Computer Works Ltd. UK) and video device access software in Matlab (VFM) (University of East Anglia, UK) were used to obtain live image input.

2.2.1. Off-line Tests

The results of off-line tests could easily be compared since the network could be challenged repeatedly with the same visual images. In the following off-line tests, we used recorded video clips to test and compare the collision detection systems.

Tests with a Simulated Background

Background detail in the absence of a colliding object could sometimes cause excitations. We used computer simulations consisting of random dots to alter background detail systematically.

A video clip with 90 frames was recorded using a robot approaching a block at speed 5.6cm/s and frame rate at 24Hz, example frames are shown in Figure 5 (a). Isolated excitations were generated using random values added to the input images. For example, at frame f, the random values were

$$L_f^n(x,y) = \begin{cases} k_n\, rand(1), & (x = 1,3,5,\cdots 100)\, AND (y = 1,3,5,\cdots,130) \\ 0, & otherwise \end{cases} \tag{19}$$

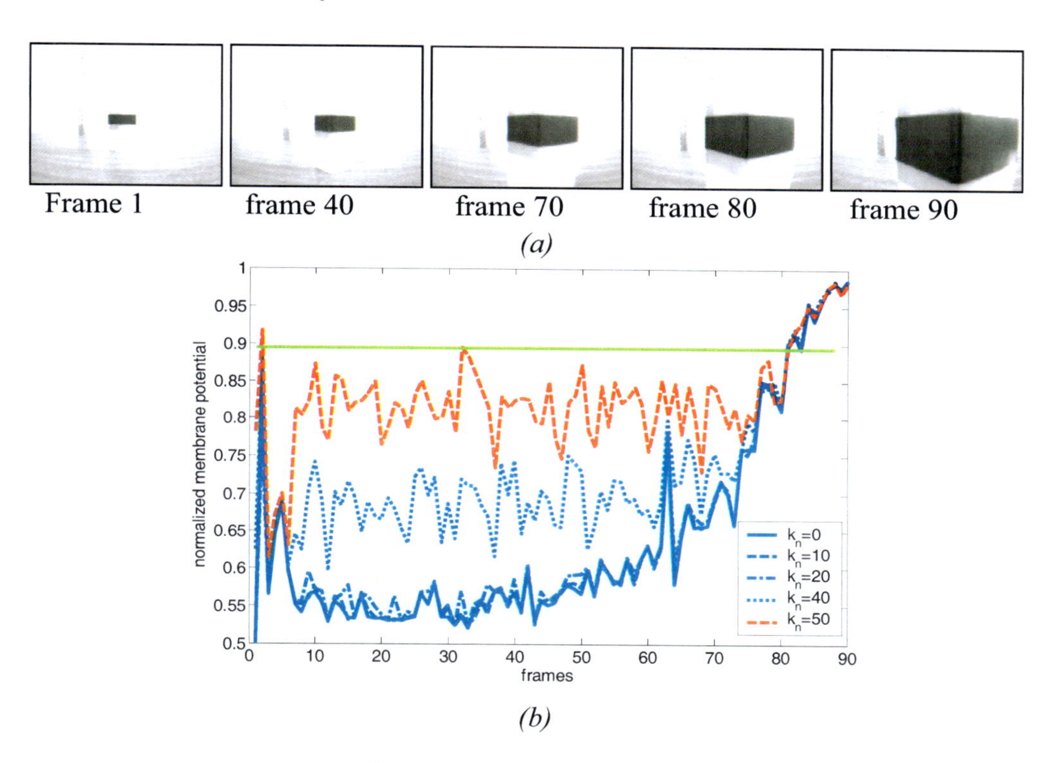

Figure 5. Continued on next page.

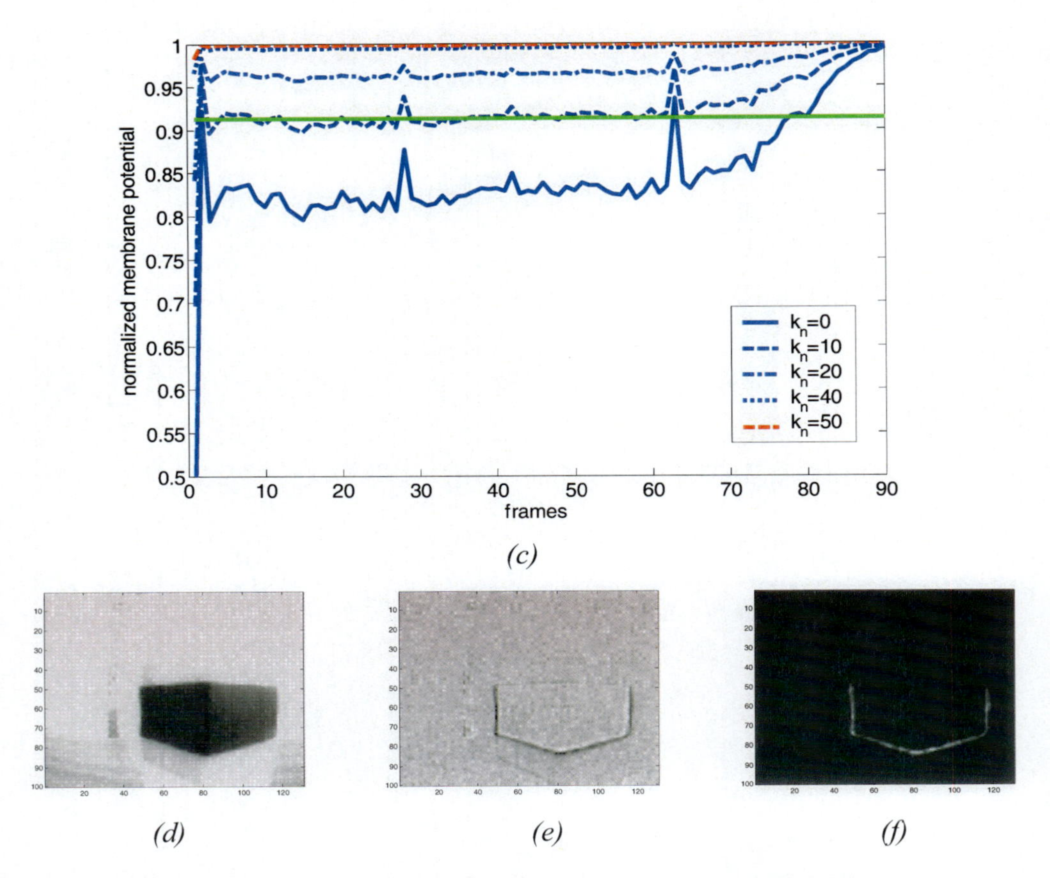

Figure 5. Off-line comparison when approaching an object. (a), the sample frames of the video clip to test the network. The video clip was taken with a robot approaching the block at speed 5.6cm/s and 24 frames per second. The collision happened at frame 91. (b), the sigmoid membrane potential at different noise level, with grouped excitation and decay (GD) processing; collisions were detected at frame 84 when noise level was less than or equal 50. The straight horizontal line around 0.89 membrane potential was the threshold. (c), the sigmoid membrane potential at different noise levels, without GD processing; the system would give a false collision signal if noise slightly exceeded 10. The thresholds have been reset to ensure both networks detect collision at frame 84, when no image noise was present. (d), the image with added noise (level k_n=50) at frame 84. (e), the image before GD processing at frame 84. (f), the image after GD processing.

where κ_n was a scale representing the level of background induced image noise (shortened to noise hereafter) and was set from 0 to 60 units with a step increase of 10, function *rand(1)* generated a uniform distribution value between (0,1). Noise at different pixels and in successive frames was independent. An element at *(x,y)* in the input image to the LGMD neural network was given by,

$$\widetilde{L}_f^{\ n}(x,y) = L_f(x,y) + L_f^n(x,y) \tag{20}$$

The averaged value of the simulated noise (averaged over 100 frames) in each isolated L cell (or pixel) was around (3.32, 6.60, 9.88, 13.20, 16.52, 19.80) for noise levels of (10, 20, 30, 40, 50, 60) respectively.

The LGMD based neural networks were challenged with the video clip and results are shown and compared in Figure 5. The network with GD processing worked well when κ_n was less or equal to 50; the collisions were consistently detected at frame 84 (Figure 5, b). The curves climbed up sharply when collision was imminent. As a comparison, a similar network but without GD processing was also tested with the same video sequence and the results shown in Figure 5 c; the threshold has been reset to ensure both networks detect collision at frame 84, when no image noise was present. However, with this threshold the network without GD processing would inevitably fail when noise, i.e. complex background, was presented as shown in the figure. The GD processing has significantly improved the network's robustness in these cases.

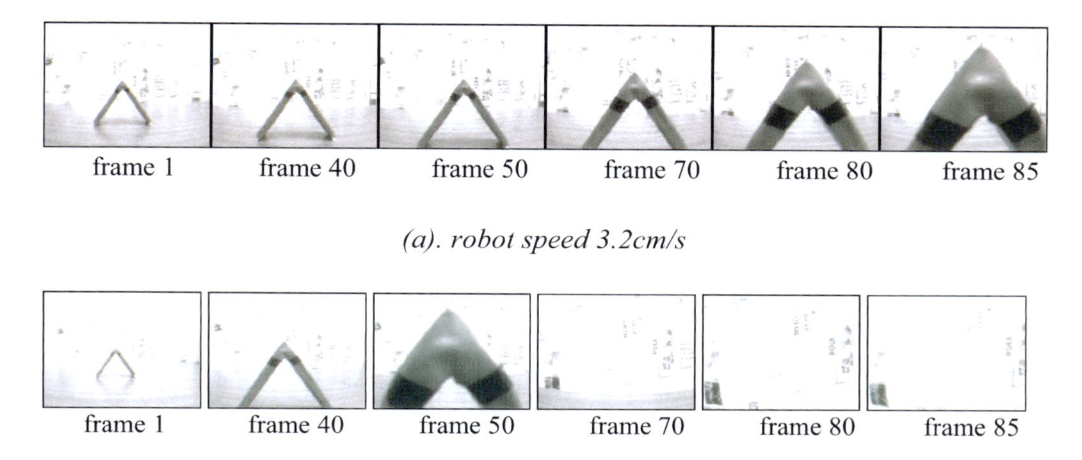

(a). robot speed 3.2cm/s

(b). robot speed 9.6cm/s

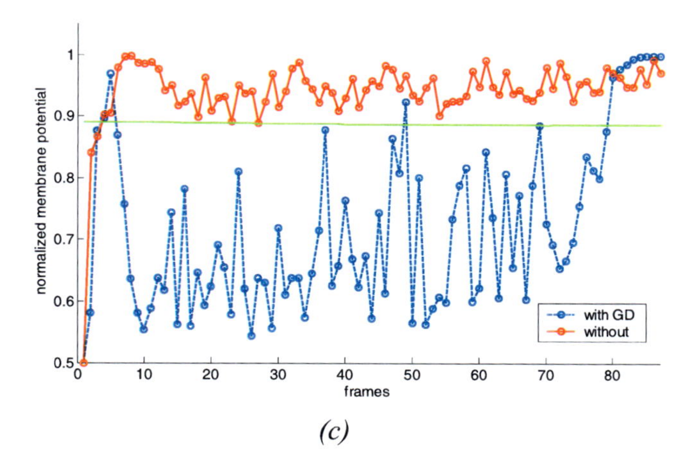

(c)

Figure 6. Continued on next page.

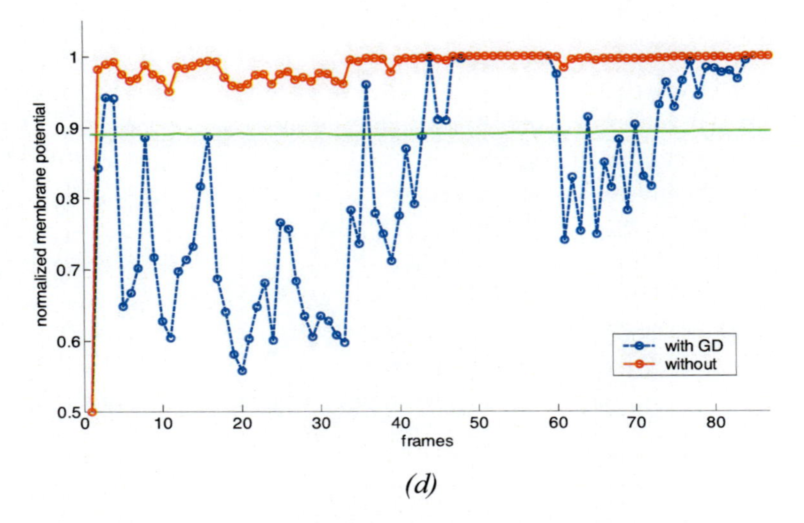

(d)

Figure 6. Comparison of the networks with/without grouped excitation and decay (GD) processing at different approaching speeds. (a), the sample frames of the video clips, which were taken by a robot's camera when the robot was approaching an inverted V shaped object at a speed of 3.2cm/s) at 24 frames per second. Collisions happened at frame 87. (b), the sample frames of the video clips with robot speed at 9.6cm/s. Collision happened at frame 51. (c), the response of the membrane potential to the 3.2cm/s video. Collision detected at frame 83 with GD processing. (d), the response of the membrane potential to the video at an approach speed of 9.6cm/s. Collision was detected at frame 47 with GD processing. The horizontal straight line around membrane potential value of 0.89 was the threshold for the network with GD processing.

Tests with Real Background Generated Image Noise

The network was also challenged with clips captured by the robot approaching an inverted V shaped object in front of a complex background at different speeds (Figure 6 a, b). The V shaped object was placed on a table 50cm away from the wall which was decorated with textured slow drifting stripes of paper. The object was 7.5cm high. Sample frames of the two video clips are shown in Figure 6 (a) and (b), the approaching speeds were 3.2cm/s and 9.6cm/s respectively. Again, the networks with GD and without GD processing were compared. The results are shown in Figure 6 (c-d).

With GD processing in the G layer, the neural network responded differently to non-colliding scenes and imminent colliding objects compared with the original network; it detected the colliding object at different approaching speeds, i.e., at frame 83 for the lower speed video and at frame 47 for the high speed video. However, without GD processing, the network was unable to detect collision at all as shown in the Figure 6 (c) and (d). These off-line tests suggested that with edge enhancement, the network could detect imminent collision robustly, especially with an open, complex background.

2.2.2. Real Time Tests

In the off-line tests, the objects and scenes were known in advance or remaining unchanged for each new test. Moving autonomously within different arenas might cause the robot to face new and unpredictable situations. To test if the collision detection system worked reliably, the best way was to challenge it with a variety of situations in real time.

Environmental Set up

In the real time experiments, the LGMD based neural network (with GD processing if not indicated otherwise) together with FFM was integrated with a Khepera robot (K-team, Switzerland) (Figure 7). The robot was controlled by a motor control unit, which could be triggered by several (4 in this study) successive spikes from the LGMD cell and outputs two commands to the left and right wheels to control the robot's turning behaviour. The luminance intensity was deliberately not controlled; however, the light from above the arenas was measured in the experiments and varied from 86μw/cm^2 to 130μw/cm^2 if not stated otherwise.

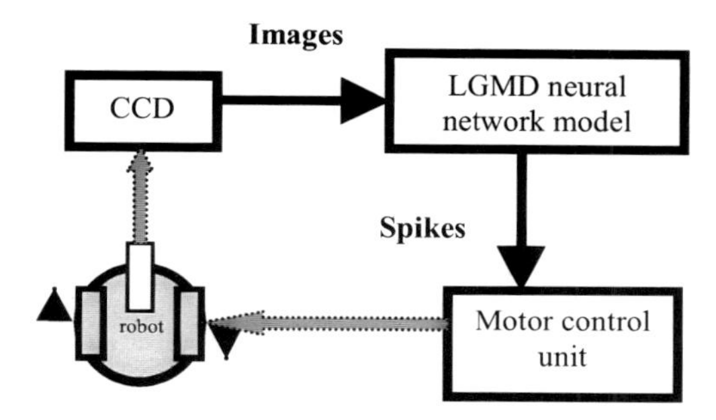

Figure 7. A diagram of the Khepera robot's (K-team, Switzerland) visual collision detection system and its connection with the motor control unit. The images are captured by a CCD camera mounted on the robot and are fed to the LGMD based neural network. Spikes are generated by the neural network and are passed to a motor control unit which has only one pattern of reaction. The motor control unit can be activated by four successive spikes. It controls the two wheels of the Khepera robot and makes the robot turn in one direction, clockwise. The turning angles are generated randomly within lower and upper limits. In the real time experiments, further background of the environment was a typical lab not excluded.

As mentioned in the introduction, previous experiments [Blanchard et. al. 2000] demonstrated the ability of a LGMD neural network to avoid collision in an arena with a simple background. Since the presented LGMD based collision detection system was aiming to tackle challenges in a more complex situation, all the arenas used in the experiments will have complex backgrounds.

In the autonomous navigating experiments, the robot was allowed to move at a speed within an arena for a period of time, for example, 15 seconds. Once it detected an imminent collision, it stopped and turned before continuing its straight line path. The turning speed was set to be 3.2cm/s for the left wheel and -3.2cm/s for the right. The turning angle was controlled by time t_α and

$$t_\alpha = t_{con} + \lambda\, rand(1) \tag{21}$$

where t_{con} was a constant and set to 0.7 second in the experiment, λ was a scale and set to 0.25 and *rand(1)* was a function to generate uniform random numbers between 0 to 1. Therefore, a turn took between 0.7~0.95 seconds.

In the following experiments, unless otherwise stated, the visual input to the LGMD neural network was turned off during a turn; the LGMD based neural network used had GD processing. The trajectories of the robot were recorded with a webcam (Trust 380 USB 2.0 SPACEC@M, http://www.trust.com) hanging above the arena and were extracted via an off-line trace extracting programme written in Matlab. In the trace extracting programme, a template matching method [Duda and Hart 1973] was used to locate the robots position in each frame.

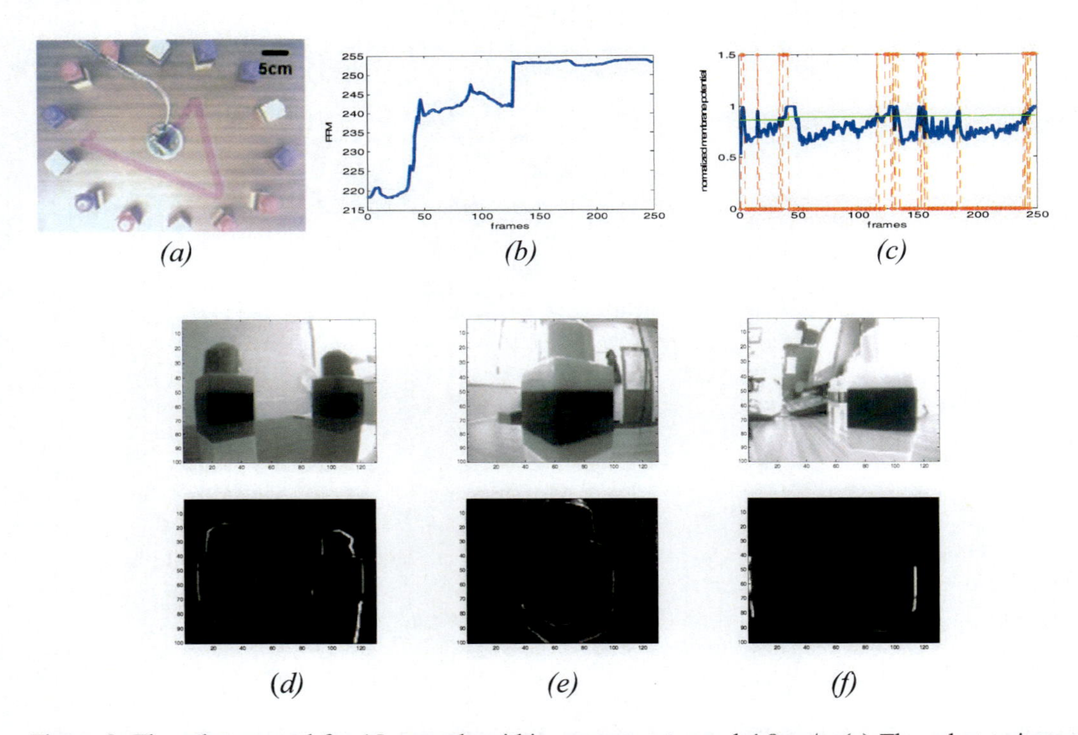

Figure 8. The robot moved for 15 seconds within an arena at speed 4.8cm/s. (a) The robot trajectory within the arena; the trajectory is indicated by bold lines. Three times of collision were detected by the LGMD based collision detection system and three turns were conducted. (b) The LGMD sigmoid membrane potential (bold red solid line), spikes (dashed line with stars at the peaks) and the threshold (solid horizontal straight line). The three collisions detected were indicated with 4 successive spikes. (c).The FFM during the movement. It reflects the changed contrast intensity after each turn. (d). The scene of the first turn and the image after GD processing. (e) The scene of the second turn and the image after GD processing. (f) The scene of the third turn and the image after GD processing.

Experiments and Results

A short robot movement (15 seconds, speed at 4.8cm/s) was made to investigate the collision detection system. Results are shown in Figure 8. Three imminent collisions were successfully detected (Figure 8 a) at frame no.44, no.127 and no.250 respectively (Figure 8, b). The membrane potential threshold jumped to a higher level when the robot turned and faced a new scene because of the change in FFM (Figure 8, b). The three scenes and processed images, just when the collisions were detected, were also shown in the figure. The stripes on the table, wires and other small objects were filtered out by the GD processing. Only the expanding edges, the feature which is used by the detection system, remained. This made the neural network more robust as it only concentrated on colliding objects.

The FFM can sometimes affect the detection moment in several frames, as shown in the results (Figure 9) from an experiment in which two collisions were detected one after the other; the collision avoidance behaviour would be either several frames later for the first colliding object or several frames earlier for the second one without FFM.

Further experiments were carried out to show how the robot behaved within an arena at different speeds (Figure 10). With the speed of 6.4cm/s, the robot sometimes was quite close to the object before it detected collision and turned (Figure 10, a) as less excitation occurred due to smaller changes in successive images. It was found that at higher speeds (9.6cm/s and 12.8cm/s) the robot increased its distance from the blocks (Figure 10, b, c). This was because more excitation was caused by higher speeds as the change in image between successive frames was larger.

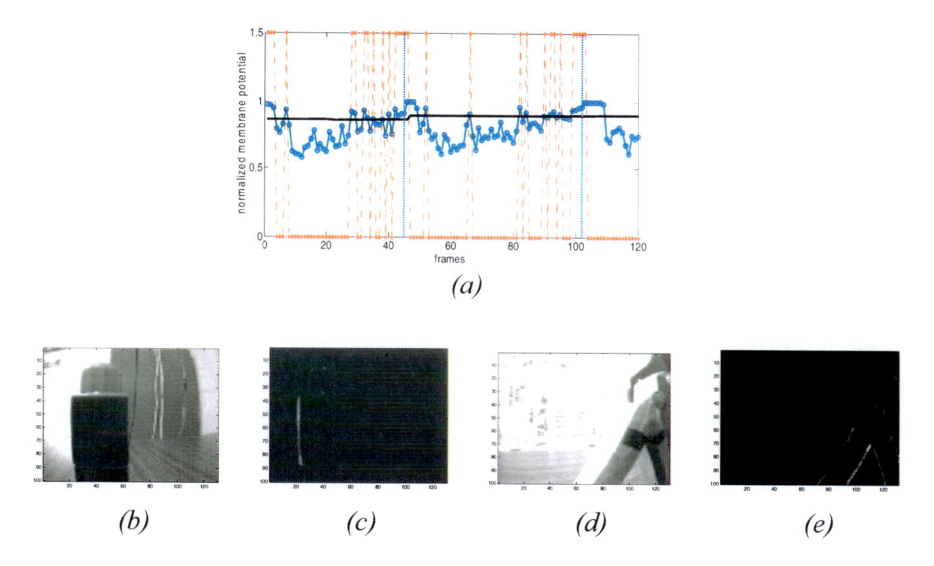

Figure 9. (a), the effects of the FFM regulated threshold. The threshold is shown as a solid straight line, sigmoid membrane potential is indicated by circles for each frame and spikes represented by upper stars. Collisions were detected at frame 45 and frame 102, and indicated by dotted vertical lines. (b), the image at frame 45. (c), the image after GD processing at frame 45. (d), the image at frame 102. (e), the image after GD processing at frame 102. In the images, white represented the highest value and black represented the lowest.

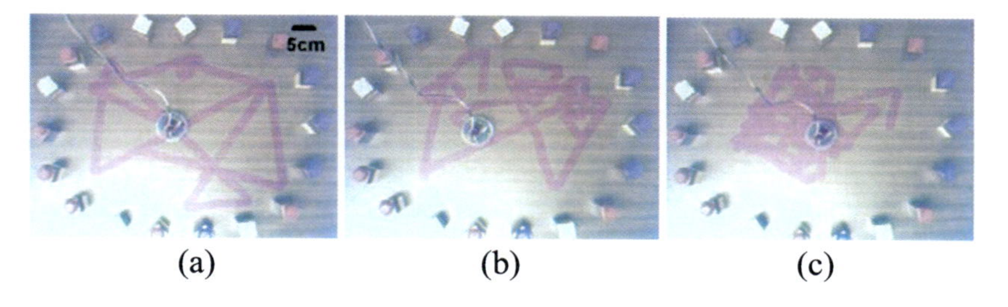

Figure 10. The robot moved autonomously within an arena at different speeds for 60 seconds. (a) robot speedwais 6.4cm/s; (b) robot speed was 9.6cm/s; (c) robot speed was 12.8cm/s.

When some of the blocks were changed to other types of objects, the neural network also worked quite well; navigating a course about 60 seconds, 12 collisions were detected and avoided as indicated in the figures (Figure 11, a and b). Mugs, strange shaped blocks and curved paper were all successfully detected by the system.

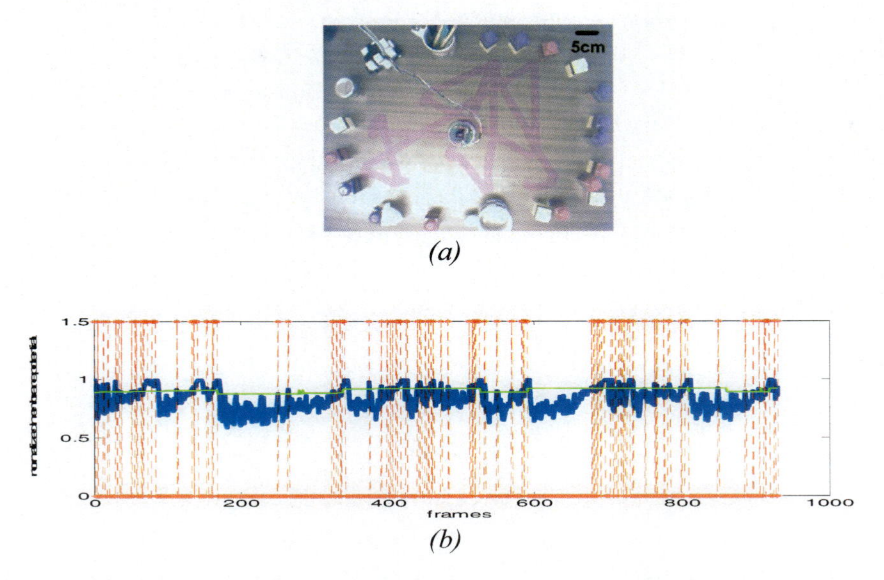

Figure 11. The robot has moved for 60 seconds within an arena surrounded by different (by shape, colour and material) objects at speed 6.4cm/s. (a) The robot trajectory within the arena. The trajectory is indicated by bold lines. 12 times of collision were detected by the LGMD based collision detection system and 12 turns were conducted. (b) Some of the scenes when imminent collisions were detected and their corresponded images after GD processing.

We next wanted to see if the system worked in extreme conditions. As shown in Figure 12, three experiments were done to test the system in three scenarios: (1) very dark with luminance intensity at 17μw/cm^2 (Ealing Electro-Optics, Holliston, MA, USA), (2) partly in sunlight (917~1,274μw/cm^2), and (3) very bright sunlight (3,240μw/cm^2) with long shadows. The experiments showed that the robot can still detect collision without any difficulty in these situations. Interestingly, it detected collision very early when facing long shadows (Figure 13, c) because the system was detecting objects relying on their contrast.

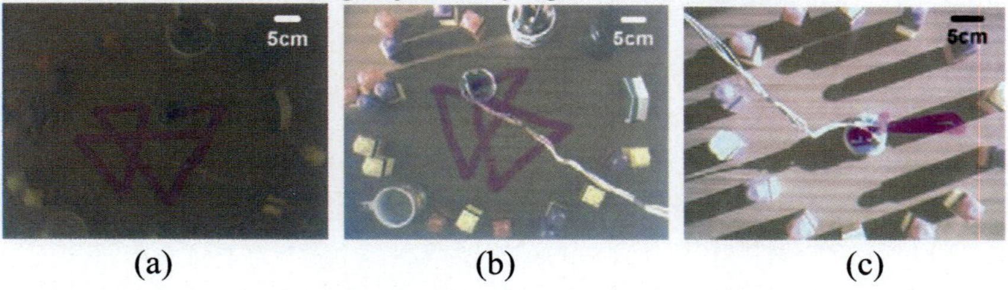

(a) (b) (c)

Figure 12. The performance of the robot with the LGMD based collision detection system in extreme conditions. (a) extremely dark (17μw/cm^2), robot speed was 6.4cm/s; (b) partly in mild sunlight

(917~1,274μw/cm^2), robot speed was 6.4cm/s; (c) in bright sunlight (3,240μw/cm^2) with long shadows, robot speed was 4.8cm/s.

2.3. Further Discussion

In the above sections, the LGMD based collision detection system was tested using both off-line and real time tests. Grouped excitation and decay (GD) processing enhanced the key features of a colliding object. The collision detection system detected collision reliably in different situations with complex backgrounds, regardless of the shape, material or colour of the colliding objects. The system allowed the robot to navigate in an unstructured, complex environment without intensive computing cost.

For many species of animal, vision plays a key role in their survival. Various visual based navigation methods have also been proposed for autonomous robots (for example, [Olson et. al. 2003, Fiala and Basu 2004]). Nowadays, visual sensors are becoming very cheap and reliable. This makes it possible for many mobile machines (e.g., mobile robots, cars, boats, planes and some toys) equipped with visual sensors and visual-based navigation systems to avoid unwanted collision automatically in the real world.

However, the locust LGMD is only one example of a visual feature detector that has evolved within the visual system of insects. In the insects' brain, there are numerous interneurons related to vision working together to extract the plentiful visual cues simultaneously. To separate/extract other visual cues from the dynamic scenes simultaneously, other specialized neurons need to be integrated into the system in the future. For example, directional selective neurons (e.g., [Rind 1990a, 1990b] in locust, [Borst and Haag 2002] in fly, [Stasheff and Masland 2002] in rabbit) could be used to detect high speed big translating objects which project over large areas of the retina. The further ongoing investigations of the LGMD and its postsynaptic interneuron in the locust also provide new ideas and alternative ways to further improve the system (for example, [Rind, 1984; Gabbiani et. al. 2004, 2006, Santer et. al. 2004, 2005]).

Although the LGMD based collision detection system's robustness was shown in the experiments, it still needs to be noted that the detection system was entirely reliant on an objects' contrast against the background. The system does fail to detect a colliding object if the object has no contrast with its background as seen by the robot.

3. Directional Selective Neural Networks (DSNN) for Collision Detection

The synthetic vision system DSNN was based on four whole-field direction-selective neurons (Figure 13 *L* (left), *R* (right), *U* (up), *D* (down) see also [Yue and Rind 2006b, 2006c] for additional details). Each of these direction selective neurons had one specified inhibitory direction generated by asymmetrically spreading lateral inhibition. Details of the synthetic vision system are given in the following section 3.1. The DSNN was trained by using evolution process which is detailed in the section 3.2 and took place in two different environments. The best trained agents representing the DSNN were then tested in their training environments.

3.1. The Synthetic Vision System

The synthetic vision system was based on direction-selective neurons, not the LGMD neuron. There are many ways to form a computational model of a direction-selective neuron (for example, [Marshall 1990, Borst and Haag 2002, Tversky and Miikkulainen 2002, Fried et. al. 2002]). As shown in Figure 13, the direction selectiveness of the whole-field neuron in this study was achieved by asymmetric lateral inhibition which spread in only one direction. As an example, the *L* neuron responded to moving edges in all directions except the inhibited left direction. The four direction-selective neurons *L, R, U* and *D* cells were then organised by several neurons (*a b c d* and *e*) and their connections (Figure 13b). The excitation and inhibition described below was indicated by a value in arbitrary units.

The Direction Selective Neuron L

P layer The first layer of the neural network was the same as the first layer of the LGMD network, which meant it consisted of photoreceptor *P* cells arranged in a matrix form; the luminance L_f of each pixel in the input image was captured by its corresponding photoreceptor cell, the change of luminance P_f between two successive frames of the image sequence was calculated and formed the output of this layer. The output of a cell in this layer was defined by the equation:

$$P_f(x,y) = \sum_i p_i P_{f-i}(x,y) + abs(L_f(x,y) - L_{f-1}(x,y)) \tag{22}$$

where $P_f(x,y)$ was the change of luminance corresponding to pixel (x, y) at frame f, x and y were pixel coordinates, L_f and L_{f-1} were the luminance, subscript f denoted the current frame and $f-1$ denoted the previous frame, the persistence coefficients p_i could be varied between 0-1.0. To increase the computing efficiency of the system, p_i was set to zero in the following evolution process. The absolute value operation in the above equation meant the differences between successive images were all positive. If there was no difference between successive images, the *P* cells were not excited.

S layer The output of the *P* cells formed the two types of inputs to the cells in the next layer- *S* layer (Figure 13). One type is excitatory and the other one is inhibitory. Unlike the LGMD network , the excitatory output of a *P* cell passed directly to its retinotopic counterpart in the S layer whereas the inhibitory output from a *P* cell was delayed by one image frame and then fed in one direction to some of its retinotopical counterpart's neighbouring cells. In the case shown in Figure 1, inhibition extended leftward. The inhibition could extend over a maximum of eight cells in any one direction. The total inhibitory input to a cell in this *S* layer was

$$I_f(x,y) = \sum_{i=1}^{N_{lay}} P_{f-1}(x+i,y) w_I(i) \tag{23}$$

where $I_f(x,y)$ was the inhibition to the cell in *S* layer at position (x, y) at frame f, N_{lay} was the number of cells that feed lateral inhibition to this cell and $w_I(i)$ was the local inhibition weight. The value of the local inhibition weight was set to be 5.5 based on previous trials to ensure direction selectivity. With this strong inhibition from the right side with one frame delay, the excitation caused by left moving edges could be weakened sharply. Inhibition with more than one frame delay would increase computing cost and its effect on the overall direction selectiveness would be negligible with the current strong inhibitory weights. With the spread of lateral inhibition N_{lay} set to 8, the system could cope with image motion slower than 8 pixels per frame, which was equivalent to 120 degrees per second angular velocity using a camera working at 25 frames per second and with a 100X80 pixel 60 degrees field of view[↓]. At the edge of a layer, the inhibitory input to a cell may be limited because there may be no neighbouring cell on its outer side. The excitatory flow gathered in the S cell would be

$$E_f(x,y) = P_f(x,y) - I_f(x,y)W_I \tag{24}$$

where W_I was the coefficient controlling the global strength of inhibition and this was set to be 1.5 based on a previous study.

L cell The excitations once they exceeded a threshold (set to be 12 based on previous trials) in the S cells were summed by the left inhibitory cell L and the excitations below the threshold were ignored in the summing operation. The summed excitation of the left inhibitory cell L was,

$$S_f^L = \sum_{x=1}^{n_k} \sum_{y=1}^{n_l} \Theta(E_f(x,y)) \tag{25}$$

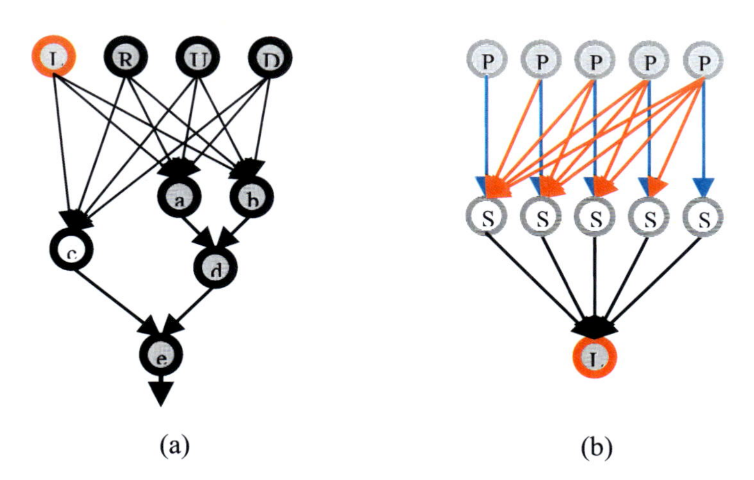

Figure 13. Schematic illustration of the synthetic vision system for collision detection. (a), A direction selective neural network with left sided lateral inhibition. It had three types of cells, i.e., photoreceptor cells P, summing cells S and direction-selective cells. Photoreceptor cells in the P layer perceived the luminance change; fed excitation directly to its retinotopically corresponding cell in the summing cell (S) layer; and fed inhibition, delayed by one frame, to its left side neighbours in the S layer. Summing cells in S layer summed excitation and inhibition and passed excitation/inhibition to different cells in the next layer. The Direction selective cell, in this example, is a left (L) cell. The L cell was one of four

[↓] The angular velocity is calculated in this way: 60(25/(100/8)) =120 degrees per second

different cells, selective for leftward (L), rightward (R), upward (U) or downward (D) motion. The connectivity and weights of a direction-selective cell did not change in the evolutionary process. Matched to the size of the input images, there were a total of 8,000 P cells, 8,000 S cells and 1 L cell in this left inhibitory neural network. (b), The organisation of the synthetic vision system consisted of a combination of four direction-sensitive neural networks each with asymmetric inhibition projecting to the left (L), right(R), up(U) or down(D). The L, R, U and D neurons passed their excitation/inhibition to all the cells in the next layer, that is, cells a, b and c. Cell a and b then fed their information to cell d; the final cell e received excitation/inhibition from cell c and d; spikes would be produced in the e cell if its threshold was exceeded. Five successive spikes means a collision was recognised by the network. Note in the figures, circles filled with grey means that excitation and inhibition were summed and an absolute value taken. In this example all but the S neurons had their outputs rectified. There were a total of 80 X 100 P cells shared by the 4 direction- selective neural networks, 4 X 80 X 100 S cells, 4 direction-selective neurons, 4 intermediate cells and 1 spiking cell in this synthetic vision system.

Since cells in this layer were also organised in a matrix, n_h was the number of cells counted in a row , n_v was the number of cells in a column and Θ denoted the threshold operation:

$$\Theta = \begin{cases} 1 & if\ E_f(x,y) > 12 \\ 0 & otherwise \end{cases} \tag{26}$$

The excitation of the *L* cell was then sigmoid as,

$$s_f^L = (1 + e^{-S_f^L n_c^{-1}})^{-1} \tag{27}$$

where n_c was the total number of the cells in *S* or *P* layer. The number of cells in *S* layer was the same as that in *P* layer. Since each cell in P or S layer corresponded to a pixel in an input image, the total number of cells n_c was determined by the size of input images. For this paper the input images were 100 by 80 pixels. Therefore, n_c was 8,000 in *P* or *S* layer. Since $S^L{}_f$ was greater than or equal to zero according to equation (25), the sigmoid excitation $s^L{}_f$ varied within (0.5~1). The L cell had one inhibitory direction, left, and was not expected to respond to left translating movements that occurred within a certain velocity range. An L cell and its presynaptic networks could be made to respond to only one specified direction with inhibition that spread to all directions apart from the specified one; however, it would be computationally very expensive. The robustness of these direction-selective cells has been demonstrated previously [Yue and Rind 2006b].

Combination of L, R, U and D Outputs

The four direction selective neurons: *L, R, U* and *D* were further organised as shown in Figure 13 (b) [Yue and Rind 2006c]. The cells *a, b* and *c* gathered information from the four neurons,

$$\begin{pmatrix} E_f^a \\ E_f^b \\ E_f^c \end{pmatrix} = \begin{pmatrix} w_{L-a} & w_{R-a} & w_{U-a} & w_{D-a} \\ w_{L-b} & w_{R-b} & w_{U-b} & w_{D-b} \\ w_{L-c} & w_{R-c} & w_{U-c} & w_{D-c} \end{pmatrix} (s_f^L\ s_f^R\ s_f^U\ s_f^D)^T \tag{28}$$

where E^a_f was the excitation in the a cell at frame f, w_{L-a} was the weight of the connection between the cell L and the cell a, and the other symbols were named in a similar way. The weights were allowed to vary within specified domains in this study, as listed below in the matrix,

$$\begin{pmatrix} w_{L-a} & w_{R-a} & w_{U-a} & w_{D-a} \\ w_{L-b} & w_{R-b} & w_{U-b} & w_{D-b} \\ w_{L-c} & w_{R-c} & w_{U-c} & w_{D-c} \end{pmatrix} \sim \begin{pmatrix} (0.5 \sim 1.5) & (-1.5 \sim -0.5) & (-0.5 \sim 0.5) & (-0.5 \sim 0.5) \\ (-0.5 \sim 0.5) & (-0.5 \sim 0.5) & (0.5 \sim 1.5) & (-1.5 \sim -0.5) \\ (0.5 \sim 1.5) & (0.5 \sim 1.5) & (0.5 \sim 1.5) & (0.5 \sim 1.5) \end{pmatrix} \quad (29)$$

With these predefined adaptable weights in equation (29), cell c gathered excitation from the direction selective cells, and cell a and b compared the weighted excitations they got from the direction selective cells.

The cell d compared information from the cells a and b,

$$E_f^d = abs(w_{a-d} abs(E_f^a) + w_{b-d} abs(E_f^b)) \quad (30)$$

where the weights were limited within: $w_{a-d} \in [0.5 \sim 1.5]$, $w_{b-d} \in [-1.5 \sim -0.5]$. Finally, the "spiking cell" e compared information from the cells c and d,

$$E_f^e = abs(w_{d-e} E_f^d + w_{c-e} E_f^c) \quad (31)$$

The weights were bounded in the similar way: $w_{d-e} \in [0.5 \sim 1.5]$ and $w_{c-e} \in [-1.5 \sim -0.5]$. Once the excitation in cell e exceeded a given threshold, a spike was produced. If spikes were produced in all of five successive frames then this was defined as detection of an imminent collision. The optimal spike number for collision decision making has been investigated previously [Yue et. al. 2006]; using multiple consecutive spikes instead of one can avoid false alarms triggered by visual perturbations such as camera shake. The threshold for cell e was allowed to vary within the range from 0.0 to 4.0.

The connectivity and the values of the weights of the input of the direction selective neurons (L, R, U and D) were fixed (see Figure 13). In contrast the weights of their output connections (see Figure 1, network on right hand side) could be altered during the evolutionary process. With the current structure, the different visual cues extracted by opposite direction selective neurons were compared by cell a and b, and further compared by cell d; the overall excitation of the four direction selective neurons were summed by cell c; the outputs from cell c and cell d were then compared by cell e to form the final outputs of the synthetic vision system.

In the whole synthetic vision system, there were 8000 P cells shared by the 4 direction selective neural networks, 32000 S cells in total (because each of the direction selective neural networks has 8000 S cells), 4 whole-field direction-selective neurons and 5 further organised cells (namely, a b c d and e).

3.2. Evolutionary Training

Genetic Algorithms [Holland 1975, Goldenberg 1989, Chipperfield and Fleming 1995, Yue et. al. 2006] are adaptive heuristic search algorithms based on the evolutionary ideas of natural selection and genetic mutation and have been proved to be efficient in tuning visual neural networks [e.g.,. Yue et. al. 2006]. In this chapter, we also explored the overall performance of the bio-inspired synthetic vision system via an evolutionary training process.

Population The population of 40 agents in each generation representing the synthetic vision system were processed via a genetic algorithm [Chipperfield and Fleming 1995]. The first generation had one recommended agent with predefined connection weights/threshold (Table 2, Start) and 39 agents with randomly produced connection weights/threshold equally distributed in the corresponding bounding intervals. The connection weights/threshold of each agent were then coded in binary numbers as its chromosome. The precision of the binary representation of each weights/threshold was 10 bits. To form a new generation, the worst performing agents (20% of the whole population in a generation) were replaced, i.e., the generation refreshing rate was 0.2. New agents (20% of a whole population) were produced by the best performing parents in the last generation through crossover. The stochastic universal sampling method was used in selecting parents for breeding [8]. Mutations were made to the binary chromosomes of these newly produced agents with a mutation rate 0.2. This meant 20% of these binary bits in the chromosomes were randomly chosen to flip, either from 1 to 0 or from 0 to 1, during a mutation process. The generation refresh rate and mutation rate were set based on our previous experiments and studies [Yue et. al. 2006].

$$F_k = \left(1 - \frac{\sum_{i=1}^{N_v} f_{event}^i}{M_{nb}}\right) \times 100\% \tag{32}$$

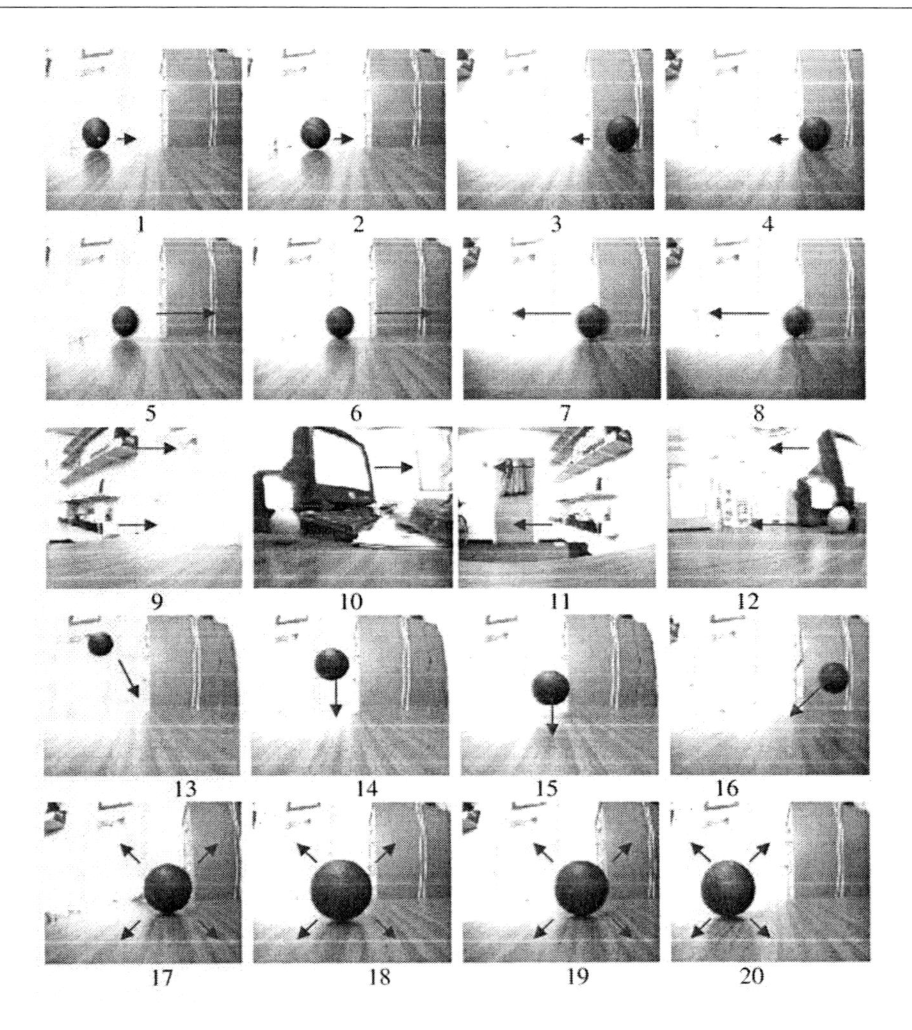

Figure 14. Samples from video sequences making up the robotic laboratory environments in which the agents evolved. The number under each image was the number of its corresponding video sequence. The arrows in the images were added to show the visual motion direction. The black ball was 95mm in diameter. In video sequences 1 and 2, the ball was moving across the field of view from left to right at intermediate speed, taking 18 and 27 frames respectively; in video sequences 3 and 4, the ball was moving across the field of view from right to left at an intermediate speed, taking 19 and 20 frames respectively; in video sequences 5 and 6, the ball was moving rapidly across the field of view from left to right, taking 11 and 14 frames respectively; in the video sequences 7 and 8, the ball was moving fast across the field of view from right to left, taking 11 and 13 frames respectively; in video sequences 9 and 10, the robot was turning anticlockwise at about 50°/s while moving forwards, at 3.2cm/s; in video sequences 11 and 12, the robot was turning clockwise at about 50°/s while moving forwards, at 3.2cm/s; in video sequences 13, the ball was bouncing to the right; in video sequence 14 and 15, the ball was bouncing up and down; in video sequences 16, the ball was bouncing to the left; in video sequences 17, the ball was approaching the robot at 0.4~0.5m/s from right side; in video sequence 18 and 19, the ball was approaching the robot at 0.4~0.5m/s from the central area; in video sequence 20, the ball was approaching the robot at 0.4~0.5m/s from left side. There were 60 frames in each video sequence. Sequences 17~20 were collision sequences.. The robot's field of view was 60°.

Fitness The Agents' fitness value, namely weighted success rate, was evaluated based on their behaviour. In each generation, an agent that responded to all visual events correctly, i.e. recognised imminent collisions and made no mistakes with translating scenes or other challenges, scored a fitness value (success rate) of 100%; an agent that failed in all events scored a fitness value of 0%; an agent that failed in a non-colliding challenge scored a lowered fitness value (reduced success rate); an agent that failed in a colliding event got a sharp reduction in fitness value since a collision event was much more important in scoring than a non-collision one.In the robotic laboratory, one undetected collision event received a penalty equivalent to four false alarmed non-collision events; in the car driving environment, one undetected collision received a penalty equivalent to eight false alarms in non-collision events. The fitness of an agent could be formulated as the following, where F_k was the fitness value of the kth agent in the population, f^i_{event} was the score for the ith event in the total N_v events, M_{nb} was the summation of all the possible high scores, and f^i_{event} was different for a collision and a non-collision event, and failure and success,

$$f^i_{event} = \begin{cases} K_{col} & \text{if collision event}, \quad \text{failure} \\ 0 & \text{if collision event}, \quad \text{success} \\ K_{non} & \text{if non-collision event}, \quad \text{failure} \\ 0 & \text{if non-collision event}, \quad \text{success} \end{cases} \tag{33}$$

Where K_{col} was the score for failure in a collision event, K_{non} was the score for failure in a non-collision event. For a collision event, failure meant no collision signal was sent out by the agent 30~3 frames before collision. K_{col} was several times bigger than K_{non} , so that an agent that failed in all collision events and an agent that failed in all non-collision events will have the same fitness value: 50%. In the robotic laboratory, $N_v = 20$ (including 4 collision events), K_{col} was 4 and K_{non} was 1; in the driving environment, $N_v = 18$ (including 2 collision events), K_{col} is 8 and K_{non} was 1. Therefore, M_{nb} was 32 in both of the environments.

Figure 15. Sample images from video sequences making up the car driving environment in which the agents evolved. The number under each image was the number of its corresponding video sequence. In video sequences 1~6, the agent faced fast translating cars or figures, while driving at very low speed; in video sequences 7~8, there were road symbols, while driving at high speed; in video sequences 9~11, another car was cutting in or approaching but was not going to collide, while driving at normal speed on a motor way; in video sequences 12~13, the agent faced slowly walking pedestrians, while driving at low speed; in video sequences 14~16, turning and driving at low speed; in video sequences 17~18, collision at low and high speed respectively. The collision sequences are 17 and 18. All the sequences were recorded at 25 frames per second.

Evolution environments As illustrated in Figure 14 and 15, the two different environments were represented by two groups of video sequences. Each sequence represented one type of scene that could cause strong excitation in the photoreceptor layer. One of the environments was the output of a video camera mounted on the turret of a Khepera robot (http://www.k-team.com) interacting with a black ball in a robotic laboratory. The ball was either moving/bouncing at a velocity of 0.3~1.25m/s across the robot's field of view in a translating way or approaching it on direct collision course at different speeds and from different sides. In the fast translating events, the ball was moving across the field of view within 10~14 frames, corresponding to 0.40~0.56 seconds. The robotic scenes also included turning as this often happened as one of the robot's motor activities. A total of 20 of these robotic scenes were used to construct a virtual environment for the agents to evolve in, (sample images are shown in Figure 14). The robot moved only in the turning scenes. The second environment consisted of car driving scenes which were more complex, and include normal driving on a highway, collision with a car, turning, pedestrians, road symbols and high speed translating cars/vans, (sample images are shown in Figure 15). The car used for collision was actually an inflatable car. A total of 18 car driving video sequences were used in this second evolution environment.

All the input images to the synthetic vision system were resized to 100 pixels horizontally by 80 pixels vertically; images were grey scale ranging in intensity from 0 to 255. The video images were all taken at 25 frames per second (fps).

Training results During evolutions agents in the driving environment were shown images videoed outdoors from a moving vehicle; agents in the robotic environment were shown images generated by a mobile Khepera robot navigating in a simplified laboratory environment. Three evolutions were done for each evolving environment and the best agents in driving and robotics laboratory were selected for further testing.

Table 2. The fitnesses (success rates) of the evolved best agents per generation, over 400 generations

Generation	Best agent evolved in driving environment	Best agent evolved in robotic laboratory
(Start guess)	*(41%)*	*(59%)*
1	88%	91%
25	94%	91%
100	97%	94%
400	100%	100%

The performances of the best agents in their training environment, together with the performance of the start guess agent, are listed in Table 2. The weights of the connections between cells of the best robotics agents and the driving agent after 400 generations are shown in the table 3. The start guess agent with a balanced weight value scored very low success rates- 41% and 59%. However, the best agents' performance improved in subsequent generations during the evolution processes. By the 400th generation, the best agents in each environment had reached a 100% success rate (Table 2).

Table 3. The weights of the connections between cells of the best robotic agent and the driving agent after 400 generations

Connection	Start	Rob. lab	Driving	Connection	Start	Rob. lab	Driving
L-a	1	1.1080	0.5225	a-d	1	1.3915	1.4013
R-a	-1	-1.4892	-1.0015	b-d	-1	-0.6945	-1.4951
U-a	0	0.0582	0.3573	L-c	1	1.4687	1.1403
D-a	0	0.4961	-0.0963	R-c	1	1.4140	1.2038
L-b	0	0.4277	-0.0367	U-c	1	0.5225	0.5958
R-b	0	-0.2615	0.1872	D-c	1	0.7356	0.5518
U-b	1	1.0132	1.4765	d-e	1	1.4208	1.2146
D-b	-1	-1.0005	-0.8226	c-e	-1	-0.8324	-1.1276
				e threshold	3.0	2.3268	3.0873

3.3. Tests

Test sequences Two groups of sequences for testing the evolved agents are shown in Figure 16 (a~b). These test sequences included some challenging scenes. For the agent in the robotic laboratory, the test included: ball looming, translating nearby, translating at high speed, bouncing from left to right, near miss (ball approaches the robot but missed it), and robot turning. For the agent in the driving environment, the test included:

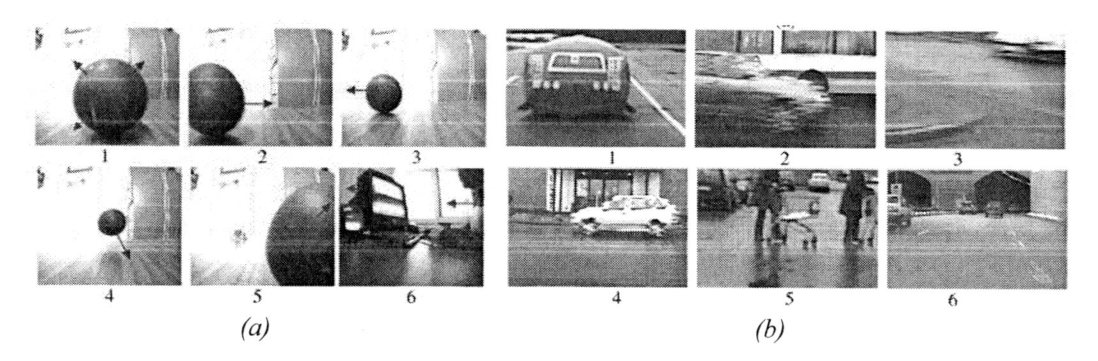

Figure 16. Sample images from all test sequences that were used for testing the agents. For each of the two evolution environments and the computer generated visual stimuli "environment" six corresponding sequences were used. The number under each sample image is its corresponding video sequence number. (a), video sequences in robotic laboratory. 1: looming ball at about 0.4m/s; 2: translating ball nearby, taking 16 frames to move across the view field; 3: fast translating ball from right to left, taking 17 frames to move across; 4: bouncing ball from left to right; 5: looming ball near missing at about 0.45m/s; 6: robot turning at about 50°/s. Ball size: 95cm in diameter. The arrows are added for schematically indicating the visual motion direction. (b), driving scenes. 1: colliding with a balloon car; 2: fast translating car nearby; 3: turning; 4: translating car; 5: translating figures with trolleys; 6: getting into and out of tunnel. The arrows indicated the direction of motion of the visual stimuli.

collision with a car, high speed translating car at different distances, turning, translating human figures and driving into and out of a tunnel. The best evolved agents were not only tested by using sequences from the environment they evolved in but also tested by using sequences from new environments.

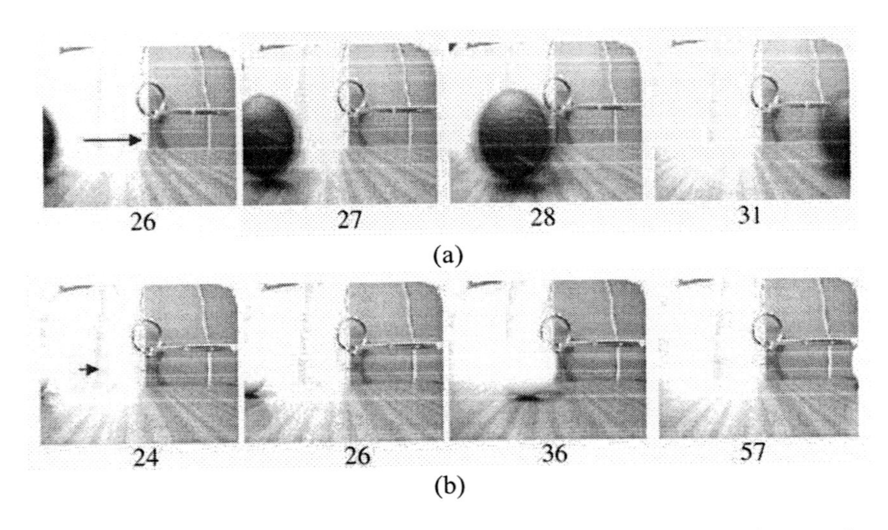

Figure 17. Samples of the extreme test sequences from the robotic laboratory used to test the capability of the direction selective neurons and the evolved agents. The numbers under the sampled images are the frame numbers. (a), video sequence 1: black translating ball at close range, taking 6 frames to move across the view field, i.e., from frame no.26 to frame no.31; equivalent angular velocity is 250°/s. (b), video sequence 2: white translating ball against a white background, taking 34 frames to move across the view field, i.e., from frame no.24 to frame no.57; equivalent angular velocity 44°/s. The black ball and the white ball were 95cm in diameter. The arrows were added to give an indication of the overall motion direction.

Special test sequences To test the capability of the directional selective neurons and the best agents, two special test sequences from the robotic laboratory were also recorded as shown in Figure 17. One of the sequences was a rapidly translating black ball, which crossed the field of view within six frames; its speed was equivalent to an angular velocity of 250°/s, as shown in Figure 17 (a). This high angular velocity visual scene is quite rare in driving scenarios. However, in the robotic laboratory, it could be achieved by deliberately throwing the ball near the robot's camera. Another sequence was a white slowly translating ball against white and/or grey background. The ball was not resolved when moving on a white background until frame 36 as shown in Figure 17 (b).

Test results The best agent from each evolution environment was then challenged with the three groups of test sequences and results are shown in the Table 4. In the test sequences, as shown in the Table 4, the synthetic vision system achieved a 100% success rate in its familiar environment.

Details of the excitation levels and spikes produced by the two best agents evolved in the two environments when challenged with some of the test sequences are compared in Figures 18~22. The evolutionary changes have resulted in the two best agents behaving differently from each another when challenged with the test stimuli. For example, when challenged by a fast translating ball (Figure 16 (a), video sequence 2), one responded with peaks in excitation and sparse spikes, whereas the other responded with continued high levels of excitation, although excitation remained below the threshold for spike production (horizontal dotted line), as shown in Figures 19.

Table 4. The performance of the two best agents (evolved in robotic laboratory and evolved in driving environment respectively) challenged with the two groups of test sequences

	Best agent evolved in driving environment		Best agent evolved in robotic laboratory	
Test sequences	Driving scenes	Robotic lab.	Driving scenes	Robotic lab.
1	√	×	√	√
2	√	√	√	√
3	√	√	√	√
4	√	√	√	√
5	√	√	√	√
6	√	√	√	√

Note: Sequence No.1 was the collision event in the three testing groups; for the computer generated test sequences, No.2 was also a collision event. √ indicates correctly detected, i.e., detected collision for collision events and did not signal collision for non-collision events; X indicates failure in an event.

The patterns of output excitation of the two best agents were of similar form when challenged with collision events; however, the output behaviour of the synthetic vision systems or agents was often quite different as shown in Figure 18 and Figure 21. One reason for this lay in the differences in their thresholds for spike production (as shown in Table 3, one was 2.3268 and the other was 3.0873). It was also found that the best agent evolved using images from the robotic laboratory produced trains of spikes in response to an accelerated translating bar (Figure 21(b)). This response was similar to that shown by the locust LGMD neuron to a moving bar that accelerated as it moved. The LGMD also responded strongly and persistently to an accelerating edge but only responded briefly to one that moves with a constant speed [Rind and Simmons 1992].

The best agents were also capable of distinguishing real collisions from those of 'collision-like' (Figure 16 (b), sequence no.6) or near miss (Figure 16 (a), sequence no.5) visual scenes as detailed in Figure 20. Driving into a tunnel was a typical 'collision-like' scenario (Figure 16 (b), sequence no.6) that may have confused the agents. When driving into a tunnel, however, the best agent evolved in the robotic laboratory responded with sparse spikes and the best agent evolved in driving responded with no spikes at all. Since the physical size of the tunnel was big, its expanding pattern in visual field was quite different to that of a colliding car, i.e., the expansion rate of the tunnel was slow and excitation passing through inhibition layer was not enough to exceed threshold. The evolved synthetic vision system has made use of these differences. When challenged with the ball on a near miss trajectory (Figure 16 (a), sequence no.5), both of the evolved best agents dealt with it correctly though their excitation patterns were slightly different as shown in Figure 20.

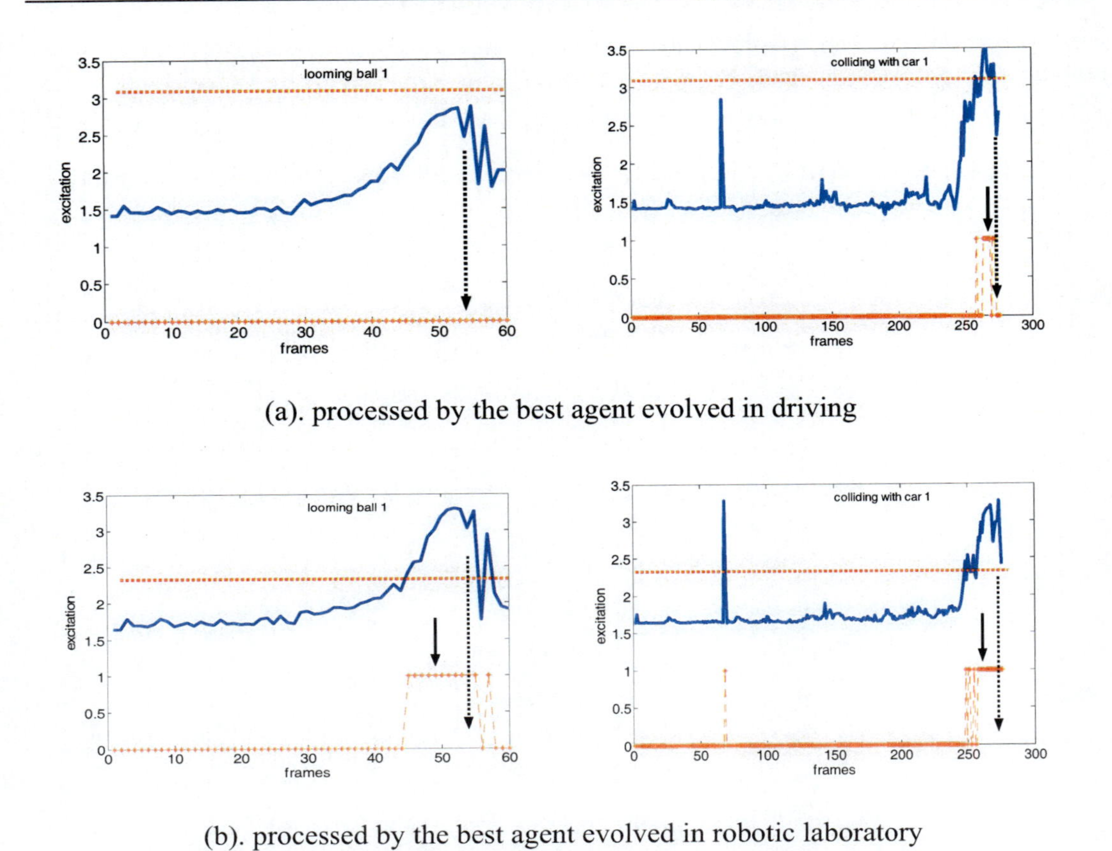

Figure 18. The excitation levels of the two best agents when processing the collision events in the different training environments, i.e. looming ball (Figure 16 (a), video sequence 1) and the car collision scenes (Figure 16 (b), video sequence 1). For the looming ball, the collision was actually going to occur at frame 54 as indicated by dotted arrows; the best agent evolved in the robotic laboratory recognised the imminent collision at frame 49 as indicated by a solid arrow; no collisions were detected by the driving environment evolved best agent. For the collision with the car, the collision was actually going to occur at frame 273 as indicated by dotted arrows; the best agent evolved in the robotic lab recognised the imminent collision at frame 261 as indicated by a solid arrow; the best agent evolved in the driving environment recognised the imminent collision at frame 267 as indicated by a solid arrow. The thresholds are indicated with red dotted lines. A collision was detected when five successive spikes occurred, as indicated by five elevated asterisks in the bottom trace.

The success of the best agents relied on the responses of the direction selective neurons. When visual movement was extremely fast and beyond the direction selective neurons ability to respond selectively, the best agents would not be capable of a correct response. For example, the four direction selective neurons all responded similarly to the rightward rapidly translating ball (Figure 17, (a)) when the angular speed of the ball was extremely high (250°/s in this case); the best agents evolved in the robotic laboratory made an incorrect decision (Figure 20, right). Currently asymmetrical lateral inhibition only spread 8 cells away, therefore, the system could only cope with visual motion slower than 120°/s (with current input images and camera setting: 60° field of view, 100 pixels in horizontal and 25 frames per second). The ability of these direction selective neurons also relied on contrast in the input images. For the white ball on a white background (Figure 17, (b)), the four direction selective neurons did not respond to the moving white ball until after frame no.36 when contrast became available (Figure 20, left).

The two evolved best agents have also been challenged with collision scenes with a wide range of size, shape, texture and speed of colliding objects [Yue and Rind 2006c]. Although the size, shape texture and approach speed of the colliding objects differed, the robotic agent detected these imminent collisions several frames in advance. These tests taken together, have demonstrated the reliability of the evolved best agents to detect different imminent collisions in a familiar environment.

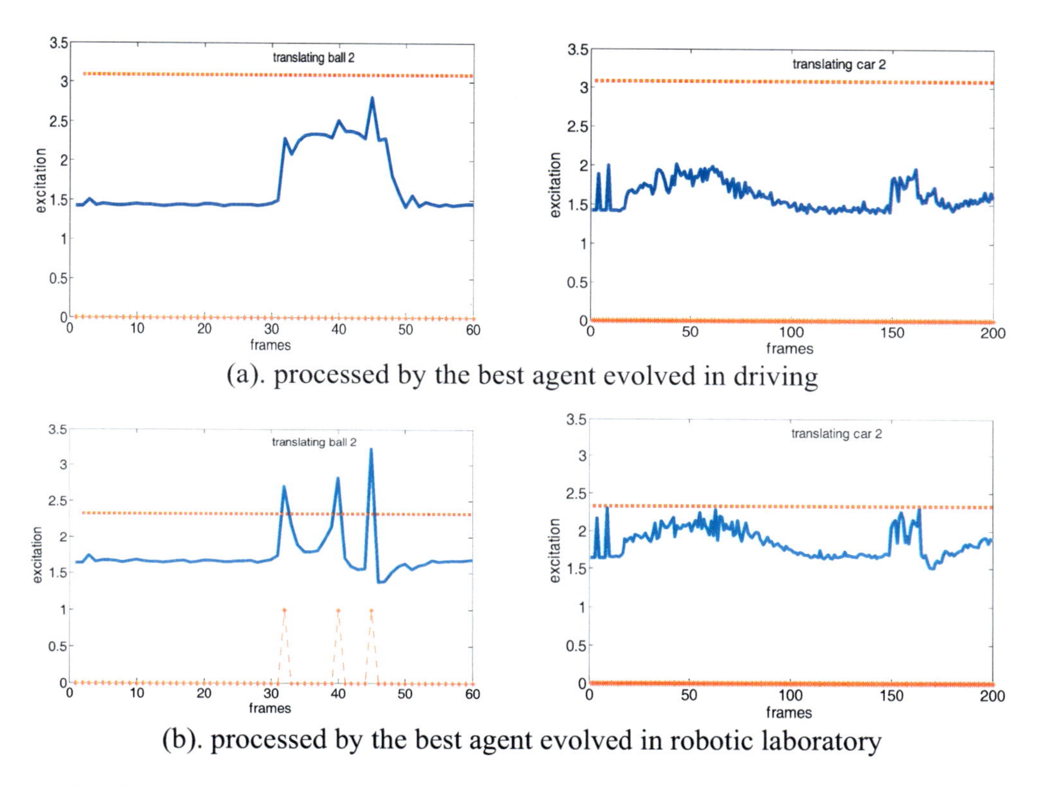

Figure 19. The excitation level of the two best agents when processing translating events in different environments, i.e., a translating ball (Figure 16 (a), video sequence 2) and a translating car (Figure 16(b), video sequence 2). The thresholds are indicated with red dotted lines. The spikes are indicated with elevated asterisks in the bottom trace. Both of the two best agents made correct decisions for the translating ball though patterns of the response excitation were quite different. The patterns of the response excitation of the two best agents for the translating car were similar when challenged with the translating car.

To understand the effect of different approaching speeds on the robotic agent's ability to detect a collision, we tested the agent with video clips in which the robot approached the textured white ball at different speeds. Results are shown in Figure 22. The robotic agent detected these collisions 1~7 frame before they occurred; no obvious correlation was observed between the approach speed and the number of frames before collision when collision was first detected. This means when a collision was detected, the distance from the camera to the approaching object was further if the approach speed is greater. Similar responses have also been observed in an LGMD model [Yue and Rind 2005, Yue and Rind 2006b].

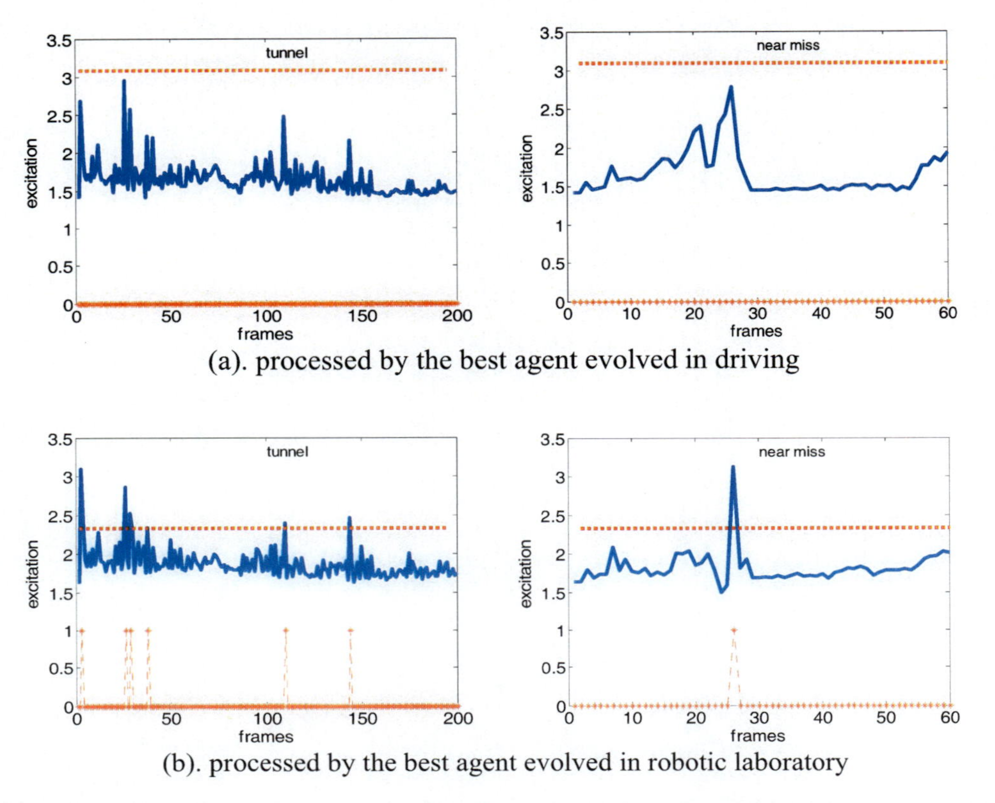

Figure 20. The excitation level of the two best agents when processing other looming but not colliding events in different environments, i.e., a looming ball near missing (Figure 16 (a), video sequence 5) and driving into a tunnel (Figure 16 (b), video sequence 6). The thresholds are indicated with red dotted lines. The spikes are indicated with elevated asterisks in the bottom trace. Both of the two best agents responded to the scenes of driving into the tunnel in similar excitation pattern, however one generated sparse spikes and the other did not produce any spike since they have different thresholds. For the ball on a near miss trajectory, both of the two best agents responded with sharp impulse-like excitation and made the correct decision.

A further test was carried out to see if its response was selective for approaching objects that were on a collision course. We challenged the robotic agent with a ball rolling towards it at 370 mm/s along an approximately 1,000 mm long track. The approach angle of the ball was systematically moved from a direct collision course to *5.7* degrees off a collision course, as shown in Figure 23 (a). For each approach angle the excitation levels of the final spiking cells in the best robotic agent were plotted as before (Figure 23 (b)). A line shows the time at which the image of the approaching ball began to move out of the robot's field of view. In the frame prior to this time excitation levels were lower as the deflection angles became larger although more trials would be required to quantify this effect. The robotic agent was able to correctly predict the collisions in the first two video clips. This experiment demonstrated the ability of the robotic agent to distinguish real collisions from near-misses.

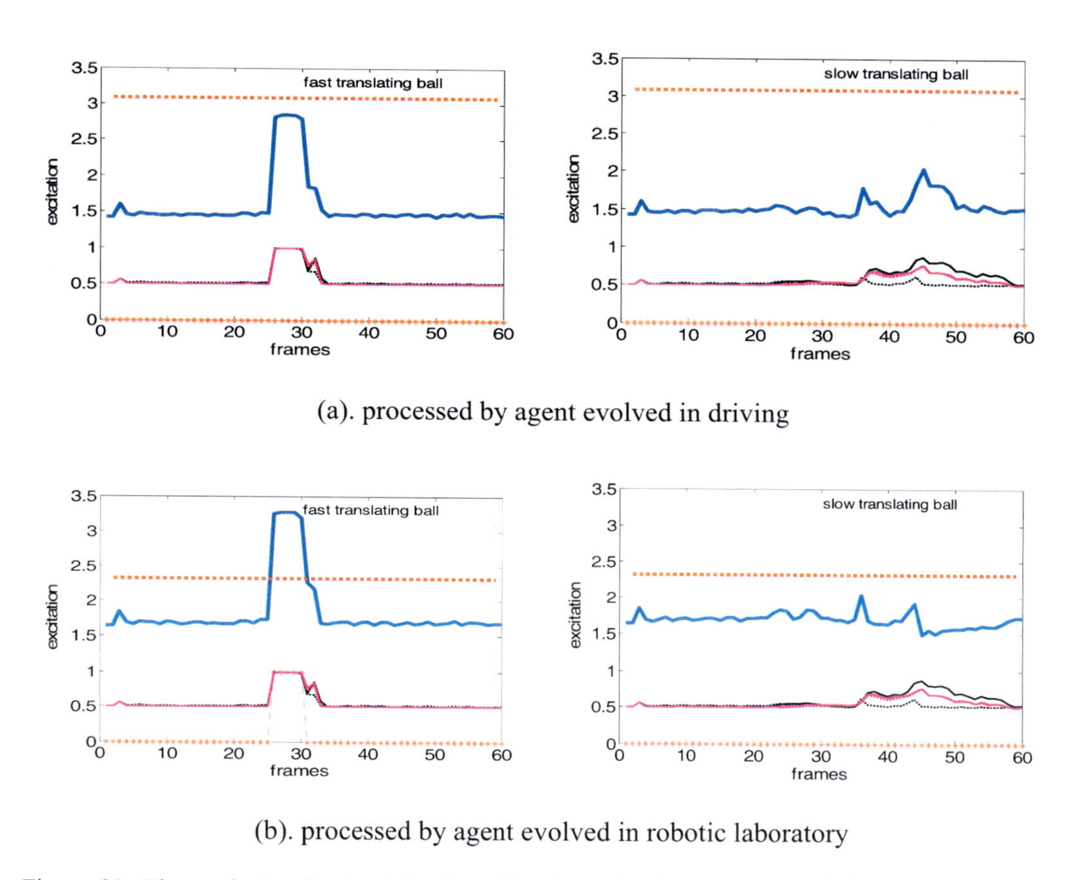

Figure 21. The excitation level of the four direction selective neurons and the two best agents in processing the two extraordinary visual scenes: a fast translating black ball (Figure 17, video sequence (a)) and a slow translating white ball on a white background (Figure 17, video sequence (b)). The thresholds are indicated with red dotted lines. The spikes are indicated with elevated asterisks. The excitation of the four direction selective neurons are indicated with solid and dashed lines fluctuating between (0.5~1.0). The responses of the four direction selective neurons to the fast translating ball were very similar. Even the best agent was not able to detect the visual motion patterns in this case. The current asymmetrical lateral inhibition only spread 8 cells away. Therefore, the system could only cope with visual motion slower than an angular velocity of 120°/s (with current input images and camera setting: 60° field of view angle, 100 pixels in horizontal and 25 frames per seconds). In the case of white ball on a white background, the four direction selective neurons did not respond to the moving white ball until frame no.36 when contrast became available.

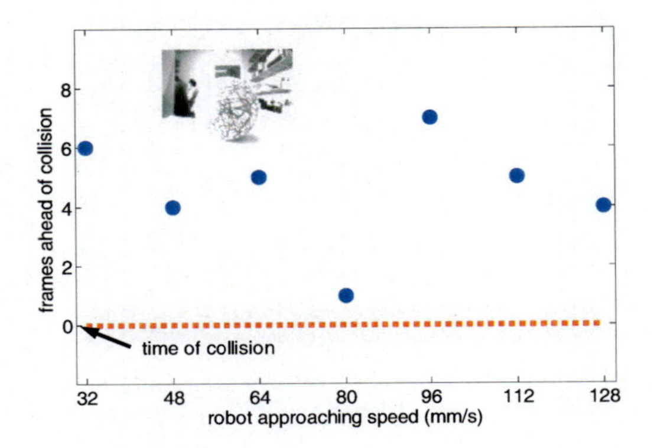

Figure 22. At different approach speeds, the evolved best robotic agent detected imminent collision 1~7 frames before it happened. The y-axis is the number of frames before collision when it was detected. The exact frame numbers before collision are indicated by filled circles. The ball used in the experiment is the same textured ball as described in Figure 5.2, Figure 5.3 and Figure 13. Time of collision for each case was the time when the robot touched the ball and was aligned to zero on the y-axis.

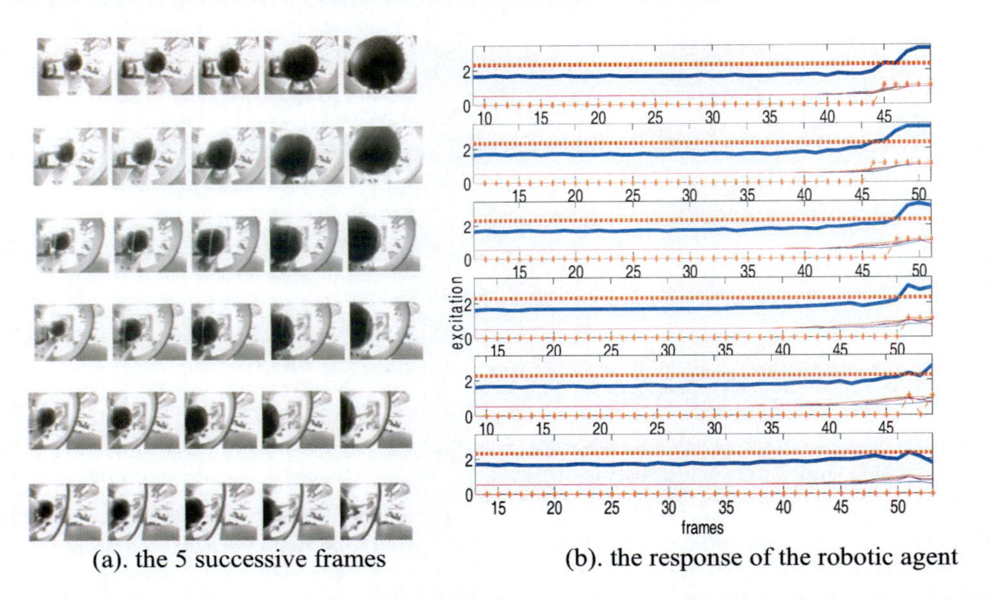

Figure 23. At different approach angles, the evolved best robotic agent responded to collision and non-collision scenes correctly. The ball used in the experiment is black 65mm in diameter. The ball was delivered on a 1000mm long 70mm high track and gained speed due to gravity. Two pieces of transparent monofilament at the lower end of the track stopped the ball from contacting the robot. The distance between the robot's body centre and the high end of the track was maintained at 1075mm in each recording. The centre of the image was fixed to the high end of the track and calibrated during the recording with a mini TV. In the 1st and 2nd video clips, the ball would hit the robot if the monofilament was not present; in the 3rd clips, the ball would graze one side of the robot; in the others, the ball is not on a collision course with the robot. From top to bottom, the deflection angle of the approaching ball became bigger each time. (a), the last five successive frames before/when the ball touched the strings are shown. A line marks the time at which the image of the approaching object moved out of the view of the robot (b), the corresponding responses from the best robotic agent; excitation is shown relative to the last frame at which the ball contacted the monofilament.

3.4. Further Discussions

From these results, we found that the synthetic vision system based on whole-field direction-selective neural networks could evolve to recognise imminent collision in simple or very complex environments. This study suggested that asymmetric lateral inhibition based whole field direction-selective neurons can be organised and adapted to play an important role in tasks such as collision detection particularly in complex dynamic scenes.

The overall organisation of these direction selective neurons in animals have not been fully understood although these neurons have been found in many different animal species [Barlow and Levick 1965, Barlow and Hill 1963, Borst and Haag 2002, Fried et. al. 2002, Hassenstein and Reichardt 1956, Livingstone 2005, Priebe and Ferster 2005, Rind and Bramwell 1996, Rind and Simmons 1992, Stasheff and Masland 2002, Vaney and Taylor 2002] and recent studies in fly shows that directionally selective motion detectors from each eye can interact via electrical synapses between the axons [Farrow et. al 2006]. Our research has provided an alternative way of investigating the roles of these neurons in animals. As shown and discussed above, these direction selective neurons could be easily organised to cope with very complex visual tasks. Our modelling results may suggest that interactions between these neurons could also occur in animals driven by complex visual tasks. On the other hand, the evolved best synthetic vision system could be a functional sensor for artificial intelligent robots/agents.

In the synthetic vision system, the visual motion patterns were extracted simultaneously from complex backgrounds by the early image processing neurons, the direction-selective neurons. Since these neurons were robust in extracting visual cues [Yue and Rind 2006b], the whole vision system was reliable. The synthetic vision system could process video sequences at 30 frames per second in a 2.80 GH Dell INSPIRON 5150 computer in the experiments. This system could therefore be used in robots for real time autonomous navigation in complex environments, where conventional robotic vision techniques are difficult to apply [DeSouza and Kak 2005].

In the above experiments, the agents were challenged with recorded video images in which visual events were the same for every agent. It will be interesting to see how the synthetic vision system evolves if the agents were put into a real/virtual 3D environment and allowed to move around. Additionally, to cope with visual events in different environments, it may be necessary to extract additional visual motion cues, such as diagonal motion. In the future, other specialized structures may be integrated into the synthetic vision system so that the agents can interact with different environments efficiently.

4. Conclusion

In the above subchapters, two different types of bio-inspired visual neural networks have been presented. First a LGMD based neural network as a real time robust collision detector with edge enhancement, especially for collision detection in complex backgrounds. Experiments showed that the LGMD based neural network as described, worked reliably in the situations with complex backgrounds. Integrated with a robot, the collision detection system demonstrated its reliability in detecting imminent collisions with only visual input over a range of speeds of movement and in a variety of situations enabling the robot to avoid collision autonomously within these arenas.

The second were the the asymmetric inhibition based DSNNs which were organized by artificial evolution into a synthetic vision system for collision detection in simple or complex environments. The synthetic vision system had significant potential for adapting to different environments and recognized collision reliably. We showed that the evolved agents with the proposed vision system could recognize imminent collision reliably in their particular environments. Further systematic tests demonstrated the reliability of the best DSNN agent, evolved in a simple environment, in detecting different colliding objects. This study suggests that whole-field direction-selective neurons, with selectivity based on asymmetric lateral inhibition, can be organised into a synthetic vision system which can then be adapted to play an important role in collision detection in complex dynamic scenes.

Although research on specialized neurons in biological systems is still going on, the properties of these neurons have shown surprising computing efficiencies in processing images for relevant visual cues. Both of the neural networks operated successfully in real time using a PC. We hope more visual processing systems will be inspired by biological visual systems in animals in the future.

Acknowledgement

This work is supported by EU IST-2001-38097. We thank M. Soininen of Volvo Car Corporation for providing the driving video footage used in this chapter and our colleagues for their helpful discussions.

References

Adams, M. D. Sensor modelling, design and data processing for autonomous navigation. River Edge, NJ, World Scientific, 1998.

Barlow, H. B. & Hill, R. M. Selective sensitivity to direction of movement in ganglion cells of rabbit retina. *Science*, vol.139, 412–414, 1963.

Barlow, H.B. & Levick, W.R. Mechanism of directionally selective units in rabbits retina. *J. Physiol. (Lond.),* vol.178, 477-504, 1965.

Blanchard, M. Verschure, P. F. M. J. & Rind, F. C. Using a mobile robot to study locust collision avoidance responses. *Int. Journal of Neural Systems*, vol. 9, 405-410, 1999.

Blanchard, M. Rind F.C. & Verschure, P.F.M.J. Collision avoidance using a model of the locust LGMD neuron. *Robotics and Automonous Systems*, vol.30, 17-38. 2000.

Borst, A. & Haag, J. Neural networks in the cockpit of the fly. *J. Comp. Physiology*, vol.188, 419-437, 2002.

Chipperfield, A.J. & Fleming, P.J. The Matlab genetic algorithm toolbox. *IEE Colloqium on Applied Control Techniques Using MATLAB*, Digest No.1995/014, 26, Jan,1995

DeSouza, G.N. & Kak, A.C. Vision for mobile robot navigation: a survey. *IEEE Transactions on Pattern Analysis and Machine Intelligence*, vol.24 (2), 237-67, 2002.

Duda, R.O. & Hart, P.E. Pattern classification and scene analysis. John Willy and Sons Inc. New York, 1973.

Everett, H.R. Sensors for Mobile Robots: Theory and Application. AK Peters, Wellesley, MA, 1995.

Farrow, K. Haag, J. & Borst, A. Nonliear, binocular interactions underlying flow field selectivity of a motion-sensitive neuron. *Nature Neuroscience*, vol.9, 1312-1320, 2006.

Fiala, M. & Basu, A. Robot navigation using panoramic tracking. *Pattern Recognition*, vol.37, 2195-2215, 2004.

Fried, S.I. Muench, T.A. & Werblin, F.S. Mechanisms and circuitry underlying direction selectivity in the retina. *Nature*, vol.420, 28 Nov. 2002, 411-414.

Gabbiani, F., Krapp, H.G., Hatsopoulos, N., Mo, C-H., Koch, C., & Laurent, G. Multiplication and stimulus invariance in a looming-sensitive neuron. J. Physiology-Paris, vol.98, 19-34, 2004.

Gabbiani, F., & Krapp, H.G., Spike-frequency adaptation and intrinisic properties of an identified, looming-sensitive neuron. J. Neurophysiology, vol.96(6), 2951-2962, 2006.

Goldenberg, D.E. Genetic Algorithms in search, optimization and machine learning. Addison-Wesley, Reading, Mass., 1989

Harrison, R.R., & Koch, C. A silicon implementation of the fly's optomotor control system. *Neural Computation*, vol.12, 2291-2304, 2000.

Hassenstein, B. & Reichardt, W. Systemtheorische analyse der Zeit-, Reihenfolgen- und Vorzeichenauswertung bei der Bewegungsperzeption des Rüsselkäfers Chlorophanus. *Zeitschrift für Naturforschung*, vol.11b, 513–524, 1956.

Holland, J.H. Adaptation in natural and artificial systems. Ann Arbor: The University of Michigan Press. 1975

Horridge, G.A. What can engineers learn from insect vision? *Phil. Trans. R. Soc. Lond.* 1992, B, vol.337, 271-282.

Huber, S.A. Franz M.O. & Buelthoff, H.H. On robots and flies: modelling the visual orientating behaviour of flies. *Robotics and Autonomous Systems*, vol.29, 227-242, 1999.

Iida, F. Biologically Inspired Visual Odometer for Navigation of a Flying Robot. *Robotics and Autonomous Systems*, vol.44/3-4, 201-208, 2003.

Indiveri, G. & Douglas, R. Neuromorphic vision sensors. *Science*, vol.288, 1189-1190, 2000.

Livingstone, M.S. Direction inhibition: a new slant on an old question. *Neuron*, vol.45, 5-7, 2005.

Manduchi, R. Castano, A. Talukder A. & Matthies, L. Obstacle detection and terrain classification for autonomous off-road navigation. *Autonomous Robots*, vol.18, 81-102, 2005.

Marshall, J. A. Self-organizing neural networks for perception of visual motion. *Neural Networks*, vol.3, 45-74, 1990.

Olson, C.F., Matthies, L.H. Schoppers, M. & Maimone, M.W. Rover navigation using stereo ego-motion. *Robotics and Autonomous Systems*, vol.43, 215-229, 2003.

O'Shea, M. Rowell, C.H.F., Williams, J.L.D. The anatomy of a locust visual interneurone: The descending contralateral movement detector. *Journal of Exp. Biology*, vol.60, 1–12, 1974.

Priebe, N.J. & Ferster, D. Direction selectivity of excitation and inhibition in simple cells of the cat primary visual cortex. *Neuron*, vol.45, 133-145, 2005.

Rind, F.C. A chemical synapse between two motion detecting neurones in the locust brain. J. Exp. Biol., vol.110, 143-167, 1984.

Rind, F.C. A directionly selective motion-detecting neurone in the brain of the locust: physiological and morphological characterization. *J. Exp. Biol.,* vol.149, 1-19, 1990a.

Rind, F.C. Identification of directionly selective motion-detecting neurones in the locust lobula and their synaptic connections with an identified descending neurone. *J. Exp. Biol.* Vol.149, 21-43, 1990b.

Rind, F.C. & Bramwell, D.I. Neural network based on the input organization of an identified neuron signaling impending collision. *Journal of Neurophysiology*, vol.75, 967– 985, 1996.

Rind, F.C. Simmons, P.J. Orthopteran DCMD neuron: A reevaluation of responses to moving objects. I. Selective responses to approaching objects. *Journal of Neurophysiology*, vol.68, 1654–1666, 1992.

Rind, F.C. & Simmons, P.J. Seeing what is coming: Building collision sensitive neurons. *Trends in Neurosciences*, vol.22, 215-220, 1999.

Rind, F.C. Motion detectors in the locust visual system: from biology to robot sensors. *Microscopy Research and Technique*, vol.56, 256-269, 2002.

Rind, F.C. Santer, R.D.J., Blanchard, M. & Verschure, P.F.M.J. Locust's looming detectors for robot sensors. *Sensors and Sensing in Biology and Engineering*, FG Barth, JAC Humphrey, and TW Secomb (Eds.), Spinger-Verlag, Wien, New York, 2003.

Rind, F.C., Stafford, R. & Yue, S. Technical Report D11: Biological Model Report, Project IST-2001-38097, LOCUST: Life-like object detection for collision avoidance using spatiotemporal image processing, 2004. http://www.imse.cnm.es/locust/main.html

Rind, F.C. Bioinspired sensors: from insect eyes to robot vision. *Frontiers in Neuroscience: Methods in Insect Sensory Neuroscience*, Christensen T.A. (Eds.), CRC Press Boca Raton, London, New York, 2005

Rowell, C.H.F. O'Shea, M. Williams, J.L. The neuronal basis of a sensory analyser, the acridid movement detector system .IV. The preference for small field stimuli. *J. Experimental Biology*, vol.68, 157-185, 1977.

Ruichek, Y. Multilevel- and neural-network-based stereo-matching method for real-time obstacle detection using linear cameras. *IEEE Transactions on Intelligent Transportation Systems*, vol.6 (1), 2005, 54-62.

Santer, R. D., Stafford R. & Rind, F. C. Retinally-Generated Saccadic Suppression of a Locust Looming Detector Neuron: Investigations Using a Robot Locust. *J.R. Soc. Lond. Interface*, vol.1, 61-77, 2004.

Santer,R.D., Simmons,P.J. & Rind, F.C. Gliding behaviour elicited by lateral looming stimuli in flying locusts. *Journal of Comparative Physiology*, vol.191, 61-73, 2005.

Schlotterer, G.R. Response of the locust descending contralateral movement detector neuron to rapidly approaching and withdrawing visual stimuli. *Canadian Journal of Zoology*, vol.55, 1372–1376, 1977.

Simmons, P.J., Rind, F.C. Orthopteran DCMD neuron: A reevaluation of responses to moving objects. II. Critical cues for detecting approaching objects. *Journal of Neurophysiology*, vol.68, 1667–1682, 1992.

Stafford, R., Santer R.D. & Rind, F.C. A bio-inspired visual collision detection mechanism for cars: combining insect inspired neurons to create a robust system. *Bio Systems: Sixth International Workshop on Information Processing in Cells and Tissues (IPCT2005)* (in press)

Stasheff, S.F. & Masland, R.H. Functional inhibition in direction-selective retinal ganglion cells: spatiotemporal extent and intralaminar interactions. *J. Neurophysiology*, vol.88, 1026-1039, 2002.

Tversky, T. & Miikkulainen, R. Modeling direction selectivity using self-organizing delay-adaptation maps. *Neurocomputing*, vol.44-46, 679-684, 2002.

Vaney, D.I. & Taylor, W.R. Direction selectivity in the retina. *Current Opinion in Neurobiology,* vol.12, 405-410, 2002.

Vahidi, A. & Eskandarian, A. Research advances in intelligent collision avoidance and adaptive cruise control. *IEEE Transactions on Intelligent Transportation Systems*, vol.4(3), 143-153, 2003.

Webb, B. & Reeve, R. Reafferent or redundant: integration of phonotaxis and optomotor behaviour in crickets and robots. *Adaptive behaviour*, vol.11 (3), 137-158. 2003.

Wichert,G. Can robots learn to see. *Control Engineering Practice*, vol.7, 783-795, 1999.

Yue, S. & Rind, F.C. A Collision detection system for a mobile robot inspired by locust visual system. *IEEE Int. Conf. on Robotics and Automation*, Spain, Barcelona, Apr.18-21, 2005, 3843-3848.

Yue, S., Rind, F.C. Keil, M.S., Cuadri, J. & Stafford, R. A bio-inspired visual collision detection mechanism for cars: optimisation of a model of a locust neuron to a novel environment. *Neurocomputing,* August 2006, vol.69 (13-15), 1591-1598.

Yue, S. & Rind, F.C. Collision detection in complex dynamic scenes using a LGMD based visual neural network with feature enhancement. *IEEE Transactions on Neural Networks*, May, 2006a, vol.17(3), 705-716.

Yue, S. & Rind, F.C. Visual motion pattern extraction and fusion for collision detection in complex dynamic scenes. *Computer Vision and Image Understanding*, 2006b, vol.104 (1), 48-60.

Yue, S. & Rind, F.C. A synthetic vision system using directionally selective motion detectors to recognize collision. *Artificial Life*, 2006c, (in press).

In: Pattern Recognition in Biology
Editor: Marsha S. Corrigan, pp. 63-103

ISBN 978-1-60021-716-6

Chapter 3

ADVANCES AND CHALLENGES IN 3D AND 2D+3D HUMAN FACE RECOGNITION

Shalini Gupta[1]***, Mia. K. Markey***[2] ***and Alan C. Bovik***[1]
[1]Department of Electrical and Computer Engineering,
The University of Texas at Austin, TX 78712, USA
[2] The University of Texas Department of Biomedical Engineering,
Austin, TX 78712, USA

Abstract

Automated human face recognition is required in numerous applications. While considerable progress has been made in color/two dimensional (2D) face recognition, three dimensional (3D) face recognition technology is much less developed. 3D face recognition approaches based on the appearance of range images and geometric properties of the facial surface have been proposed. Methods that combine 2D and 3D modalities also exist. These innovations have advanced the field and have created novel areas of investigation. The purpose of this chapter is to provide a summary and critical analysis of the progress in 3D and 2D+3D face recognition. The chapter also identifies open problems and directions for future work in the area.

1. Introduction

Robust and accurate identification of humans is required for numerous tasks. Over the years, a number of scientific approaches have been investigated to identify individuals. A widely explored approach is biometrics, the measurement of anatomical, physiological, and behavioral characteristics believed to be unique to individuals. The measurement of anatomical attributes of human beings is also called anthropometry. Phillips *et al.* outlined the characteristics of an ideal biometric based identification system: a) all members of the population should posses the biometric; b) the biometric signature of each individual should differ from that of other individuals in a controlled population; c) the biometric signature of each individual should not vary under the conditions in which it is recorded; and d) the system should resist countermeasures [87].

The first truly scientific approach to establish an individual's identity based on anthropometry was the Bertillon system introduced in France in the nineteenth century [95]. In this system, anthropometric measurements, *e.g.*, the skull width, foot length, cubit, trunk, along with hair color, eye color, and front and side view photographs were recorded. Bertillonage, as it was called, remained in vogue in police departments around the world for a number of years during the latter part of the nineteenth century. It was eventually abandoned due to questions regarding its infallibility as an accurate identification system, and was replaced by more reliable systems based on fingerprints. Fingerprint matching was considered the most reliable method for person identification for a large part of the twentieth century, and was later augmented with DNA matching.

With the availability of computers, a natural step forward has been to automate the task of human identification. In the recent years, due to the availability of computers with greater speeds and memory, automated biometric systems have become a topic of considerable interest. They have numerous applications including secured access to ATM machines and buildings, automatic surveillance, forensic analysis, retrieval of images from mugshot databases in police departments, automatic identification of patients in hospitals, checking for fraud or identity theft, and human computer interaction. Automated biometric systems expedite processing of large volumes of data, and in some cases can work in realtime. Currently, personal identification numbers, access codes/cards, bar codes, and radio frequency ID tags are popularly employed for identification. However, these are susceptible to loss or theft. Identification numbers and access codes/cards also require substantial user involvement. Hence, they are of limited utility for identifying very young children and seriously ill persons.

Many biometric techniques have been explored for automated human identification including face, iris, retina, fingerprint, palmprint, hand, gait, voice, and handwriting recognition. Of these, iris and fingerprint systems are reported to be highly accurate [93], but they require substantial subject cooperation. They are difficult to deploy in realtime screening and surveillance applications, where minimal user cooperation is desired, or where the system is to be operated covertly. Face recognition as a biometric modality requires less subject cooperation, is amenable to surveillance applications, and can be developed using relatively low cost components.

Although human beings are highly skilled at recognizing faces, there are also deficiencies in the face recognition abilities of humans. For example, a study of DNA exonerations reported that 84% of wrongful convictions were due in part to false recognition by eyewitnesses or victims [99]. Researchers believe that the face cognition abilities of humans are influenced by cross-racial effects [68] and other biases. Furthermore, many psychological aspects of human face processing and cognition are not well understood.

As another example in the criminal justice domain, consider the construction of a lineup, in which eyewitnesses or victims are presented with six face images to inspect. One of these is of the suspect. It is recommended that besides the suspect's facial image, the other images be of individuals close in appearance to the suspect [116]. In order to automatically construct effective lineups, it is necessary to define quantitative measures of similarity of facial appearance. Such measures can also be useful for content-based image retrieval from facial databases. Researchers in the cognitive sciences emphasize the need to define objective, quantitative measures of similarity of human faces [92]. They believe that insights gained from the field of computer vision could aid in enhancing the understanding of the mechanisms driving human face processing and cognition.

Considerable research attention has been directed, over the past two decades, towards developing reliable automatic face recognition systems. For the most part, research efforts have concentrated primarily on intensity/color/two dimensional (2D) images, and commercial systems are now available for this task [115].

Two dimensional face recognition systems are easy to construct with relatively cheap off-the-shelf components, but they are inadequate for robust face recognition. The Face Recognition Vendors Test was conducted in the year 2002 (FRVT 2002) to establish performance metrics for fully automatic 2D face recognition algorithms [91]. It was reported that the performance of the three best algorithms dropped nearly in half for facial images captured under varying ambient illumination conditions, or varying facial poses. Synthetic 2D frontal face images generated by employing three dimensional (3D) morphable models [13], greatly improved recognition results for faces with large pose variations. Hence at the FRVT 2002, using 3D face models for pose correction was identified as a potential solution to the pose problem in face recognition.

Three dimensional face recognition technology has emerged in the recent years, in part, due to the availability of improved 3D image acquisition devices and processing algorithms. For 3D face recognition algorithms, 3D facial models are employed either by themselves (3D algorithms) or in conjunction with 2D facial images (2D+3D algorithms).

In this chapter, we present a comprehensive, up-to-date literature review of the existing 3D and 3D+2D face recognition techniques. We focus primarily on recognition techniques developed for facial models captured in realtime using 3D acquisition devices. While, techniques for 3D and 2D+3D face recognition have also been reviewed previously [3, 15, 100], we present a more comprehensive and up-to-date literature survey.

We first outline the broad categories of tasks that are performed by automatic recognition systems and quantitative measures to assess their performance. We then discuss the main 2D face recognition algorithms that have influenced the development of analogous techniques for 3D face recognition, and have been employed in numerous 2D+3D approaches. A detailed discussion of existing approaches for 3D and 2D+3D face recognition follows. We conclude by enumerating open problems in area and identifying potential directions for future work.

2. Face Recognition Tasks

The two main tasks performed by any automatic human recognition system are verification/authentication and identification [88]. These are discussed in the following sections.

2.1. Verification

Verification/authentication is a one-to-one matching task wherein a person claims to be a specific entity known to the system (Figure 1). The database of people known to the system is referred to as the 'gallery'. The individual whose identity is verified/authenticated by the system is referred to as the 'probe'. The facial representation of the probe is compared against the gallery representation of the claimed entity. If the similarity score between the two is greater than a predefined threshold, the individual is identified as the claimed entity; otherwise, he or she is rejected as an imposter. An example of a verification scenario is

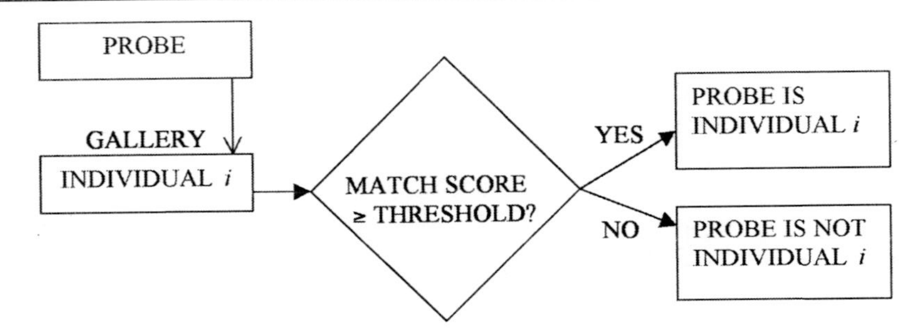

Figure 1. A schematic diagram of an automatic biometric verification system.

automated secure access to a building.

The performance of verification systems is reported in terms of a receiver operating characteristic (ROC) curve [94]. A ROC curve is a plot of the relationship between the false acceptance rate (FAR) and the false rejection rate (FRR). The verification scenario really can be thought of as a two class problem where match scores are either classified as intra-intity scores or inter-entity scores. FAR is defined as the proportion of comparisons between two different individuals that are falsely accepted by the system [101]. FRR is the proportion of comparisons between two instances of the same individual that are falsely rejected by the system [101]. Both FAR and FRR vary as the system's decision threshold is varied [33]. A single performance metric typically reported for verification systems is the equal error rate (EER), where FAR = FRR. For an ideal system EER=0%. The area under the ROC curve is also sometimes reported as a measure of performance. This area ranges from zero for an ideal system to 0.5 for a system with chance performance. Figure 2 presents a typical ROC curve. Note that different, but equivalent, formulations of ROC curves are used in evaluating classifications systems in other disciplines (*e.g.*, in medical imaging [71]).

ROC curves are also closely related to precision-recall curves frequently employed for evaluating the performance of general information retrieval systems [31]. Recall is equal to the true positive fraction, which measures the fraction of intra-entity scores that are correctly labeled by the system. Precision on the other hand measures the fraction of match scores that are labeled by the system as intra-entity scores and are truly so. Precision-recall curves are generally regarded superior than ROC curves for evaluating the performance of two-class decision systems with highly skewed data sets. Data sets employed to evaluate face verification systems are also highly skewed, as the number of intra-entity match scores is usually much less than the number of inter-entity match scores. Yet, currently only ROC curves are employed to evaluate their performance. Hence, investigations into the utility of precision-recall curves for the task are warranted.

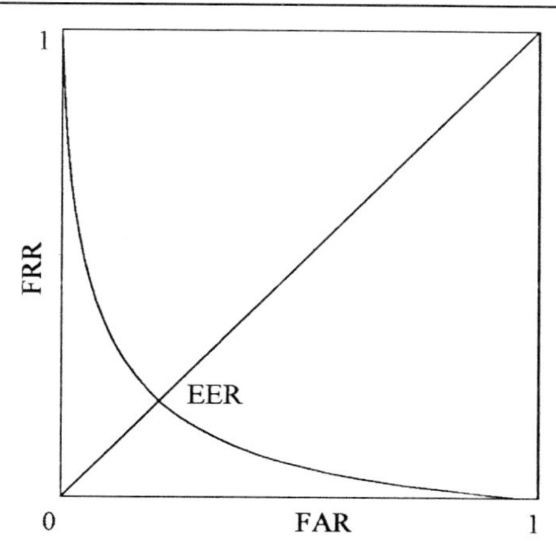

Figure 2. A Receiver Operating Characteristic Curve.

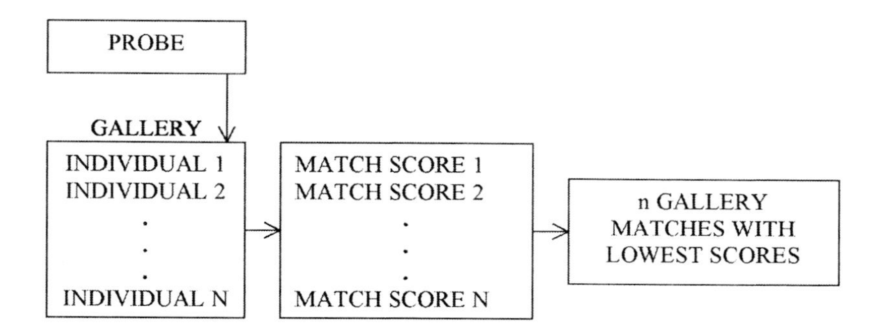

Figure 3. A schematic diagram of an automatic biometric identification system.

2.2. Identification

Identification is a one-to-many matching task wherein an unknown individual's identity is established by comparing his/her biometric against a database of known individuals. The closest matches in the gallery are found. For example a person may be identified by comparing against a database of mugshots in a police department. A generalization of the identification task is the watchlist task. In a watchlist task, each probe is compared against signatures of all entities known to the system and entities resulting in the highest n similarity scores that are also above a predefined threshold value are considered matches. Setting a predefined threshold value seems to be is an arbitrary protection against selecting gallery individuals among the top n matches that are not close in appearance to the probe.

The performance of an identification system can be evaluated in terms of a cumulative match characteristic (CMC) curve [88, 89]. This formulation assumes a 'closed universe' model where all individuals that query the system are present in the gallery. The CMC curve is a plot of the recognition rate (RR) versus the top n database matches considered. It is the ratio of the number of probes for which the correct gallery match is present among the top n matches to the total number of probes that query the system. If the closed universe assumption is false, *i.e.*, a probe is not present in the gallery, then the maximum RR achieved is less than 100%.

Verification, identification, and watchlist tasks present different design challenges. For example, imposters attempting to fool a verification access control system would disguise themselves as someone known to the system. Thus, it is important that a high security access control system have a low FAR so as not to falsely allow access to an imposter. By comparison, persons attempting to evade identification merely need to disguise themselves as anyone other than themselves. Hence, the FRR of, *e.g.*, airport screening systems should be low, so that individuals with disguises are correctly identified. Hence, while designing biometric systems it is necessary to keep in mind the final intended application. A system optimized for one application may not perform acceptably for another.

3. 2D Face Recognition Algorithms

2D face recognition technology has developed considerably over the past two decades. Broadly speaking, the two main approaches employed for 2D face recognition are: (a) based on the appearance of the whole face or (b) based on local facial features/geometric templates. Comprehensive survey papers detailing the progress of 2D face recognition algorithms have been written [26, 124].

Over the last decade, a number of independent tests have also been administered to compare the performance of 2D face recognition algorithms. These include a series of Facial Recognition Technology (FERET) tests [89, 90], and the Face Recognition Vendors Tests [118]. In such evaluations, a few 2D face recognition algorithms have consistently demonstrated superior performance. These include algorithms based on principal component analysis (PCA), linear discriminant analysis (LDA), local feature analysis (LFA), and elastic bunch graph matching (EBGM).

Commercials systems based on mature 2D face recognition algorithms include those by Cognitec Systems GmbH, Dresden, Germany; Identix®, Minnetonka, MN; and Eyematic Interfaces Inc., Inglewood, California now called Neven Vision, Santa Monica, California. These systems were among the top three performers at the FRVT 2002. The system by Identix® was based on LFA, and that by Eyematic Interfaces Inc./Neven Vision was based on EBGM. All these systems performed well for frontal or nearly frontal 2D face images captured indoors, but their performance decreased with variable illumination conditions and head rotations.

These top ranking 2D face recognition algorithms have also inspired similar algorithms for 3D face recognition. Furthermore, many of these 2D techniques have also been combined with 3D techniques to form 2D+3D face recognition algorithms. While, the focus of this paper is not 2D face recognition, the significant ones are discussed here since they form the basis for many 3D and 2D+3D strategies.

3.1. Holistic Appearance Based Techniques

Among the holistic 2D face recognition techniques, based on the appearance of facial grayscale/color images, are a number of subspace projection methods. These include algorithms that employ principal component analysis (PCA), and linear discriminant analysis

(LDA). In these methods, a facial image is regarded as an instance in N dimensional feature space, where N is the number of pixels in the image. All human face images are modeled to lie on a linear/non-linear subspace of the N dimensional feature space. Statistical techniques are employed to learn the subspace using an ensemble of facial images. All facial images are projected onto the subspace before classification.

Holistic appearance based methods generally use information from the entire image. They do not employ high level knowledge about human faces to segment parts of the face. Hence, they tend to be less robust to outliers, cluttered backgrounds, and occlusions. Furthermore in order to reliably learn the facial subspace, they require a large number of training images of many subjects under diverse imaging conditions. Two of the important holistic 2D face recognition techniques, PCA and LDA, are discussed here.

3.1.1. Principal Component Analysis

PCA or eigenfaces was one of the first successful techniques developed for 2D face recognition [51, 110]. In this technique, a set of orthogonal basis vectors, that maximize the variance of facial image data, are obtained by eigen decomposition of the scatter matrix of facial images [32]. The eigenvectors are referred to as eigenfaces. PCA results in compact representations of high dimensional data, which are optimal in the mean squared sense. That is, PCA minimizes the mean squared error between the original image, and the corresponding image reconstructed from the eigen directions. If an image is reconstructed from all the eigen directions, the mean squared error between the original image and its reconstructed version is zero.

For face recognition, the top M ($M \leq N$) eigen vectors, which account for most of the variation of the data, are retained. All gallery and probe facial images are projected onto the M directions. Faces in the transformed space are compared by means of a suitable distance metric. PCA is advantageous in that it possesses a closed form solution. It has resulted in effective 2D face recognition algorithms and is regarded as a benchmark, against which many others are compared [89]. Despite its success with 2D face recognition, it is not intuitively obvious as to what discriminatory information about human faces is encoded in the different eigen directions. Furthermore, its performance degrades with facial variations including expression, pose and illumination changes.

3.1.2. Linear Discriminant Analysis

Another successful approach for 2D face recognition is based on Fisher's linear discriminant analysis [5, 123]. This technique is also called fisherfaces. In this technique, high dimensional data are linearly projected onto $C - 1$ LDA directions, where C is the number of classes, such that the ratio of the between class scatter to the within class scatter is maximized. LDA can successfully discriminate between linearly separable classes [34]. For face recognition, the dimensionality of the data is first reduced via PCA or some other technique, before applying LDA. This is done to ensure that the within class scatter matrix, which is involved in the LDA calculations, is non-singular. Better results for 2D face recognition have been reported with LDA than PCA [5]. This could be explained by the fact while LDA projects data onto novel directions, such that the classes are most separated, PCA is not specifically tailored towards classification problems.

3.2. Local Feature Based Techniques

Two dimensional face recognition techniques based on local facial features/geometric templates, employ characteristics of localized regions of the face as features for recognition. Such techniques require an additional step of automatically locating specific parts of the face using flexible geometric templates or intensity/texture characteristics of specific facial features. The performance of feature based approaches depends on the accurate localization of facial landmarks. Segmentation of facial features is a non-trivial task. Segmentation techniques need to be specific enough to locate only the desired facial features, yet general enough to do so for a diverse variety of facial images. These are contrasting goals and thus it is difficult to achieve both.

If facial features can be reliably segmented, effective face recognition techniques based on local facial features can be developed. They are likely to be more robust to changes in facial pose, expressions and illumination, holes, occlusions, and the presence of noise than holistic appearance based approaches. This is because some facial features derived from localized facial regions would remain unchanged on varying other conditions. Furthermore, the information encoded in face recognition techniques based on local facial features is easier to interpret. For example, the discrimination ability of various sub-parts of the face can be evaluated independently of others.

In this section we discuss two successful 2D face recognition techniques that are based on local facial properties, namely LFA and EBGM.

3.2.1. Local Feature Analysis

It is argued that PCA does not exploit the inherent correlations and redundancies between neighboring pixels of an image. Eigen directions are also not topographical in that it is not understood how the various eigenfaces relate to each other. In order to overcome some of these deficiencies of PCA, the local feature analysis approach to face recognition, was developed [84]. Being a proprietary software of Identix®, Minnetonka, MN, the exact details of how this technique is applied to face recognition are not available. It is known however, that LFA is local in the sense that LFA kernels capture variations in sub-regions of the face. However, like PCA and LDA, it is applied directly to whole face regions without first segmenting different parts of the face. LFA kernels are statistically derived local and topographical sparse representations of the face. Such representations can provide information about the discrimination ability of the different parts of the face. LFA is reported to perform well with 2D frontal faces captured with constant illumination [91].

3.2.2. Elastic Bunch Graph Matching

Wiskott *et al.* developed another successful technique for 2D face recognition called elastic bunch graph matching [117]. EBGM is based on Gabor filter coefficients [14] extracted from specific facial fiducial points. In this technique, fiducial points are manually located on gallery images. From each point, a 'jet', comprising of forty coefficients for Gabor filters at 5 spatial frequencies and 8 orientations, is extracted. Jets from all gallery images are concatenated to form a data structure called a 'bunch'. A flexible 'face graph' is also constructed by connecting fiducial points by straight lines. The face graph and the bunch together form a data structure called an 'elastic bunch graph' (EBG). EBGs are employed for both automatic localization of fiducial points on probe faces, as well as for facial recognition. It is reported that EBGM works well for varying facial expressions and illumination conditions. However, it performs poorly for faces with large pose variations.

4. 3D Facial Models

In this section we describe the techniques used to acquire 3D facial models in realtime and also data structures employed to represent them for face 3D face recognition algorithms.

4.1. 3D Facial Model Acquisition

3D facial models are acquired using both active and passive techniques. Besl provides a summary of the different 3D imaging techniques [7]. The most widely employed active 3D acquisition technique is based on laser range finders [23,37,44,59,60,65,72,73,85,96,102]. A range finder projects light from a laser source onto a scene and records its reflection. The depth of the surface closest to the camera is determined by triangulation. Laser range finders produce dense and accurate 3D models, but require longer acquisition times than passive techniques. Furthermore, they require that the human subject be perfectly still during image capture [69]. Thus, laser range finders are unsuitable for high throughput screening applications. Another concern is the intrusive nature of laser light for human eyes.

Passive techniques employed for 3D facial image acquisition include stereo imaging [41–43, 55, 55, 70, 81, 122], and approaches based on structured light [2, 6, 10, 16, 36, 67, 78, 107, 119]. In stereo imaging systems, multiple cameras simultaneously capture a face from different view points. Depth information is resolved using camera calibration parameters and disparity information which is obtained from the different view points. To register points in the different images, a random light pattern is sometimes projected onto the scene [81].

In the structured light approach, a standard light pattern, *e.g.*, a light stripe pattern, is projected onto the scene [10]. Deformation of the light pattern and camera calibration parameters are employed to resolve the depth at each point in the scene by a process of triangulation. While passive techniques are faster, safer, and cheaper than laser range finders, they are typically less accurate and contain more missing data.

Attempts have also been made to recover the 3D shape of faces from one or more 2D images by morphing generic 3D facial models [13, 53, 57, 74, 75, 121]. Many such techniques involve variants of the 'shape from X' algorithms, but the recovery of 3D shape from a single texture image is an ill-posed problem. Such techniques can be useful for generating 3D facial models of subjects whose 2D images are available but their 3D images cannot be captured in realtime.

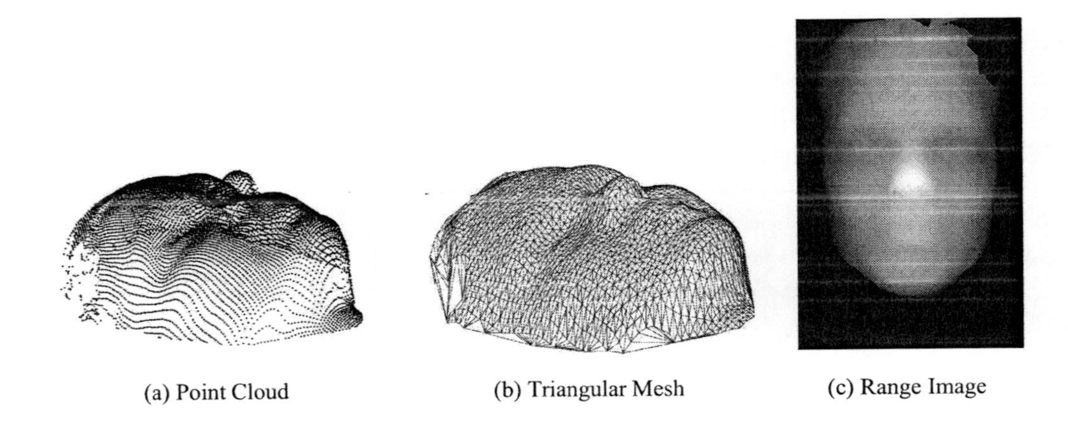

(a) Point Cloud (b) Triangular Mesh (c) Range Image

Figure 4. This figure shows the different 3D representations of a facial surface. In Fig 4(a), the face is represented as a 3D point cloud. In Fig. 4(b) the 3D points are joined to form a triangular mesh representation. In Fig 4(c) the 3D face is represented as a range image where each value in the matrix corresponds to the depth at that point.

4.2. Facial Surface Representation

Point clouds, triangulated surface meshes, or range images are employed to represent facial surfaces. The point cloud representation contains the (x, y, z) coordinates of a set of points on the facial surface (Figure 4(a)). These points can be connected to their nearest neighbors via straight lines resulting in a triangulated mesh representation (Figure 4(b)). Compact point cloud or surface mesh representations can be obtained by sampling surfaces densely in regions containing detailed information, and sparsely in relatively smooth regions. The 3D points in these representations are usually unstructured and thus they require relatively involved algorithms for processing.

A range image, also referred to as a 2.5D surface or depth map, consists of (x, y) points on a regular rectangular grid. Each (x, y) point is associated with a z value of the point on the surface closest to the acquisition device (Figure 4(c)). Range images can be produced by orthographic projection of surface meshes or 3D point clouds. Three dimensional acquisition devices that capture range images directly are also available. As points in a range image are placed along a regular rectangular grid, they can be processed via relatively straightforward image processing algorithms.

5. 3D Face Recognition Algorithms

In recent years, numerous 3D face recognition techniques have been developed. In some respects, 3D face recognition techniques are advantageous relative to 2D techniques. The pose of 3D face models can be relatively easily corrected by rigid rotation and translation. This substantially alleviates the pose problem. Three dimensional models are normally in real-world dimensions and hence do not need to be rescaled prior to processing. Theoretically speaking, the shape of a face is independent of external factors such as ambient lighting during image acquisition. Knowledge of the shape of a face, and the direction and intensity of ambient illumination during capture, enables the calculation of facial surface reflectance properties. Surface reflectance is an intrinsic property of the facial surface. It can be employed to synthetically generate images of the face under different illumination conditions, which also helps to alleviate the facial illumination problem.

Concerns have been raised with regards to the illumination invariant nature of 3D face recognition algorithms [15]. It is argued that since intensity images are employed to construct 3D models in passive 3D image acquisition techniques, variations in illumination can alter the shape of the model constructed. However, in a recent study it was reported that the performance of 3D face recognition algorithms did not change significantly when 3D models were acquired under varying illumination conditions using a stereo imaging device [54]. It has also been observed that while the performance of 2D face recognition algorithms improves significantly when images are compensated for illumination variations, no significant improvement in performance of 2D+3D algorithms is observed [67, 107]. This evidence suggests that 3D face recognition algorithms may be effected less by changes in illumination conditions than 2D algorithms.

The two main categories of 3D face recognition techniques are: (a) based on the appearance of facial range images ('appearance based'), and (b) based on the geometric properties of 'free form' facial surfaces ("free form' based'). Campbell *et al.* provide an excellent overview of computer vision techniques for 3D object representation and recognition in general [18]. In the following sections, appearance (statistical learning) based and 'free form' based 3D face recognition approaches are discussed.

Although a number of 3D and 2D+3D face recognition algorithms have been proposed, most have been evaluated on small private data sets not accessible to other researchers. This makes it impossible to directly compare their performances. A few studies report the performance of multiple 3D face recognition algorithms on a common data set [6,10,12,36,41,43,45,49,50,80,83,102,119]. However, for most such studies, hypothesis testing procedures have not been conducted to confirm the statistical significance of the observed differences.

Three dimensional and 2D+3D face recognition studies also vary considerably with regards to certain aspects of their experimental design and evaluation protocol. For example, different research groups have employed different sized data sets. Hence, their results should be interpreted, bearing in mind that the performance of face recognition algorithms decreases on increasing the number of subjects in the database [91]. Furthermore, in different studies, different number of images of each person have been included in the gallery. It should be noted that increasing the number of images of each person in the database generally improves the performance of recognition algorithms [23].

There are also inconsistencies in the manner in which results have been reported by different researchers. For example, for the recognition performance, many research groups report only the rank n RR value instead of the rank 1 RR value. Similarly, for the verification performance, instead of reporting both the FAR and the FRR for a particular operating point, in many studies only one out of the two values has been reported. Hence, in order to assess and compare the performance of sate-of-the-art 3D and 2D+3D face recognition algorithms, it is necessary to test them on a large common data set using a fixed evaluation protocol. Results of statistical comparisons between the different algorithms should also be reported.

The Face Recognition Grand Challenge organized in the year 2005 (FRGC 2005) was a move in this direction [85]. As a part of the challenge, the performance of a few 3D and 2D+3D face recognition algorithms was evaluated using a standardized protocol on the two large data sets called FRGC v0.1 and FRGC v0.2 (Table 10). The performance of the 3D PCA algorithms was considered as the baseline [85, 86]. All other algorithms were compared against it. Some limitations of the FRGC data sets have also been noted. These include motion artifacts in range images, inconsistent expressions in the 2D and 3D images because of time delay between their capture, and occlusion problems due to facial hair [69].

5.1. 3D Appearance Based Techniques

Three dimensional face recognition techniques based on the appearance of facial range images are similar to 2D holistic appearance based techniques. The only difference being that they employ range images instead of intensity images. For the most part, they are straight forward extensions of techniques that have been successful with 2D facial images.

A number of preprocessing and normalization steps are usually required in these algorithms. Their purpose is to localize and segment the human head; remove spike noise and holes (regions of missing data); align heads to a canonical position; and to generate range images in that position. Three or more points on the face are manually or automatically located to determine the head pose. For most algorithms, the canonical position is the frontal pose with the tip of the nose located at the center of the image.

The appearance based methods that have been investigated for 3D face recognition include PCA, LDA, LFA, independent component analysis (ICA), hidden Markov models (HMM), and optimal component analysis (OCA) (Tables 1, 2, and 3). These are discussed in detail in the following sections.

5.1.1. Principal Component Analysis

PCA is the most widely explored appearance based technique for 3D face recognition (Tables 1, 2, and 10). The PCA algorithm is regarded as a baseline for evaluating the performance of other 3D face recognition algorithms [86]. It has also been applied to 3D facial detection to distinguish between face and non-face regions in range images [30]. For face recognition via 3D PCA, eigensurfaces, similar to 2D eigenfaces, are computed. The optimal spatial resolutions for 3D PCA are reported to be up to 0.98 mm per pixel along the x and y directions, and resolutions ranging from 0.5 to 3 mm per pixel along the z direction [20]. This implies that higher resolutions may not be required for 3D PCA.

Overall, rank one recognition rates ranging from 100% to 68% have been reported for 3D PCA for databases of varying sizes and complexity (Tables 1 and 2). Although earlier experiments with smaller databases resulted in high performance for 3D PCA, more realistic performance estimates can be had from recent studies with larger and more complex databases.

Numerous algorithms have been proposed in which PCA is applied to depth values of range images [2, 20, 21, 23, 24, 35, 44, 45, 108], to (x, y, z) values of 3D facial point clouds [13], to geometry preserving isometric sphere representations of facial surfaces [77], to horizontal gradients, vertical gradients and depth curvature representations [42]. It is not clear as to which of these representations is most effective for use with PCA. Heseltine *et al.* reported that PCA on horizontal gradients of range images was more effective than PCA applied to depth or curvature representations [42]. However, the authors did not report whether these differences were statistically significant. Pan *et al.* also reported improved results for the case when PCA was applied to isometric spherical representations of facial surfaces instead of applying it directly to range images [77]. They evaluated both algorithms on the FRGC 2005 v0.1 data set.

Furthermore, 3D PCA has been applied either to entire range images [2, 108] or only to regions of range images that contain the main facial features [13, 20, 21, 23, 24, 35, 44]. It is likely that eliminating regions of the range image other than the main facial features may be advantageous as it would remove undesirable noise from hairstyles, clothing, and cluttered backgrounds.

The performances of 2D grayscale PCA and 3D PCA have been compared in a number of studies [20, 21, 23, 24, 35, 108] (Table 2). However, it has not been conclusively established as to which of the two modalities results in superior performance. In [21] 2D PCA and 3D PCA were not observed to be significantly different. Three dimensional PCA was observed to be significantly superior to 2D PCA in [20]. However, employing color images instead of grayscale images for 2D PCA has been reported to perform better than 3D PCA [35, 108].

In all these studies it was observed that combinations of the 2D and 3D modalities resulted in significantly superior performance relative to either of them. Chang *et al.* argue that this increase may be merely due to the availability of multiple images of a subject in the gallery for multi-modality approaches [23]. They did, however, observe that the performance for a multi-modality single image 2D+3D PCA approach was superior to a single modality multiple images 2D+2D PCA approach.

5.1.2. Independent Component Analysis

One study has been reported where independent component analysis was applied to facial range images for recognition [45]. ICA considers not only the linear relationships between pixels in a facial image, but also higher order relationships. It projects data linearly onto a set of new basis vectors that are as statistically independent as possible [48]. Analogous to the results reported for 2D face recognition [4], the performance of 3D ICA has been reported to be superior to 3D PCA [45]. Hence, it is worthwhile to further investigate the potential of ICA for 3D face recognition.

5.1.3. Linear Discriminant Analysis

Linear discriminant analysis techniques have also been investigated for 3D face recognition (Tables 1 and 3). LDA has been applied to various 3D representations including range images, and horizontal and vertical gradients of range images. In all studies, the gradient representations are reported to yield the best recognition performance [6, 41, 43]. However, it is unclear as to which of horizontal or vertical gradient representations, is better for 3D LDA. In one analysis the horizontal gradient representation was found to be optimal [41, 43], while in another the vertical gradient representation was reported as the superior of the two [6].

Consistent with results reported for 2D face recognition [5], 3D LDA has also been reported to be better than 3D PCA [36,41,43]. Results also suggest 3D LDA as being superior to 2D LDA, and 2D+3D LDA as being superior to either of the individual modalities [6].

5.1.4. Local Feature Analysis

A single study has been reported that investigates the utility of the LFA technique for 3D face recognition [6]. In this study, the authors employed the 2D LFA FaceIt ® package available from Identix®, Minnetonka, MN for 3D face recognition. They observed that the vertical gradient representation was optimal for LFA. They further observed that the 2D LFA algorithm was significantly superior to 3D LFA, and that the 2D+3D LFA algorithm was superior to either of them.

5.1.5. Optimal Component Analysis

Srivastava *et al.* explored a technique called optimal component analysis for 3D face recognition [102]. They derived optimal linear projections of data in a lower dimensional space such that the performance of the nearest neighbor classifier was maximized. The technique involved a computationally expensive iterative procedure to learn the basis vectors. The authors reported that the method was superior to PCA, LDA, or ICA for 3D face recognition and was capable of handling variable facial expressions. However, they also did not report results of hypothesis testing for comparing the performances of these algorithms.

5.1.6. Hidden Markov Models

Two dimensional face recognition techniques based on hidden Markov models (HMM) [98] have also been applied to range images for 3D face recognition [2]. These techniques exploit the fact that facial features naturally occur in a fixed order from top to bottom and from left to right, irrespective of changes in illumination, pose, and facial expression. Different facial components are modeled as states in Markov models that are learnt from an ensemble of facial images. Achermann *et al.* observed the 3D HMM technique to be superior to 3D PCA [2], but statistical analyses were not conducted to confirm whether this difference was significant. Since a small database was employed in the study (Table 1), the difference is less likely to be statistically significant.

Methods based on embedded hidden Markov models (EHMM) [76] have also been studied for face recognition using color, grayscale intensity, and range images, and for combinations of these modalities [66, 67, 105–107] (Table 3). The 3D EHMM technique was reported to perform poorly relative to the grayscale EHMM and color EHMM techniques [105–107]. Like other appearance based techniques, the combined grayscale+3D EHMM and color+3D EHMM approaches are reported to perform better than the individual modalities.

Results of the studies based on EHMM also suggest that while changes in facial pose significantly alter the performance of both the 2D and 3D modalities, 3D algorithms may be effected less than 2D algorithms, by changes in illumination conditions during image acquisition. This can be concluded from the fact that when 2D+3D EHMM techniques were compensated for varying facial pose, by either augmenting the gallery with synthetic facial images with varying facial poses, [105–107], or by transforming all images to a frontal canonical pose before recognition [66, 67], significant improvement in their performances was observed. On the other hand, when pose compensated images were also compensated for varying facial illumination by either augmenting the database with images lit synthetically from various directions [105–107], or by re-lighting all images to frontal illumination [66, 67], the performance of 2D+3D EHMM algorithms did not change substantially.

The pose and illumination compensation technique where all faces were rotated to frontal pose and re-illuminated to frontal lighting [66, 67], was observed to be better than the database enrichment technique [105–107]. For the database enrichment technique it is difficult to ascertain *a priori* the optimal number and types of images of each individual that should be added to the gallery.

5.2. 'Free form' 3D Face Recognition Algorithms

A 3D object that cannot be recognized as either a planar or a naturally quadric surface is referred to as a 'free form' object [18]. The human face is an example of a 'free form' object. Numerous 3D face recognition techniques based on 'free form' object descriptions have been proposed. These descriptions may be associated with individual points on a surface, one dimensional surface curves, or two dimensional surface patches. The 'free form' 3D object recognition techniques that have been investigated for human face recognition are discussed in the following sections.

5.2.1. Facial Surface Matching

A number of techniques for 3D face recognition have been developed where the shapes of two facial surfaces are compared directly (Tables 3, 4, 8, 9, and 10). The general philosophy of these approaches is to register two facial surfaces as closely as possible in 3D space and to compare them by means of a suitable metric. Different versions of this basic technique have been investigated by varying the registration procedure employed and/or the metric employed for calculating the distance between the two surfaces.

Coarse normalization, and coarse normalization followed by fine normalization are the two main approaches that have been employed for registering facial surfaces. In course normalization [1, 37, 55, 59, 60], the gross pose and position of each facial surface in 3D space is computed and they are transformed to canonical frontal positions. The two surfaces, in this orientation, are compared by means of either the mean squared error (MSE) metric [37, 55, 59], or the Hausdorff distance (HD) metric [1, 60] proposed by Huttenlocher *et al.* [47]. This normalization technique however is not adequate as the accuracy of these distance metrics is dependent on precise alignment of the two surfaces.

For most 3D face recognition algorithms that employ matching of facial surfaces, coarse normalization is followed by fine normalization [22, 49, 52, 61, 62, 64, 65, 69, 70, 79–81, 96, 97]. The most frequently used algorithm for fine normalization is the Iterative Closest Point (ICP) algorithm [9]. In ICP, one 3D surface is iteratively transformed rigidly in 3D space to place it as close as possible to the other surface. Coarse normalization, in this case, helps the fine normalization iterative optimization procedure to converge correctly and prevents it from being trapped in a local minima.

Mean squared error, between pairs of closest points on the two surfaces, has been often employed as the objective criterion to be minimized in the iterative procedure [22,49,69,70]. In other studies the HD or the partial HD metric [62, 79, 80, 96], a combination of the MSE and HD metrics [52, 97], or a combination of the MSE between the pairs of closest 3D points and the MSE between closest points and surfaces [62, 64, 65] , as proposed by Chen and Medioni [27], have also been employed as the objective function. The partial HD metric is known to be more robust than the MSE metric to small mis-alignmemts between surfaces, and to the presence of holes, noise, and occlusions. It is also known that employing MSE based on point-to-surface distances [27] instead of MSE based on point-to-point distances [9] in the ICP algorithm, makes it less likely to be trapped in a local minimum. Russ *et al.* incorporated a multi-resolution approach with the ICP surface matching algorithm and reported high recognition rates with images from the FRGC v0.1 data set [97]. Slightly lower performance was reported on the same set of images when only a random set of points from the nasal region were employed for facial surface matching [52].

The ICP algorithm and its variants have yielded fairly successful algorithms for 3D and 2D+3D face recognition (Tables 3, 4, 8, 9, and 10). Face recognition algorithms based on ICP have been reported to be robust to changes in facial poses [65], and varying illumination conditions during 3D image acquisition [54]. They have also been reported to perform better than 3D PCA [49, 80], color LFA [70], and to 2D LDA [65] approaches.

It is not clear whether the addition of 2D grayscale/color information to the 3D ICP algorithm results in an increase in its performance. Lu *et al.* reported better performance for a combined 2D LDA+3D ICP algorithm relative to either of the individual modalities [65]. Similarly, Maurer *et al.* noted that the performance of the 3D ICP algorithm improved when it was combined with the 2D EBGM algorithm [69]. On the other hand, significant improvement in performance was not reported for the case when grayscale information and (x, y, z) co-ordinate values were employed for 4D ICP relative to when only the spatial co-oridnates were employed [81].

One major limitation of iterative surface matching based 3D face recognition algorithms is that they are computationally very expensive. For example, it has been reported that a single comparison between a pair of 3D facial surfaces required 20 seconds on an average Pentium 42.8 Ghz CPU [65]. This expense can become prohibitive for searching large databases. This computational expense stems from the two factors. First, the calculation of MSE between pairs of closest points on the two surfaces, or the HD metric, requires computations of the order $O(MN)$, where M is the number of 3D points on one surface and the N is the number of points on the other surface. Second, each matching procedure additionally requires a slow iterative optimization procedure to align a pair of facial surfaces. The iteration can also sometimes undesirably terminate in a local minimum. Recently 3D face recognition algorithms based on facial surface matching have been developed that employ the complex-wavelet structural similarity metric (CW-SSIM) [40]. This metric is known to be robust to small translational and rotational mis-alignments [114]. Matching facial surfaces using the CW-SSIM metric was shown to be more accurate, robust and efficient than employing the MSE or the HD metric.

Another limitation of 3D face recognition algorithms that employ ICP is that they are not robust to changes in facial expression [64, 65, 69, 81]. Employing only regions of the face that are relatively rigid and are not altered significantly by changes in facial expression, has been investigated as a solution to this problem [22].

5.2.2. Surface Normal Orientation Statistics

Attempts have also been made to match facial surfaces using the statistics of the orientation of facial surface normals (Table 5). Specifically, the extended Gaussian image approach [56, 103, 104] and the phase Fourier transforms [25, 82] have been investigated. The underlying idea of these techniques is to obtain a unique signature of each facial surface in terms of the distribution of the orientation of the surface normals. Facial surfaces are then matched by comparing their corresponding signatures.

Such techniques for facial surface matching are advantageous relative to the iterative surface matching techniques in some respects. They do not require an iterative procedure for aligning surfaces. Furthermore they are independent of the scale of 3D models [25], and have been shown to be robust to small rotations [25]. Yet, such techniques have not been explored extensively or rigorously. The few studies that exist, have employed small data sets, and none of them report face recognition performance estimates including the RR and EER values (Table 5). Hence, further investigation into such techniques is warranted in order to establish their true utility for 3D face recognition.

5.2.3. Profile Matching

One dimensional profile curves are can be easily obtained from 3D facial models by intersecting them with planes along various orientations, relative to 2D facial images. A number of studies have been reported where the shapes of the different facial profile curves have been compared in order to recognize faces (Tables 3, 6, and 8).

Interestingly, many studies have reported the central vertical facial profile curve to be the most discriminatory of all profile curves for 3D face recognition. Nagamine *et al.* [73] found vertical profiles located in the central region of the face to be the most discriminatory, followed by annular shaped curves centered at a point 40 mm above the tip of the nose. They also found horizontal facial profiles to be the least discriminatory of the three. Beumier *et al.* reported similar results. They observed a marginal decrease in performance of face recognition algorithms when 15 vertical profiles that spanned the entire facial surface were replaced by only 3 central profiles [10, 12]. Zhang *et al.* found that between the central vertical profile and two horizontal facial profiles (one passing through the forehead and the other through cheeks), the central vertical profile was the most discriminatory [122].

Consequently, a number of techniques have also been investigated to automatically locate the central vertical profile or the natural axis of bilateral symmetry on the human facial surface. Cartoux *et al.* automatically detected the axis of bilateral symmetry by iteratively minimizing the difference between principal curvatures on opposite sides of the facial surface [19]. Others have proposed a method of locating the axis of bilateral symmetry by iteratively aligning a facial surface to its mirror image [78–80, 122].

Consistent with other face recognition algorithms, Beumier *et al.* found that fusion of grayscale and shape information of the central and lateral vertical profiles results in improved performance, relative to either of the two individual modalities [11].

Three dimensional face recognition algorithms based on facial profiles are advantageous in that, they can provide information about the discriminatory information contained in the different sub-regions of the facial surface. However, they require an additional step of reliably locating the profiles. Furthermore, the overall accuracy of the recognition algorithms depends on accurate localization of profiles. The performance of profile based approaches for 3D face recognition lowers with the presence of variable facial expressions [122].

5.2.4. 3D Local Geometric Features

Studies that have considered local geometric features of 3D facial surfaces for recognition are fewer in number, relative to holistic techniques that do not require automatic segmentation of facial landmarks (Tables 7, 8, 9, and 10).

A number of methods for automatically locating facial landmarks that are purely based on facial surface characteristics have been proposed. The most widely used method is to locate facial landmarks is to exploit either their unique curvature characteristics [22, 28–30, 37, 38, 46, 65] using the H-K segmentation algorithm [8], or to employ the curvature properties of facial profile curves [58, 122]. Another approach is to align 3D faces rigidly [119] or non-rigidly [49] to generic face templates for which the locations of fiducial points/facial features are known *a priori*. The tip of the nose has been detected as the most prominent point of 3D facial models in a canonical frontal position [59, 73]. However, this heuristic method is not successful for arbitrary facial poses. Yacoob and Davis labeled components of the facial surface based on contextual reasoning [120]. When color/texture information is also available along with the 3D shape information, it has also been employed to automatically locate facial landmarks [46, 55, 109, 111–113].

The local geometric features of facial surfaces that have been employed as features for face recognition include position, surface area and curvatures properties of facial landmarks/fiducial points, and 3D Euclidean distances, ratios of distances, geodesic distances and angles between them. The local shape of facial landmarks/regions about fiducial points have been quantified by means of Gaussian curvature values [38, 72], Guassian-Hermite moments [119], 'point signatures' [28, 113], and by 2D and 3D Gabor filter coefficients of facial range images [46, 111]. In order to quantify the relationships between the facial landmarks, 3D Euclidean distances [38, 46], angles and 3D Euclidean distances [58, 72], or geodesic and 3D Euclidean distances have been employed [39].

The method based on 'point signatures' [28] has also been reported to be superior to methods based on PCA applied to the 3D point clouds, profile matching, and PCA applied to range images [49]. Its performance was equivalent to an approach where the positions of only 10 fiducial points were compared after iteratively aligning a pair of facial models. Another approach based on the shape characteristics of local facial features, and surface matching has also been reported to perform better than profile matching [119].

A technique, where the facial sub-regions of the eyes, nose and mouth by fitting with subdivisional surfaces, and are matched separately, has been evaluated on the FRGC v0.1 and the FRGC v0.2 data sets [50, 83]. It performed better than the baseline 3D PCA algorithm. Another 2D+3D face recognition technique based on EGBM applied to range and grayscale images performed on par with this technique on the FRGC v0.2 data set [52]. In this study, the 2D EBGM algorithm performed better than the 3D EBGM algorithm. This could be explained by the fact that Gabor filter coefficients that are employed to quantify texture in grayscale images, might not encode discriminatory information for facial range images that do not contain as much texture variations (Figure 4(c)).

Face recognition techniques based on local facial features present several advantages relative to holistic techniques. First, instead of an ad hoc set of local facial features, their selection can be based on domain knowledge about the structural diversity of human faces. Such an approach has been demonstrated to result in effective 3D face recognition algorithms [39]. Second, techniques based on local facial features are less affected by global changes in the appearance of facial images including changes facial expressions [28, 39, 113].

Nonetheless, techniques for 3D face recognition based on local facial features have received little attention relative to holistic techniques. This may be due to the fact that such techniques require an additional step of reliably locating facial landmarks, which may effect their overall performance. Nonetheless, if facial landmarks can be reliably located, evidence from the literature on both 2D and 3D face recognition suggests that powerful techniques for 3D face recognition can be developed. Hence, there is a need to further explore the potential of 3D face recognition algorithms that employ features from facial sub-regions, and to find ways to combine them with techniques for robust facial feature detection.

5.3. Ensemble Approaches

A number of modular and ensemble methods consisting of combinations of multiple 'free form' 3D face recognition techniques have also been investigated (Table 8). For a majority of such algorithms, it has been observed that the combined approach results in superior performance, relative to the individual constituent approaches. Gökberk *et al.* observed this for the case when scores of 3D face recognition algorithms based on LDA, surface normals, and profiles were combined in parallel using a nonlinear rank-sum method [36]. Significant improvement in performance was also reported for hierarchical combinations of a surface matching classifier with 3D LDA [36], or 2D LDA [65]. A combination of features from the whole face for surface matching, and features from the mouth, nose and orbital regions also resulted in superior performance, relative to the individual sets of features [119]. Pan *et al.* observed superior performance when they combined outputs of a facial surface matching algorithm and a central profile matching classifier using the MAX rule [78, 79].

6. Expression Invariant Approaches

Achieving invariance to changes in facial expressions is one of the major open problems of automatic 2D and 3D facial recognition. While the pose of 3D facial models can be easily corrected, changes in facial expression, which are non-rigid transformations of the facial surface are not as trivial to eliminate. Numerous studied have demonstrated that the performance of 2D, 3D and 2D+3D techniques is considerably reduced when facial images with arbitrary expressions are employed [50, 65, 69, 81, 83, 111, 122]. Gallery enrichment has been the most common approach employed in 2D face recognition to introducing some degree of invariance to facial expressions. The idea, is to include multiple images of an individual in the gallery with different facial expressions. There are two problems with this technique. First, it increases computational costs. Second, there is no reliable way to decide on the number and type of facial expressions to add to the gallery, which would, in principle, encompass all facial expressions of an individual.

With the introduction of 3D face recognition techniques, a few other methods for achieving expression invariance have also been explored (Table 9). For 3D expression invariant face recognition, one approach has been to obtain an invariant representation of the face. The other has been to employ regions of the face that are known to remain relatively rigid under facial expression changes. These techniques are discussed in detail in this section.

A number of expression invariant representations of the human facial surface have been proposed. Bronstein *et al.* modeled changes in facial expressions as isometric deformations of the facial surface [16, 17]. For such deformations, the intrinsic properties of surfaces including geodesic distances between all pairs of points remain unchanged. The authors' approach was to isometrically embed facial surfaces into expression invariant canonical forms before recognition using multidimensional scaling. Similarly, invariance to facial expression changes was also observed for a algorithm that employed geodesic and 3D Euclidean distances between anthropometric facial fiducial points as features [39].

Wang and Chua employed 2D/3D Gabor coefficients calculated from range images at specific facial fiducial points to generate expression invariant representations of human face [111]. Their technique met with moderate success. Lu and Jain proposed a technique where 3D models were matched rigidly using ICP then non-rigidly deformed using thin-plate spline deformation to generate a displacement vector image [63]. Each point in the displacement vector image contained a vector connecting a point on the original face before deformation to its location after deformation. These vector images were classified as intra-personal and inter-personal deformations.

The other approach of employing regions of the face that are relatively rigid to changes in facial expressions has not been investigated as extensively. Chang *et al.* investigated matching local regions of the facial surfaces including the nose and the nose bridge using ICP [22]. In another study 'point signatures' were employed to quantify the local surface curvature of fiducial points located at relatively rigid regions of the face [28, 113]. However, it is not clear whether this technique achieved expression invariance due to favorable properties of the 'point signature' representation or due the fact that facial features were derived from relatively rigid regions of the face.

Table 1. The table summarizes the studies that have investigated appearance based 3D face recognition algorithms. For some studies, the authors have not specified the number of images/subject they included in the gallery and probe data set, but instead provide only the total number of images employed (*e.g.* Heseltine [42]). Missing information is indicated by '–'. The decrease in performance of the 3D PCA algorithm with the increase in the number of subjects in the database is evident from this table.

Author	Method	Data Set			Performance
		# Subjects	# Gallery Images	# Probe Images	
Achermann [2]	PCA, HMM	24	5/subject	5/subject	RR = 100%
Heseltine [42]	PCA	100	40	290	EER = 12.7%
Hesher [44]	PCA	37	5/subject	1/subject	RR = 94%
Hesher [45]	ICA	37	5/subject	1/subject	RR = 97%
Heseltine [41, 43]	LDA	230	–	–	EER = 9.3%
Srivastava [102]	OCA	67	3/subject	3/subject	RR = 99%

Table 2. This table summarizes the studies that have combined PCA classifiers applied to different imaging modalities, including grayscale, color, infra red and 3D range images. It is evident from the table that the combination of modalities consistently results in superior performance relative to the individual modalities.

Author	Method		Data Set			Performance
	2D	3D	# Subjects	# Gallery Images	# Probe Images	
Tsutsumi [109]	PCA	PCA	24	9/subject	35/subject	$FAR_{2D+3D} = 4.5\%$ $FRR_{2D+3D} = 3.7\%$
Chang [21]	PCA	PCA	166	1/subject	1/subject	$RR_{2D+3D} = 92.8\%$ $RR_{2D} = 83.1\%$ $RR_{3D} = 83.7\%$
Chang [20]	PCA	PCA	200	1/subject	1/subject	$RR_{2D+3D} = 98.5\%$ $RR_{2D} = 89\%$ $RR_{3D} = 94.5\%$
Chang [24]	PCA	PCA	127	1/subject	297	$RR_{2D+3D+IR} = 100\%$ $RR_{2D} = 90.6\%$ $RR_{3D} = 91.9\%$ $RR_{IR} = 71.0\%$
Chang [23]	PCA	PCA	198	1/subject	670	$RR_{2D+3D} = 97.5\%$
Tsalakanidou [108]	color PCA	PCA	40	2/subject 3 poses	2/subject	$RR_{color+3D} = 98.8\%$ $RR_{color} = 93.8$ $RR_{3D} = 85$
Godil [35]	color PCA	PCA	200	1/subject	1/subject	$FAR_{color+3D} = 20\%$ $FRR_{color+3D} = 1\%$ $RR_{color+3D} = 82\%$ $RR_{color} = 78\%$ $RR_{3D} = 68\%$

Table 3. The table summarizes numerous 2D+3D face recognition algorithms. Here again, it can be observed that combining 2D and 3D modalities results in better performance, relative to the individual modalities.

Author	Method		Data Set			Performance
	2D	3D	# Subjects	# Gallery Images	# Probe Images	
BenAbdelkader [6]	LFA,	LFA,	185	2/subject	2/subject	$EER_{2D+3D} = 0\%$
Tsalakanidou [106, 107]	gray EHMM	EHMM	50	5/subject	55/subject	$EER_{2D+3D} = 5.5\%$ $EER_{2D} = 9.3\%$ $EER_{3D} = 20\%$
Tsalakanidou [105]	color/gray EHMM	EHMM	20	5/subject	1402	$EER_{2D+3D} = 7.5\%$ $EER_{color} = 15.1\%$ $EER_{2D} = 13.8\%$
						$EER_{3D} = 16.7\%$
Tsalakanidou [107]	color/gray EHMM	EHMM	20	2-5/subject	2124	$EER_{2D+3D} = 7.7\%$ $EER_{2D} = 7.9\%$ $EER_{3D} = 14.8\%$ $RR_{2D+3D} = 77.3\%$
Malassiotis [66, 67]	color EHMM	EHMM	20	2-5/subject	–	$RR_{color_3D} = 99.1\%$ $RR_{color} = 90.7\%$ $RR_{3D} = 96.4\%$
Lu [62]	color	ICP, shape	18	1/subject	63	$RR_{2D+3D} = 84\%$
Lu [64, 65]	LDA	ICP	200	1/subject	598	$RR_{2D+3D} = 90\%$ $RR_{2D} = 77\%$ $RR_{3D} = 86\%$
Papatheodorou [81]	4D ICP	3D ICP	62	1/subject	5/subject	$FRR_{2D+3D} = 0\%$ $FAR_{2D+3D} = 0\%$
Beumier [11]	Profiles	Profiles	100	9/subject	–	$EER_{2D+3D} = 1.2\%$

Table 4. The table presents the existing surface matching based 3D face recognition algorithms. For a majority of them, the mean squared error and the Hausdorff distance metrics have been employed along with the ICP algorithm. This approach has been fairly successful even with large data set of 200 subjects, but is limited by its high computational cost.

Author	Method	Data Set			Performance
		# Subjects	# Gallery Images	# Probe Images	
Gordon [37]	depth MSE	8	2/subject	1/subject	RR = 100%
Lao [55]	depth MSE	10	3/subject	1/subject	RR = 96%
Lee [59]	depth μ,σ MSE	35	1/subject	1/subject	Rank 10 RR = 100%
Medioni [70]	ICP	100	6/subject	1/subject	EER < 2%
Lu [61]	ICP and shape feature	18	1/subject	113	RR = 96.5%
Irfanöglu [49]	ICP	30	2/subject	1/subject	RR = 96.7%
Achermann [1]	Partial HD	24	5/subject	5/subject	RR = 99.2%
Lee [60]	depth weighted HD	42	1/subject	1/subject	Rank 5 RR = 98%
Pan [80]	HD-ICP	30	1/subject	2/subject	EER = 3.24%
Russ [96, 97]	HD-ICP	200	1/subject	1/subject	FRR = 2% FAR = 0%
Gupta [40]	CW-SSIM	12	1/subject	29/subject	EER = 9.13% Rank 1 RR = 98.6%

Table 5. The table presents a summary of the 3D face recognition techniques based on the statistics of the orientation of surface normals. Clearly these techniques have not been greatly explored and warrant further investigations.

Author	Method	Data Set			Performance
		# Subjects	# Gallery Images	# Probe Images	
Lee [56]	EGI	6 images	–	–	–
Tanaka [103, 104]	EGI	37 images	–	–	–
Paquet [82]	PFT	24 images	–	–	–
Chang [25]	PFT	15 images	–	–	–

Table 6. This table summarizes the 3D face recognition algorithms that involve the matching of various facial profile curves. For all such techniques, the vertical central profile curve has resulted in the best performance.

Author	Method	Data Set			Performance
		# Subjects	# Gallery Images	# Probe Images	
Cartoux [19]	Profiles	5	1-3/subject	1/subject	RR = 100%
Pan [78]	HD Profile matching	30	1/subject	2/subject	EER = 2.22%
Nagamine [73]	Profiles	16	9/subject	1/subject	RR = 100%
Beumier [10, 12]	Profiles	120	1/subject	5/subject	EER = 4.75%
Zhang [122]	Profiles	166	166	32	RR = 96.9%

Table 7. This table summarizes the studies based on geometric facial features, including curvatures, angles, Euclidean and geodesic distances between facial landmarks, for 3D face recognition.

Author	Method	Data Set			Performance
		# Subjects	# Gallery Images	# Probe Images	
Gordon [38]	Curvature, Distances	8	2/subject	1/subject	RR = 100%
Moreno [72]	Curvature, Angles, Distances	60	1/subject	6/subject	RR = 78%
Lee [58]	Angles and Distances	100	–	–	RR = 96%
Gupta [39]	Euclidean and Geodesic Distances	105	1/subject	663	Rank 1 RR = 98.64% EER = 1.4%

Table 8. This table presents the ensemble approaches to 3D face recognition, where multiple 3D classifiers/representaions/fetaures are combined. For all analyses, the ensemble approach results in superior performance, relative to the individual 3D classifiers.

Author	Method	Data Set			Performance
		# Subjects	# Gallery Images	# Probe Images	
Pan [79]	Profiles and surface matching	30	1/subject	2/subject	EER = 1.11%
Xu [119]	z values and local shape	30	5/subject	1/subject	RR = 96.1%
Gökberk [36]	LDA and Normals and Profiles	106	3/subject	2-3/subject	RR = 99.1%

Table 9. The table summarizes 3D face recognition algorithms that have been reported to be robust to changes in facial expression to some degree. The two main ideologies are to to either transform the facial surface into an expression invariant form or to employ facial shape characteristics that are not significantly altered by changes in facial expression.

Author	Method		Data Set			Performance
	2D	3D	# Subjects	# Gallery Images	# Probe Images	
Chua [28]	–	Point Sign	6	4/subject	1/subject	RR = 100%
Wang [113]	2D Gabor	Point Sign	50	3/subject	3/subject	RR = 90%
Wang [111]	2D Gabor	3D Gabor	30	1/subject	3/subject	RR = 79.58%
Bronstein [17]	Isometric embedding	Isometric embedding	157	–	–	–
Bronstein [16]	–	Isometric embedding	30	65	155	EER = 1.9% RR = 100%
Chang [22]	–	ICP of rigid regions	449 FRGC v0.2	1/subject	1-9/subject	RR = 88.6%
Lu [63]	–	Spline deformation	100	1/subject	1-2/subject	FRR = 34% FAR = 1% RR = 91%

Table 10. This table summarizes the 3D face recognition algorithms that have been tested on the Face Recognition Grand Challenge data sets. The three algorithms tested on the FRGC v0.2 data set, which contains images of 466 individual, can be regarded as the current state-of-the-art in 2D+3D face recognition. Notably, the algorithm based on 2D+3D EBGM performed the best of the three.

Author	Method		Data Set			Performance
	2D	3D	# Subjects	# Gallery Images	# Probe Images	
Pan [77]	–	Isometric flattening & PCA	276 FRGC v0.1	–	–	EER = 2.83%
Russ [97]	–	MSE and HD-ICP	198 FRGC v0.1	1/subject	–	FRR = 6.5% FAR = 0.1% RR = 98.5%
Koudelka [52]	–	HD-ICP	198 FRGC v0.1	1/subject	1/subject	RR = 94%
Kakadiaris [50]	–	AFM	275 FRGC v0.1	152	608	RR = 99.3%
Passalis [83]	–	AFM	466 FRGC v0.2	ROC III	ROC III	FRR = 3.6% FAR = 0.1%
Hüsken [46]	EBGM	EBGM	466 FRGC v0.2	ROC III	ROC III	FRR = 2.7% FAR = 0.1%
Maurer [69]	EBGM	ICP	466 FRGC v0.2	4007	4007	FRR = 6.5% FAR = 0.1%

7. Conclusion

The field of 3D face recognition has grown rapidly since FRVT 2002. Both appearance based and 'free form' based 3D face recognition algorithms have been proposed. Numerous approaches to combine 3D and 2D modalities have also been investigated. While it still remains a matter of debate as to which of 2D or 3D modalities is superior, it has been conclusively established that combing the two modalities holds promise for improving face recognition performance. The availability of 3D data has considerably alleviated the pose problem, but achieving expression invariance still remains to be solved. Debate also exists about the 'illumination invariant' nature of 3D data, but cues from existing studies point towards the fact that 3D techniques may be less sensitive to changes in illumination during image capture than 2D techniques.

Considerable attention is now also being directed towards robust testing and evaluation of 3D face recognition algorithms on a common database. This will help to objectively evaluate the current sate-of-the-art in the area. A number of areas, however, still remain unexplored. Research efforts need to be directed towards developing 3D feature and profile detection algorithms. A better understanding of facial image statistics and surface geometric properties is required to understand the workings of the holistic appearance based statistical approaches for 3D face recognition. Assessing the relationship between multiple modalities and improving methods of combining them are also likely to advance the field further.

References

[1] B. Achermann and H. Bunke. Classifying range images of human faces with hausdorff distance. In *Proceedings 15th International Conference on Pattern Recognition*, volume vol.2, pages 809–813. IEEE Comput. Soc, Dept. of Comput. Sci., Bern Univ., Switzerland, 2000.

[2] B. Achermann, X. Jiang, and H. Bunke. Face recognition using range images. In *Proceedings. International Conference on Virtual Systems and MultiMedia* , pages 129–136. Int. Soc. Virtual Syst. & MultiMedia (VSMM), Inst. of Comput. Sci. & Appl. Math., Bern Univ., Switzerland, 1997.

[3] L. Akarun, B. Gokberk, and A. A. Salah. 3d face recognition for biometric applications. In *13th European Signal Processing Conference (EUSIPCO)* , Antalya, Turkey, Spetember 2005.

[4] M. S. Bartlett, J. R. Movellan, and T. J. Sejnowski. Face recognition by independent component analysis. *Neural Networks, IEEE Transactions on*, 13(6):1450–1464, 2002.

[5] P. N. Belhumeur, J. P. Hespanha, and D. J. Kriegman. Eigenfaces vs. fisherfaces: recognition using class specific linear projection. *Pattern Analysis and Machine Intelligence, IEEE Transactions on*, 19(7):711–720, 1997.

[6] C. BenAbdelkader and P. A. Griffin. Comparing and combining depth and texture cues for face recognition. *Image and Vision Computing*, 23(3):339–352, 2005.

[7] P. J. Besl. Active, optical range imaging sensors. *Machine Vision and Applications*, 1(2):127–152, 1988.

[8] P. J. Besl and R. C. Jain. Segmentation through variable-order surface fitting. *Pattern Analysis and Machine Intelligence, IEEE Transactions on*, 10(2):167–192, 1988.

[9] P. J. Besl and H. D. McKay. A method for registration of 3-d shapes. *Pattern Analysis and Machine Intelligence, IEEE Transactions on*, 14(2):239–256, 1992.

[10] C. Beumier and M. Acheroy. Automatic 3d face authentication. *Image and Vision Computing*, 18(4):315–321, 2000.

[11] C. Beumier and M. Acheroy. Face verification from 3d and grey level clues. *Pattern Recognition Letters*, 22(12):1321–1329, 2001.

[12] C. Beumier, M. Acheroy, P. H. Lewis, and M. S. Nixon. Automatic face authentication from 3d surface. In P. H. Lewis and M. S. Nixon, editors, *BMVC 98. Proceedings of the Ninth British Machine Vision Conference*, volume vol.2, pages 449–458. Univ. Southampton, Signal & Image Centre, R. Mil. Acad., Brussels, Belgium, 1998.

[13] V. Blanz and T. Vetter. Face recognition based on fitting a 3d morphable model. *Pattern Analysis and Machine Intelligence, IEEE Transactions on*, 25(9):1063–1074, 2003.

[14] A. C. Bovik, M. Clark, and W. S. Geisler. Multichannel texture analysis using localized spatial filters. *Pattern Analysis and Machine Intelligence, IEEE Transactions on*, 12(1):55–73, 1990.

[15] K. W. Bowyer, K. Chang, and P. J. Flynn. A survey of approaches and challenges in 3d and multi-modal 3d+2d face recognition. *Computer Vision and Image Understanding*, 101:1–15, 2006.

[16] A. M. Bronstein, M. M. Bronstein, and R. Kimmel. Three-dimensional face recognition. *International Journal of Computer Vision*, 64(1):5–30, 2005.

[17] A. M. Bronstein, M. M. Bronstein, R. Kimmel, J. Kittler, and M. S. Nixon. Expression-invariant 3d face recognition. In J. Kittler and M. S. Nixon, editors, *Audio- and Video-Based Biometric Person Authentication. 4th International Conference, AVBPA 2003. Proceedings (Lecture Notes in Computer Science Vol.2688)*, pages 62–69. Springer-Verlag, Dept. of Electr. Eng., Israel Inst. of Technol., Haifa, Israel, 2003.

[18] R. J. Campbell and P. J. Flynn. A survey of free-form object representation and recognition techniques. *Computer Vision and Image Understanding*, 81(2):166–210, 2001.

[19] J. Y. Cartoux, J. T. Lapreste, and M. Richetin. Face authentification or recognition by profile extraction from range images. In *Interpretation of 3D Scenes, 1989. Proceedings., Workshop on*, pages 194–199, 1989.

[20] K. I. Chang, K. W. Bowyer, and P. J. Flynn. Face recognition using 2d and 3d facial data. In *Multimodal User Authentication, Workshop in*, pages 25–32, 2003.

[21] K. I. Chang, K. W. Bowyer, and P. J. Flynn. Multimodal 2d and 3d biometrics for face recognition. In *Analysis and Modeling of Faces and Gestures, 2003. AMFG 2003. IEEE International Workshop on*, pages 187–194, 2003.

[22] K. I. Chang, K. W. Bowyer, and P. J. Flynn. Adaptive rigid multi-region selection for handling expression variation in 3d face recognition. In *Computer Vision and Pattern Recognition, 2005 IEEE Computer Society Conference on*, volume 3, pages 157–157, 2005.

[23] K. I. Chang, K. W. Bowyer, and P. J. Flynn. An evaluation of multimodal 2d+3d face biometrics. *Pattern Analysis and Machine Intelligence, IEEE Transactions on*, 27(4):619–624, 2005.

[24] K. I. Chang, K. W. Bowyer, P. J. Flynn, and X. Chen. Multi-biometrics using facial appearance, shape and temperature. In *Automatic Face and Gesture Recognition, 2004. Proceedings. Sixth IEEE International Conference on*, pages 43–48, 2004.

[25] S. Chang, M. Rioux, and C. P. Grover. Range face recognition based on the phase fourier transform. *Optics Communications*, 222(1):143–153, 2003.

[26] R. Chellappa, C. L. Wilson, and S. Sirohey. Human and machine recognition of faces: a survey. *Proceedings of the IEEE*, 83(5):705–741, 1995.

[27] Y. Chen and G. Medioni. Object modeling by registration of multiple range images. *Image and Vision Computing*, 10:145–155, 1992.

[28] C.-S Chua, F. Han, and Y.-K. Ho. 3d human face recognition using point signature. In *Automatic Face and Gesture Recognition, 2000. Proceedings. Fourth IEEE International Conference on*, pages 233–238, 2000.

[29] A. Colombo, C. Cusano, and R. Schettini. Tri-dimensional face detection and localization. In S. Santini, R. Schettini, and T. Gevers, editors, *SPIE Internet imaging VI, in Proceedings of*, volume 5670, pages 68–75, 2005.

[30] A. Colombo, C. Cusano, and R. Schettini. 3d face detection using curvature analysis. *Pattern Recognition*, 39(3):444–455, 2006.

[31] J. Davis and M. Goadrich. The relationship between precision-recall and roc curves. In *ICML '06: Proceedings of the 23rd international conference on Machine learning*, pages 233–240, New York, NY, USA, 2006. ACM Press.

[32] R. O. Duda, P. E. Hart, and D. G. Stork. *Pattern Classification*. John Wiley and Sons, New York, 2nd edition, 2001.

[33] J. P. Egan. *Signal Detection Theory and ROC Analysis*. Academic Press, New York, 1975.

[34] R. A. Fisher. The use of multiple meaures in taxonomic problems. *Annals Eugenics*, 7:179–188, 1936.

[35] A. Godil, S. Ressler, and P. Grother. Face recognition using 3d facial shape and color map information: comparison and combination. In *Proceedings of the SPIE - The International Society for Optical Engineering*, volume 5404, pages 351–361, Nat. Inst. of Stand. & Technol., Gaithersburg, MD, USA, 2004. SPIE-Int. Soc. Opt. Eng.

[36] B. Gökberk, A. A. Salah, and L. Akarun. Rank-based decision fusion fro 3d shape-based face recognition. In *LNCS, International Conference on Audio- and Video-based Biometric Person Authentication*, volume 3546, pages 1019–1028, 2005.

[37] G. G. Gordon. Face recognition based on depth maps and surface curvature. In *SPIE Geometric methods in Computer Vision*, volume 1570, pages 234–247, 1991.

[38] G. G. Gordon. Face recognition based on depth and curvature features. In *Computer Vision and Pattern Recognition, 1992. Proceedings CVPR '92., 1992 IEEE Computer Society Conference on*, pages 808–810, 1992.

[39] S. Gupta, M. K. Markey, J. K. Aggarwal, and A. C. Bovik. Three dimensional face recognition based on geodesic and euclidean distances. In *Electronic Imaging, Vision Geometry XV*, volume 6499 of *Proc. of SPIE*, San Jose, California, USA, Jan 28 - Feb 1 2007.

[40] S. Gupta, M. P. Sampat, Z. Wang, M. K. Markey, and A. C. Bovik. 3d face recognition using the complex-wavelet structural similarity metric. In *IEEE Workshop on Applications of Computer Vision*, Austin, TX, 2007.

[41] T. Heseltine, N. Pears, and J. Austin. Three-dimensional face recognition: A fisher-surface approach. In A. Campilho and M. Kamel, editors, *International Conference on Image Analysis and Recognition, In Proc. of the*, volume LNCS 3212 of *ICIAR 2004*, pages 684–691. Springer-Verlag Berlin Heidelberg, 2004.

[42] T. Heseltine, N. Pears, and J. Austin. Three-dimensional face recognition: an eigen-surface approach. In *Image Processing, 2004. ICIP '04. 2004 International Conference on*, volume 2, pages 1421–1424 Vol.2, 2004.

[43] T. Heseltine, N. Pears, and J. Austin. Three-dimensional face recognition using surface space combinations. In *British Machine Vision Conference*, 2004.

[44] C. Hesher, A. Srivastava, and G. Erlebacher. Principal component analysis of range images for facial recognition. In *International Conference on Imaging Science, Systems, and Technology (CISST), in Proceedings of*, 2002.

[45] C. Hesher, A. Srivastava, and G. Erlebacher. A novel technique for face recognition using range imaging. In *Signal Processing and Its Applications, 2003. Proceedings. Seventh International Symposium on*, volume 2, pages 201–204 vol.2, 2003.

[46] M. Hüsken, M. Brauckmann, S. Gehlen, and C. Von der Malsburg. Strategies and benefits of fusion of 2d and 3d face recognition. In *Computer Vision and Pattern Recognition, 2005 IEEE Computer Society Conference on*, volume 3, pages 174–174, 2005.

[47] D. P. Huttenlocher, G. A. Klanderman, and W. J. Rucklidge. Comparing images using the hausdorff distance. *Pattern Analysis and Machine Intelligence, IEEE Transactions on*, 15(9):850–863, 1993.

[48] A. Hyvarinen, J. Karhunen, and E. Oja. *Independent Component Analysis*. John Wiley and Sons, New York, 2001.

[49] M. O. Irfanöglu, B. Gökberk, and L. Akarun. 3d shape-based face recognition using automatically registered facial surfaces. In *Proceedings of the 17th International Coneference on Pattern Recognition*, volume Vol.4, pages 183–186. IEEE Computer Society, Dept. of Comput. Eng., Bogazici Univ., Turkey, 2004.

[50] I. A. Kakadiaris, G. Passalis, T. Theoharis, G. Toderici, I. Konstantinidis, and N. Murtuza. Multimodal face recognition: combination of geometry with physiological information. In *Computer Vision and Pattern Recognition, 2005. CVPR 2005. IEEE Computer Society Conference on*, volume 2, pages 1022–1029 vol. 2, 2005.

[51] M. Kirby and L. Sirovich. Application of the karhunen-loeve procedure for the characterization of human faces. *Pattern Analysis and Machine Intelligence, IEEE Transactions on*, 12(1):103–108, 1990.

[52] M. L. Koudelka, M. W. Koch, and T. D. Russ. A prescreener for 3d face recognition using radial symmetry and the hausdorff fraction. In *Computer Vision and Pattern Recognition, 2005 IEEE Computer Society Conference on*, volume 3, pages 168–168, 2005.

[53] A. Z. Kouzani, F. He, and K. Sammut. Towards invariant face recognition. *Information Sciences*, 123(1):75–101, 2000.

[54] E. P. Kukula, S. J. Elliott, R. Waupotitsch, and B. Pesenti. Effects of illumination changes on the performance of geometrix facevision/spl reg/ 3d frs. In *Security Technology, 2004. 38th Annual 2004 International Carnahan Conference on*, pages 331–337, 2004.

[55] S. Lao, Y. Sumi, M. Kawade, and F. Tomita. 3d template matching for pose invariant face recognition using 3d facial model built with isoluminance line based stereo vision. In *Pattern Recognition, 2000. Proceedings. 15th International Conference on*, volume 2, pages 911–916 vol.2, 2000.

[56] J. C. Lee and E. Milios. Matching range images of human faces. In *Computer Vision, 1990. Proceedings, Third International Conference on*, pages 722–726, 1990.

[57] M. W. Lee and S. Ranganath. Pose-invariant face recognition using a 3d deformable model. *Pattern Recognition*, 36(8):1835–1846, 2003.

[58] Y. Lee, H. Song, U. Yang, H. Shin, and K. Sohn. Local feature based 3d face recognition. In *Audio- and Video-based Biometric Person Authentication, 2005 International Conference on, LNCS*, volume 3546, pages 909–918, 2005.

[59] Y. Lee and T. Yi. 3d face recognition using multiple features for local depth information. In M. Grgic and S. Grgic, editors, *Proceedings EC-VIP-MC 2003. 4th EURASIP Conference focused on Video/Image Processing and Multimedia Communications (IEEE Cat. No.03EX667)*, volume vol.1, pages 429–434. Faculty of Electrical Eng. & Comput, Zagreb, 2003.

[60] Y.-H. Lee and J.-C. Shim. Curvature based human face recognition using depth weighted hausdorff distance. In *Image Processing, 2004. ICIP '04. 2004 International Conference on*, volume 3, pages 1429–1432 Vol. 3, 2004.

[61] X. Lu, D. Colbry, and A. K. Jain. Three-dimensional model based face recognition. In *Pattern Recognition, 2004. ICPR 2004. Proceedings of the 17th International Conference on*, volume 1, pages 362–366 Vol.1, 2004.

[62] X. Lu, D. Colbry, A. K. Jain, and D. Zhang. Matching 2.5d scans for face recognition. In D. Zhang and A.K. Jain, editors, *Biometric Authentication. First International Conference, ICBA 2004. Proceedings (Lecture Notes in Comput. Sci. Vol.3072)*, pages 30–36. Croucher Found., Dept. of Comput. Sci. & Eng., Michigan State Univ., East Lasning, MI, USA, 2004.

[63] X. Lu and A. K. Jain. Deformation analysis for 3d face matching. In *Applications of Computer Vision, WACV '05 7th IEEE Workshop on*, pages 99–104, 2005.

[64] X. Lu and A. K. Jain. Integrating range and texture information for 3d face recognition. In *Applications of Computer Vision, 2005. (WACV 2005). Proceedings. Sixth IEEE Workshop on*, 2005.

[65] X. Lu, A. K. Jain, and D. Colbry. Matching 2.5d face scans to 3d models. *Pattern Analysis and Machine Intelligence, IEEE Transactions on*, 28(1):31–43, 2006.

[66] S. Malassiotis and M. G. Strintzis. Pose and illumination compensation for 3d face recognition. In *Image Processing, 2004. ICIP '04. 2004 International Conference on*, volume 1, pages 91–94 Vol. 1, 2004.

[67] S. Malassiotis and M. G. Strintzis. Robust face recognition using 2d and 3d data: Pose and illumination compensation. *Pattern Recognition*, 38(12):2537–2548, 2005.

[68] R. S. Malpass and J. Kravitz. Recognition for faces of own and other race. *Journal of Personality and Social psychology*, 13:330–334, 1969.

[69] T. Maurer, D. Guigonis, I. Maslov, B. Pesenti, A. Tsaregorodtsev, D. West, and G. Medioni. Performance of geometrix activeid^{TM} 3d face recognition engine on the frgc data. In *Computer Vision and Pattern Recognition, 2005 IEEE Computer Society Conference on*, volume 3, pages 154–154, 2005.

[70] G. Medioni and R. Waupotitsch. Face modeling and recognition in 3-d. In *Analysis and Modeling of Faces and Gestures, 2003. AMFG 2003. IEEE International Workshop on*, pages 232–233, 2003.

[71] C. E. Metz. Basic principles of roc analysis. *Seminars in Nuclear Medicine*, 8(4):283–298, 1978.

[72] A. B. Moreno, A. Sanchez, J. Fco, V. Fco, and J. Diaz. Face recognition using 3d surface-extracted descriptors. In *Irish Machine Vision and Image Processing Conference (IMVIP 2003)*, Sepetember 2003.

[73] T. Nagamine, T. Uemura, and I. Masuda. 3d facial image analysis for human identification. In *Pattern Recognition, 1992 . Vol.1. Conference A: Computer Vision and Applications, Proceedings., 11th IAPR International Conference on* , pages 324–327, 1992.

[74] D. Nandy and J. Ben-Arie. Shape from recognition: a novel approach for 3-d face shape recovery. *Image Processing, IEEE Transactions on*, 10(2):206–217, 2001.

[75] C. Nastar and A. Pentland. Matching and recognition using deformable intensity surfaces. In *Computer Vision, 1995. Proceedings., International Symposium on* , pages 223–228, 1995.

[76] A. V. Nefian and III Hayes, M. H. An embedded hmm-based approach for face detection and recognition. In *Acoustics, Speech, and Signal Processing, 1999. ICASSP '99. Proceedings., 1999 IEEE International Conference on*, volume 6, pages 3553–3556 vol.6, 1999.

[77] G. Pan, S. Han, Z. Wu, and Y. Wang. 3d face recognition using mapped depth images. In *Computer Vision and Pattern Recognition, 2005 IEEE Computer Society Conference on*, volume 3, pages 175–175, 2005.

[78] G. Pan, Y. Wu, and Z. Wu. Investigating profile extracted from range data for 3d face recognition. In *Systems, Man and Cybernetics, 2003. IEEE International Conference on*, volume 2, pages 1396–1399 vol.2, 2003.

[79] G. Pan, Y. Wu, Z. Wu, and W. Liu. 3d face recognition by profile and surface matching. In *Neural Networks, 2003. Proceedings of the International Joint Conference on*, volume 3, pages 2169–2174 vol.3, 2003.

[80] G. Pan, Z. Wu, and Y. Pan. Automatic 3d face verification from range data. In *Multimedia and Expo, 2003. ICME '03. Proceedings. 2003 International Conference on*, volume 3, pages III–133–6 vol.3, 2003.

[81] T. Papatheodorou and D. Rueckert. Evaluation of automatic 4d face recognition using surface and texture registration. In *Automatic Face and Gesture Recognition, 2004. Proceedings. Sixth IEEE International Conference on* , pages 321–326, 2004.

[82] E. Paquet, H. H. Arsenault, and M. Rioux. Recognition of faces from range images by means of the phase fourier transform. *Pure and Applied Optics*, 4(6):709–721, 1995.

[83] G. Passalis, I. A. Kakadiaris, T. Theoharis, G. Toderici, and N. Murtuza. Evaluation of 3d face recognition in the presence of facial expressions: an annotated deformable model approach. In *Computer Vision and Pattern Recognition, 2005 IEEE Computer Society Conference on*, volume 3, pages 171–171, 2005.

[84] P. S. Penev and J. J. Atick. Local feature analysis: a general statistical theory for object representation. *Network: Computation in Neural Systems*, 7:477–500, 1996.

[85] P. J. Phillips, P. J. Flynn, T. Scruggs, K. W. Bowyer, J. Chang, K. Hoffman, J. Marques, J. Min, and W. Worek. Overview of the face recognition grand challenge. In *Computer Vision and Pattern Recognition, 2005. CVPR 2005. IEEE Computer Society Conference on*, volume 1, pages 947–954 vol. 1, 2005.

[86] P. J. Phillips, P. J. Flynn, T. Scruggs, K. W. Bowyer, and W. Worek. Preliminary face recognition grand challenge results. In *Automatic Face and Gesture Recognition, 2006. FGR 2006. 7th International Conference on*, pages 15–24, 2006.

[87] P. J. Phillips, A. Martin, C. L. Wilson, and M. Przybocki. An introduction to evaluating biometric systems. *Computer*, 33(2):56–63, 2000.

[88] P. J. Phillips, H. Moon, P. Rauss, and S. A. Rizvi. The feret evaluation methodology for face-recognition algorithms. In *Computer Vision and Pattern Recognition, 1997. Proceedings., 1997 IEEE Computer Society Conference on*, pages 137–143, 1997.

[89] P. J. Phillips, H. Moon, S. A. Rizvi, and P. J. Rauss. The feret evaluation methodology for face-recognition algorithms. *Pattern Analysis and Machine Intelligence, IEEE Transactions on*, 22(10):1090–1104, 2000.

[90] P. J. Phillips, H. Wechsler, J. Huang, and P. J. Rauss. The feret database and evaluation procedure for face-recognition. *Image and Vision Computing*, 16(5):295–306, 1998.

[91] P.J. Phillips, P. Grother, R. J. Micheals, D. M. Blackburn, E. Tabassi, and J. M. Bone. Frvt 2002: Overview and summary. available at www.frvt.org, March 2003.

[92] S. S. Rakover and B. Cahlon. *Face Recognition: Cognitive and Computational Processes*. John Benjamins Publishing Company, Amsterdam, 2001.

[93] N. K. Ratha, A. Senior, R. M. Boile, S. Singh, N. Murshed, and W. Kropatsch. Automated biometrics. In S. Singh, N. Murshed, and W. Kropatsch, editors, *Advances in Pattern Recognition - ICAPR 2001. Second International Conference. Proceedings (Lecture Notes in Computer Science Vol.2013)*, pages 445–453. Springer-Verlag, IBM Thomas J. Watson Res. Center, Yorktown Heights, NY, USA, 2001.

[94] S. A. Rizvi, H. Moon, and P. J. Phillips. The feret verification testing protocol for face recognition algorithms. In *Automatic Face and Gesture Recognition, 1998. Proceedings. Third IEEE International Conference on*, pages 48–53, 1998.

[95] S. L. Rogers. *The Personal Identification of Living Individuals*. Charles C Thomas, Springfield, Illinois, 1986.

[96] T. D. Russ, M. W. Koch, and C. Q. Little. 3d facial recognition: a quantitative analysis. In *Security Technology, 2004. 38th Annual 2004 International Carnahan Conference on*, pages 338–344, 2004.

[97] T. D. Russ, M. W. Koch, and C. Q. Little. A 2d range hausdorff approach for 3d face recognition. In *Computer Vision and Pattern Recognition, 2005 IEEE Computer Society Conference on*, volume 3, pages 169–169, 2005.

[98] F. Samaria and S. Young. Hmm-based architecture for face identification. *Image and Vision Computing*, 12(8):537–543, 1994.

[99] B. Scheck, P. Neufeld, and J. Dwyer. *Actual Innocence*. Doubleday, New York, 2000.

[100] A. Scheenstra, A. Ruifrok, and R. C. Veltkarnp. A survey of 3d face recognition methods. In T. Kanade, A. K. Jain, and N. K. Ratha, editors, *Audio- and Video-Based Biometric Person Authentication. 5th International Conference, AVBPA 2005. Proceedings (Lecture Notes in Computer Science Vol.3546)*, pages 891–899. Springer-Verlag, Inst. of Inf. & Comput. Sci., Utrecht Univ., Netherlands, 2005.

[101] W. Shen, M. Surette, and R. Khanna. Evaluation of automated biometrics-based identification and verification systems. *Proceedings of the IEEE*, 85(9):1464–1478, 1997.

[102] A. Srivastava, X. Liu, and C. Heher. Face recognition using optimal linear components of range images. Accepted for publishing in Image and Vision Computing, 2003.

[103] H. T. Tanaka and M. Ikeda. Curvature-based face surface recognition using spherical correlation-principal directions for curved object recognition. In *Pattern Recognition, 1996., Proceedings of the 13th International Conference on*, volume 3, pages 638–642 vol.3, 1996.

[104] H. T. Tanaka, M. Ikeda, and H. Chiaki. Curvature-based face surface recognition using spherical correlation. principal directions for curved object recognition. In *Automatic Face and Gesture Recognition, 1998. Proceedings. Third IEEE International Conference on*, pages 372–377, 1998.

[105] F. Tsalakanidou, S. Malassiotis, and M. G. Strintzis. Exploitation of 3d images for face authentication under pose and illumination variations. In *3D Data Processing, Visualization and Transmission, 2004. 3DPVT 2004. Proceedings. 2nd International Symposium on*, pages 50–57, 2004.

[106] F. Tsalakanidou, S. Malassiotis, and M. G. Strintzis. Integration of 2d and 3d images for enhanced face authentication. In *Automatic Face and Gesture Recognition, 2004. Proceedings. Sixth IEEE International Conference on*, pages 266–271, 2004.

[107] F. Tsalakanidou, S. Malassiotis, and M. G. Strintzis. Face localization and authentication using color and depth images. *Image Processing, IEEE Transactions on*, 14(2):152–168, 2005.

[108] F. Tsalakanidou, D. Tzovaras, and M. G. Strintzis. Use of depth and colour eigenfaces for face recognition. *Pattern Recognition Letters*, 24(9):1427–1435, 2003.

[109] S. Tsutsumi, S. Kikuchi, and M. Nakajima. Face identification using a 3d gray-scale image-a method for lessening restrictions on facial directions. In *Automatic Face and Gesture Recognition, 1998. Proceedings. Third IEEE International Conference on*, pages 306–311, 1998.

[110] M. Turk and A. Pentland. Eigenfaces for recognition. *Journal of Cognitive Neuroscience*, 3:7186, 1991.

[111] Y. Wang and C. Chua. Face recognition from 2d and 3d images using 3d gabor filters. *Image and Vision Computing*, 23(11):1018–1028, 2005.

[112] Y. Wang and C. Chua. Robust face recognition from 2d and 3d images using structural hausdorff distanec. *Image and Vision Computing*, 24:176–185, 2006.

[113] Y. Wang, C. Chua, and Y. Ho. Facial feature detection and face recognition from 2d and 3d images. *Pattern Recognition Letters*, 23(10):1191–1202, 2002.

[114] Z. Wang and E. P. Simoncelli. Translation insensitive image similarity in complex wavelet domain. In *Acoustics, Speech, and Signal Processing, 2005. Proceedings. (ICASSP '05). IEEE International Conference on*, volume 2, pages 573–576, 2005.

[115] H. Wechsler, P. J. Phillips, V. Bruce, F. F. Soulie, , and T. S. Huang, editors. *Face Recognition: From Theory to Applications*. Springer-Verlag, Berlin, 1998.

[116] G. L. Wells and E. P. Seelau. Eyewitness identification: Psychological research and legal policy on lineups. *Psychology, Public Policy, and Law*, 1:765–791, 1995.

[117] L. Wiskott, J.-M. Fellous, N. Kuiger, and C. von der Malsburg. Face recognition by elastic bunch graph matching. *Pattern Analysis and Machine Intelligence, IEEE Transactions on*, 19(7):775–779, 1997.

[118] www.frvt.org, 2002.

[119] C. Xu, Y. Wang, T. Tan, and L. Quan. Automatic 3d face recognition combining global geometric features with local shape variation information. In *Automatic Face and Gesture Recognition, 2004. Proceedings. Sixth IEEE International Conference on*, pages 308–313, 2004.

[120] Y. Yacoob and L. Davis. Labeling of human face components from range data. In *Computer Vision and Pattern Recognition, 1993. Proceedings CVPR '93., 1993 IEEE Computer Society Conference on*, pages 592–593, 1993.

[121] C. Zhang and F. S. Cohen. 3-d face structure extraction and recognition from images using 3-d morphing and distance mapping. *Image Processing, IEEE Transactions on*, 11(11):1249–1259, 2002.

[122] L. Zhang, A. Razdan, G. Farin, J. Femiani, M. Bae, and C. Lockwood. 3d face authentication and recognition based on bilateral symmetry analysis. *Visual Computer*, 22(1):43–55, 2006.

[123] W. Zhao, R. Chellappa, and P. Phillips. Subspace linear discriminant analysis for face recognition. Technical Report CAR-TR-914, Center for Automation Research, University of Maryland, College Park, 1999.

[124] W. Zhao, R. Chellappa, P. J. Phillips, and A. Rosenfeld. Face recognition: a literature survey. *ACM Computing Surveys*, 35(4):399–459, 2003.

In: Pattern Recognition in Biology
Editor: Marsha S. Corrigan, pp. 105-148
ISBN 978-1-60021-716-6

Chapter 4

PSYCHOLOGICAL STUDIES OF THE CRITERIA FOR 3D FREE FORM SHAPE MATCHING

Yonghuai Liu, Horst Holstein and Mark Lee
Department of Computer Science
The University of Wales, Aberystwyth
Ceredigion SY23 3DB, Wales, UK

Abstract

The traditional closest point criterion (CPC) has been widely used for 3D image registration, object recognition, and shape matching. In this chapter, we employ psychological studies to test the hypothesis whether human beings often either consciously or subconsciously apply this criterion to establish correspondences between overlapping 3D free form shapes captured from different nearby viewpoints. To this end, we design four sets of data with different tasks and invite a class of year 2 undergraduates at a university and two classes of year 7 and year 9 children at a comprehensive school respectively for test.

After experiments, the answer books were collected and carefully compared against the standard answers determined at the phase of data design. The comments on why participants perform the given tasks like that were collected and summarized. Both correlation analyses and significance tests were performed between the average marks on each question and the participants in different age groups, showing that age has no significant impact on the average performance of participants.

What the participants drew and wrote on the answer papers and commented on why they performed the given tasks like that show clearly that no matter whether the participants in the different age groups realized or not, (1) the CPC has been widely used to perform the 3D free form shape matching tasks in one way or another, (2) the CPC is more likely to be inconsistently used especially by younger participants, and (3) the CPC is often used by younger participants in its brute force form as one-to-one mapping, many-to-one mapping and one-to-many mapping. As a side effect of these studies, a number of useful criteria such as one-to-one mapping, relative motion and distance, and overlapping area maximization have been revealed. The psychological studies performed thus indeed deepen our understanding of the traditional CPC for 3D free form shape matching and are fruitful, as expected. The findings from the psychological studies described in this chapter will be useful for the development of novel automatic 3D free form shape matching algorithms.

1. Introduction

1.1. Motivation

Learning is a basic skill for human beings to live a normal life and solve complex problems in understanding the real world. The more competent a person is at learning, the more intelligent s/he is. After birth, human beings consciously or subconsciously learn from their parents, friends, teachers, and surroundings. Even though children are not taught how to perceive the 3D world, they do succeed in fostering the ability to interpret the 3D world around them and take appropriate actions accordingly. The problem is: How does this ability be developed with one's growth? To what extent is this ability gained from birth? Whether is it possible to teach the skills to perceive and interpret the real world? Do they subconsciously learn something by themselves?

Theoretically, it is not too difficult for human beings to construct a 3D model from two overlapping projective images captured by their two eyes [15, 22]. Nowadays, it is completely feasible to use laser range finders to directly capture 3D information of the objects of interest [17]. In Figure 1, the 3D perception and interpretation means: from the two 3D images captured, how does one identify which point in one image corresponds to a point in another; and from the resulting point matches, how does one estimate the motion of the viewer relative to the static 3D shape, or the motion of the 3D shape relative to the static viewer? Once the former has been solved, the latter becomes an easy task and vice versa. While the former is important for the perception of 3D geometry of the objects of interest, the latter is important for action and path planning. For example, a person walking on a highway road sees a car coming toward him, how does he decide whether to continue to walk on the road or to go away? The final decision depends on his guess of the car speed and moving direction and also the distance from him to the car. If the car has slowed down or changed direction, then he probably can continue to walk on the road. If the distance from him to the car is short, then he has to go away and take his action as soon as possible. The establishment of point matches in different images is at the heart of his guess.

1.2. Purpose

To facilitate 3D shape interpretation, different images captured from different viewpoints are often synthesised into a single coordinate system without changing the coordinates of any point [19], as demonstrated in Figure 1. This mimics such a scenario that the camera (viewer) is static and objects are in motion. Even children can recognise that 2-a, 3-b, 4-c, 5-d, 6-e are possible point matches, representing the same physical points on the 3D shape respectively, and all others representing appearing and disappearing points in different views. The problem is, how do they identify the possible point matches in different images captured from different viewpoints? Do the closest points in the Euclidean distance sense in the synthesized image represented as numbers 1 through 6 and letters a through g in Figure 1 represent correct point matches between two different overlapping images? Is there any rationale behind this closest point criterion (CPC)? In other words, under what conditions may children identify 2-b and 3-a, for example, as possible point matches? The purpose of this research is to experimentally investigate these interesting but challenging issues.

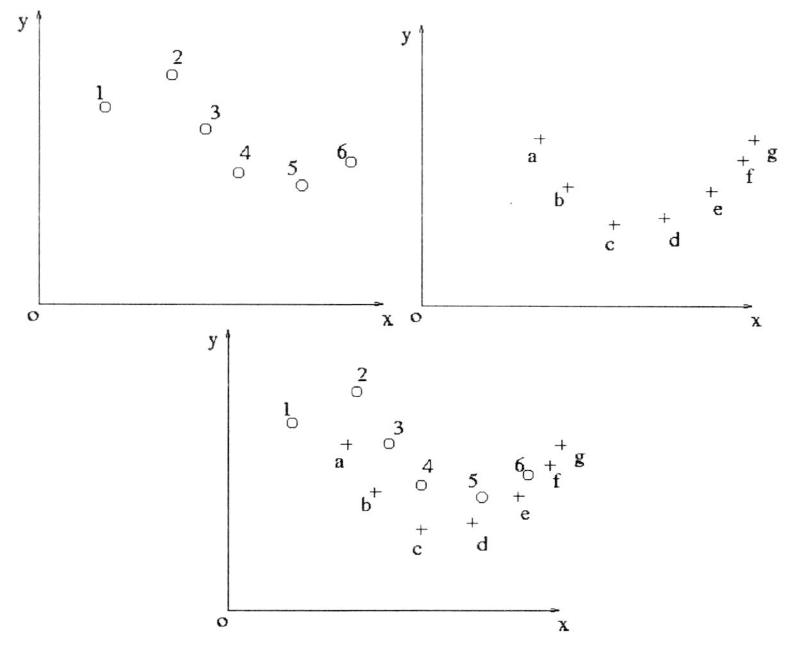

Figure 1. Left two: two images represent the same 3D shape captured from two different nearby viewpoints respectively. Right: the images captured are synthesized into a single coordinate system without changing the coordinates of any point. Here the 3D shapes are represented as discrete circles and plus signs respectively.

1.3. Significance

This research is important at least due to the following two reasons: (1) the research has a potential to reveal how human beings perceive and interpret the 3D world in different views. The implication of this research is to improve the learning efficiency like drawing skills of human beings; (2) the outcome of this research can be applied to develop novel algorithms for shape matching [8], object modelling [9, 1], 3D object recognition [10], and Internet searching [7], which find applications in machine vision and medical imaging. Thus, some patients can benefit from this research since body parts can be modelled and thus, operation process can be monitored and operation quality can ultimately be improved.

1.4. Related Work

3D free form shape matching has attracted a great deal of attention from the psychology community [12, 13, 18]. While psychological studies often use either real objects or images as stimuli, participants may subconsciously extract and store the visual, semantic and verbal representations of objects in their mind for matching against the test object [13]. For example, morph sets of textured, shaded, 3D models of familiar objects were used in [12, 13]. In [18], 2D representations of four 3D objects: an arch, a barrel, a brick, and a wedge were used. In [4], shape matching was compared in different sizes, colours, and translational motions with color matching. The shapes were represented as colored random polygons. In [3], a set of 12 rendered, simple (two-part), novel 3D objects were used. Each object in the set was depicted at two orientations in depth. In [20], each object consisted of ten solid cubes attached face-to-face to form a right-angled structure with exactly three right angled "elbows ". In [6], shaded grey-scale images of novel segmented tube-like objects were used. Even though novel objects were used in experiments [13, 6], training still enables participants to familiarize themselves with the objects and thus enables participants to be influenced with regard to subsequent matching performance. Participants may associate their life experience with such objects or images and thus facilitate them in identifying the overall shape or components of objects. In return, they show optimistic performance in the psychological tests without revealing the subconscious process taking place.

On the other hand, the data for experiments were often subject to a large motion, with a rotation angle up to 240° [13], 120° [3], and 340° [20] in depth, and as far as 130° apart [6]. The objects perceived from different viewpoints are very different. In such cases, it is understandable that matching performance is view sensitive, due to the simple yet insurmountable fact that unless all of the stimuli have been seen, complete subjective information is not assured. In order to obtain detailed description, participants have to look carefully at the objects either for the extraction of invariants [3] from each view, due to a limited field of view or occlusion, appearance and disappearance of parts of the stimuli, or for the performance of mental rotation [20]. While such large motions are not realistic at all for laser scanning systems in the process of data capturing and do not simulate human binocular vision, we consider in this chapter small motions to allow participants to draw reliable conclusions. Small motions are generally those that have rotation angles smaller than 40°, for example, around 3D rotation axes.

1.5. Our Work

Specifically, the main purpose of psychological studies described in this chapter is twofold: (1) to test the hypothesis whether the CPC is usually a default choice for human beings to interpret 3D overlapping free form shapes, and (2) to reveal what criteria are usually employed by human beings to interpret 3D overlapping free form shapes. To this end, we follow the approach proposed in [23]. The objects are first aligned, then the outcome of recognition is decided based on the goodness of the fit between the object and the model. While the alignment is often a mental process and the recognition of objects is a focus of study [20], in this chapter we attempt to employ psychological studies to reveal what criteria will be used by participants to establish correspondences between the representation of the objects perceived from two different nearby viewpoints so that the transformation for alignment can be recovered. The experimental results can be analysed from two aspects: (1) compare the drawings and the answers from participants against the references determined at the phase of experimental design, and (2) do statistics on the comments from participants. While response time and error rates of object recognition are often used to analyse the performance of subjects in experiments [3, 6], we use error rates of point and line matches for performance analysis with an attempt to reveal what novel computational model can be developed for overlapping 3D free form shape matching.

We use either points or lines (representation of features or points) to represent the 3D free form shapes to be matched. In this case, participants may have to reconstruct the object model from the points and decide how to measure the similarity between different lines. For model reconstruction, they need to consider not only the points themselves, but also the information obtained from a second viewpoint. The reconstruction process may involve complex conscious and subconscious interactions between perception, abstraction, association, remembering, discrimination, assembly, and categorization. Thus, while 3D shape matching based on point patterns is significantly more abstract and thus more difficult than that based on real objects or images [20, 3, 6, 18], it may reveal some subjective criteria for 3D free form shape matching and thus inspire the novel automatic 3D free form shape matching algorithm development.

While existing experiments often present participants with stimuli one by one [13], we present stimuli to participants once and then provide only necessary verbal explanations or visual demonstrations on point matching, but not on line matching. Since all participants have no knowledge at all about 3D point pattern matching, they have to utilize any intuition to carry out their given tasks and adjust their time and effort to respond to different tasks. Even though we limit the overall time for experiments, the time given should be sufficient for active participants to respond to the given tasks. Eventually, we can obtain reliable data and draw useful conclusions that reflect the actual activities and performance of participants.

Even though participants have no knowledge at all about point pattern matching and line similarity measurement, such a study is still feasible and desirable, since at an age of just five months, human beings begin to learn to segregate and recognize different objects using boundaries, features, depth, shape, colour, texture, and pattern information [5] and our aim is to reveal how the participants respond to and comment on the given tasks. While psychological studies show that objects may be represented as collections of specific views [6] and view sensitivity increases when similar objects need to be discriminated [12, 13], we expect that 3D shape matching of point patterns is view sensitive, since in this case, participants can hardly figure out the overall shape or components of the object that the point patterns represent. Thus it is difficult for them to use their life experience to help point pattern matching. In order to establish plausible point matches, they have to consider not only the point patterns in 3D space, occlusion, appearance and disappearance of points, but relative motion of different points in 3D space as well.

While pattern recognition is to develop computational models on how to represent and compare objects of interest, we should not forget that we are trying to simulate human beings' behaviour and cognitive intelligence. In this case, psychological studies maybe a natural choice for the pattern recognition community to reveal how human beings do the same tasks so that their behaviours can be simulated using computers. Psychological studies described in this chapter reveals that the CPC is often a default choice for human beings to interpret 3D overlapping point patterns. As a side effect of the psychological studies reported in this chapter, some criteria either consciously or subconsciously used by participants have been revealed. These criteria will be valuable for novel 3D free form shape matching algorithm development in the matching vision literature.

The rest of this chapter is structured as follows: Section 2. describes how the data are generated for the psychological test and how the psychological tests are implemented, Section 3. describes the typical comments from the participants on the psychological test, and Section 4. analyses the psychological test results, while Section 5. discusses a number of criteria revealed from the psychological test. Finally, some conclusions are drawn in

Section 6.

2. Method

2.1. Design

In the following sections, we use psychological tests to investigate whether the CPC has been either consciously or subconsciously used by human beings to interpret a 3D free form shape perceived from different nearby viewpoints. The experiments were designed in four series with different tasks: (1) Participants are asked to join points represented as plus signs and circles from two overlapping 3D point clouds that were thought to represent the same physical points on the object surface. The purpose of this series of experiments is to test what criteria are used to establish point matches between the two overlapping 3D point clouds. In order to find correct point matches, participants have to imagine both the object surface and camera motion in 3D space; (2) Given a point pattern in a coordinate system, participants are asked to duplicate this point pattern into a new coordinate system. The objective of this series of experiments is to examine whether the topology is conserved in the process of the duplication; (3) Given that each object view is represented as a feature line and various feature lines are stored in a database [7], participants are asked to identify the most similar line among the candidate ones to the query one. The objective of this series of experiments is to identify what metric (Euclidean, χ^2, or Kullback-Leibner) is used to search for the most similar line in the database to the query one; and finally (4) Given the experience that a pair of similar lines from different data sets has been found, participants are asked to select the most similar line in one set to a given line in another. The purpose of this series of experiments is to test whether lessons from past experience in identifying similar lines are learned and the topological relationship between lines is conserved in finding the most similar line in one data set to the query one in another. To this end, the generation of relative data is described as follows.

2.1.1. Point Match Establishment

A 3D shape is represented as sparse points. For the sake of easy recognition and application of 3D shape information and camera motion, the number of points is limited in the range of [8, 14]. Too many points used to represent 3D shape render the figures messy and thus make it difficult for participants to come up with sound conclusions about real point matches. For establishing possible point matches between two overlapping point clouds, it is perceivable that participants have to consider not only the point matches in the sense of closest points, but also the expected camera motion and local or even global shape information. Thus, to facilitate participants to establish correct point matches between two overlapping point clouds in 3D space, the designing of experiments must make sure that the points in the point clouds are relatively well distributed.

Point data were generated using both computer simulation and real images. In this case, the data used for psychological studies and actual 3D free form shape matching algorithm validation are the same so that the findings from the psychological studies can be readily adapted for novel 3D free form shape matching algorithm development. While points generated using computer simulation are generally well distributed and the occlusion, appearance and disappearance of points in different point clouds can be controlled, the distribution of points from sampling real images is often beyond control. Consequently, the point patterns generated using computer simulation and from sampling real image points provide varying degrees of complexity for matching.

The procedure for point pattern generation based on computer simulation is described as follows. A set of n points $\mathbf{p}_i (i = 1, 2, \cdots, n)$ were first randomly generated with uniform distribution within $[5, 105] \times [5, 105] \times [5, 105]$ which were then subject to a rotation angle θ around a rotation axis $\mathbf{h}$ (subject to normalization) randomly generated with uniform distribution within $[1, 2] \times [1, 2] \times [1, 2]$ followed by a translation vector (camera position) $\mathbf{t}$ randomly generated with uniform distribution within $[10, 110] \times [10, 110] \times [10, 110]$. Let the transformed points be $\mathbf{p}'_i (i = 1, 2, \cdots, n)$. Thus we have precise knowledge about the 3D point matches $(\mathbf{p}_i, \mathbf{p}'_i)$ $(i = 1, 2, \cdots, n | n \geq 3)$ before and after a rigid camera motion. This is used as reference for the performance evaluation of participants' point match establishment. For the simulation of real world noisy data, Gaussian random noise with mean equal to zero and standard deviation $\sigma = 0.04$ was added to the coordinates of each point. For the simulation of appearance and disappearance of points (Figure 2), the last $\alpha \times 100\%$ points in the data shape $\mathbf{P}$=$\{\mathbf{p}_1, \mathbf{p}_2, \cdots, \mathbf{p}_n\}$ and the beginning $\beta \times 100\%$ points in the model shape $\mathbf{P}'$=$\{\mathbf{p}'_1, \mathbf{p}'_2, \cdots, \mathbf{p}'_n\}$ were removed where α and β were randomly generated with uniform distribution within the interval $[0, 0.15]$ and $[0, 0.25]$ respectively. Eventually, the overlap between two point clouds can be varied from 60% to 100%. The purpose for varying the quantity of overlap is to render that no regularities in the data can be found by participants and also make the experiments more interesting. However, it is worth pointing out that since before removing points from $\mathbf{P}$ and $\mathbf{P}'$ respectively for the simulation of disappearing and appearing points, the points were not sorted in any order at all. Thus, disappearing and appearing points can appear anywhere in the point clouds, instead of just boundaries in the real images. For a small motion simulation, the rotation angle θ was chosen as either $5°$ or $10°$. The designed shapes are presented in Figure 3 with n varying from 8 to 14.

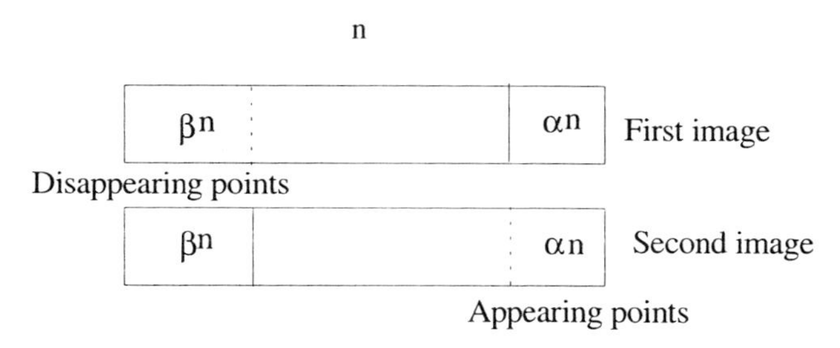

Figure 2. The simulation of the appearance and disappearance of points.

While the points in Figures 3a, 3b, 3e, and 3g well distribute, appearing and disappearing points can easily be recognised. The points in Figures 3c, 3d and 3f do not well distribute. Some points stay close together. Appearing and disappearing points appear anywhere in 3D space. Especially, the points in the middle of the figure from two point sets do not simply represent a one-to-one mapping. Thus, they are more challenging for correct point match establishment. This purpose of this series of experiments is to test whether

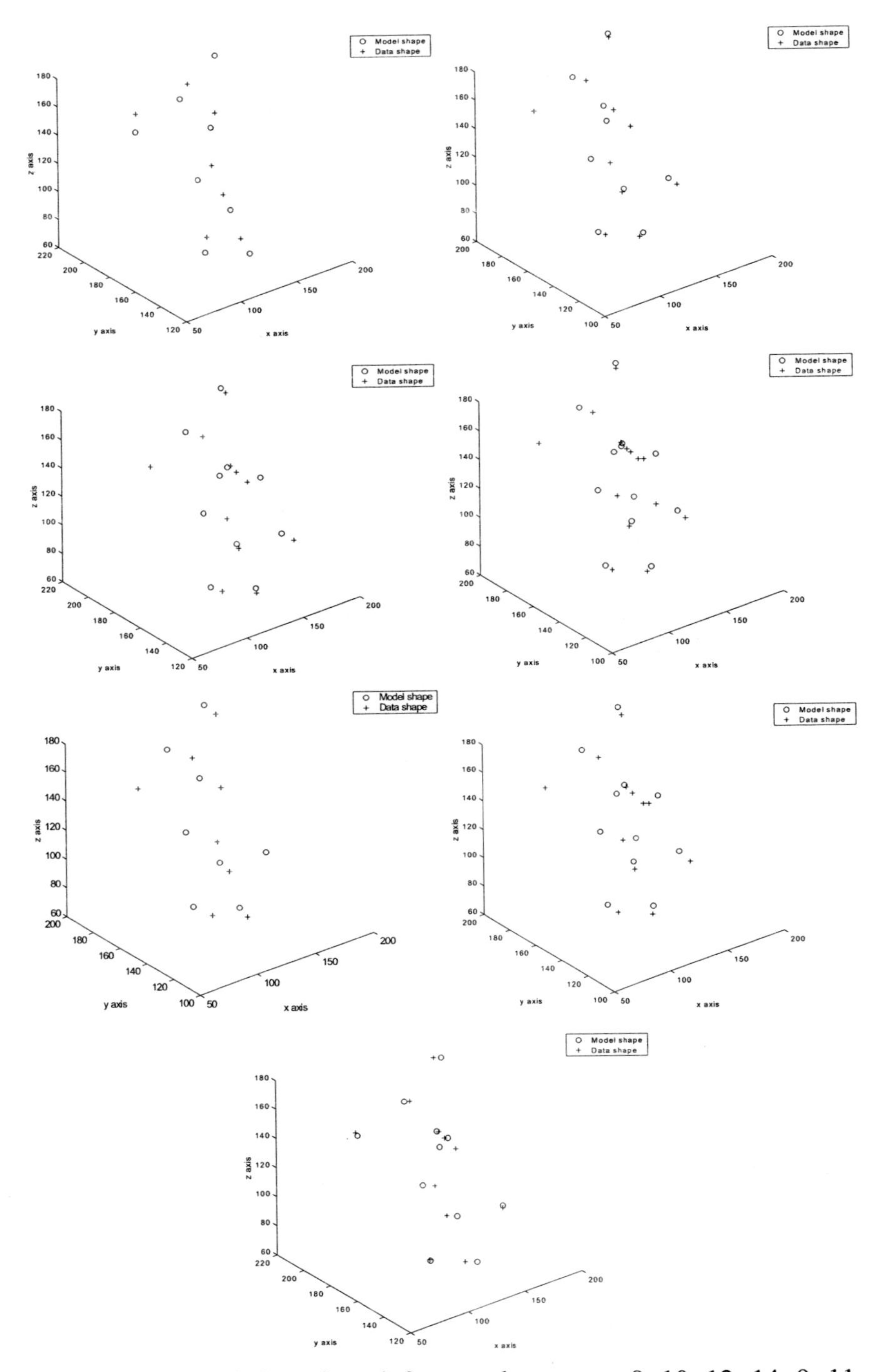

Figure 3. Computer simulation: from left top to bottom: n=8, 10, 12, 14, 9, 11, and 13. Different figures are called: a through g. *Question*: join points in the model and data shapes represented as circles and plus signs respectively that represent the same physical points in 3D space.

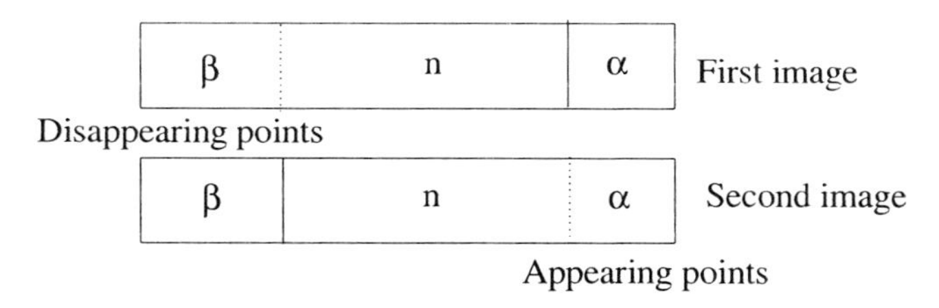

Figure 4. The simulation of appearance and disappearance of points in real images.

participants make use of the CPC in conjunction with the one-to-one mapping to correctly recognise real point matches and appearing and disappearing points between two overlapping point clouds.

The procedure for point pattern generation based on real images is described as follows. The real range images whose sizes are of 200 by 200 pixels were downloaded from a publicly available range image database currently hosted by the Signal Analysis and Machine Perception laboratory at Ohio State University and were subject to a camera motion with a rotation angle of either $20°$ or $36°$ around a 3D axis respectively. For the simulation of two overlapping 3D point clouds (Figure 4), a number of n points were first randomly selected with uniform distribution from the closest point pairs [2] between the two images to be registered where n was randomly generated with uniform distribution within the interval $[4, 10]$ and the first image was subject to a pure translational motion determined as the centroid difference of the two images. For the simulation of appearance and disappearance of points (Figure 4), a number of α points randomly selected with uniform distribution from the second image and a number of β points from the first image were then added to the point pairs obtained where α and β were randomly generated with uniform distribution within the interval $[0, n/3]$ and $[0, n/3]$ respectively. Doing so varies the percentages of appearance and disappearance of points from one experiment to another. The smallest quantity of overlap between two point clouds can be computed as: $n/(n/3 + n + n/3) = 60\%$. This overlap is typical under actual imaging conditions. The designed point sets are presented in Figures 5a through f.

From Figure 5, it can be seen that the points do not distribute well and appear messy. Some points are close together. Some points are far away from each other. Thus, it is even more challenging for correct point match establishment. The purpose of this series of experiments is to test whether the participants can make use of a number of criteria such as closest points, one-to-one mapping, relative motion, relative direction, relative distance, and overlapping area maximization to help establish correct point matches between two overlapping 3D point clouds.

The task for this experiment is to draw lines between corresponding points in the given two overlapping 3D point clouds before and after a rigid camera motion.

2.1.2. Duplicating a Figure

The objective of this series of experiments is to examine whether the topology is conserved when the figures are duplicated. This will provide evidence whether it is necessary for

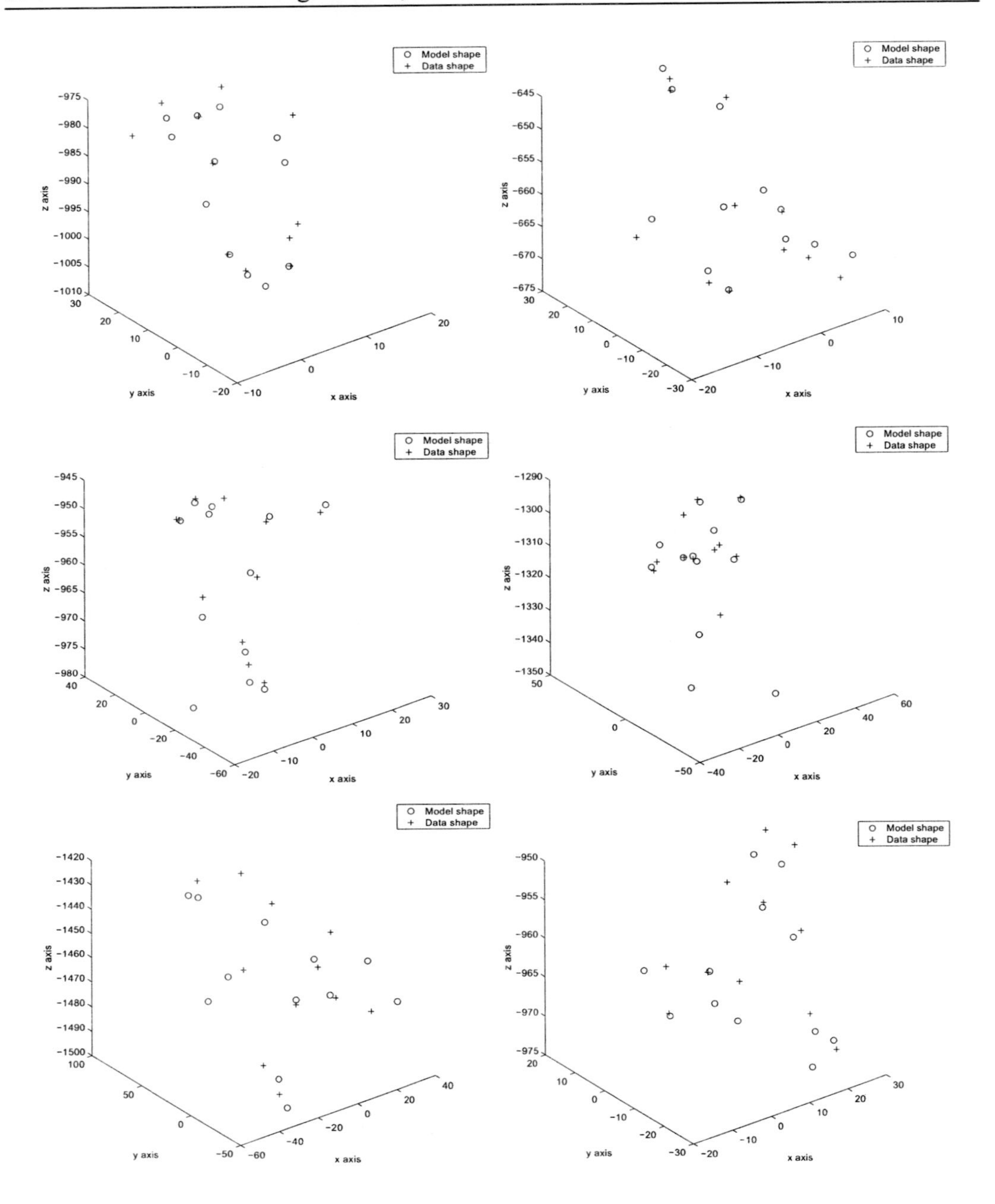

Figure 5. From left top to right bottom: a through f. *Question*: join points in the model and data shapes represented as circles and plus signs respectively that represent the same physical points in 3D space.

algorithms to conserve the topological relationship between points when establishing real point matches between different overlapping free form shapes. The experiments designed are based on real images downloaded from the same source above. The points were selected using the same method as that described above. The designed figures are presented in Figure 6. The task for this experiment is to duplicate point patterns from one grid to another.

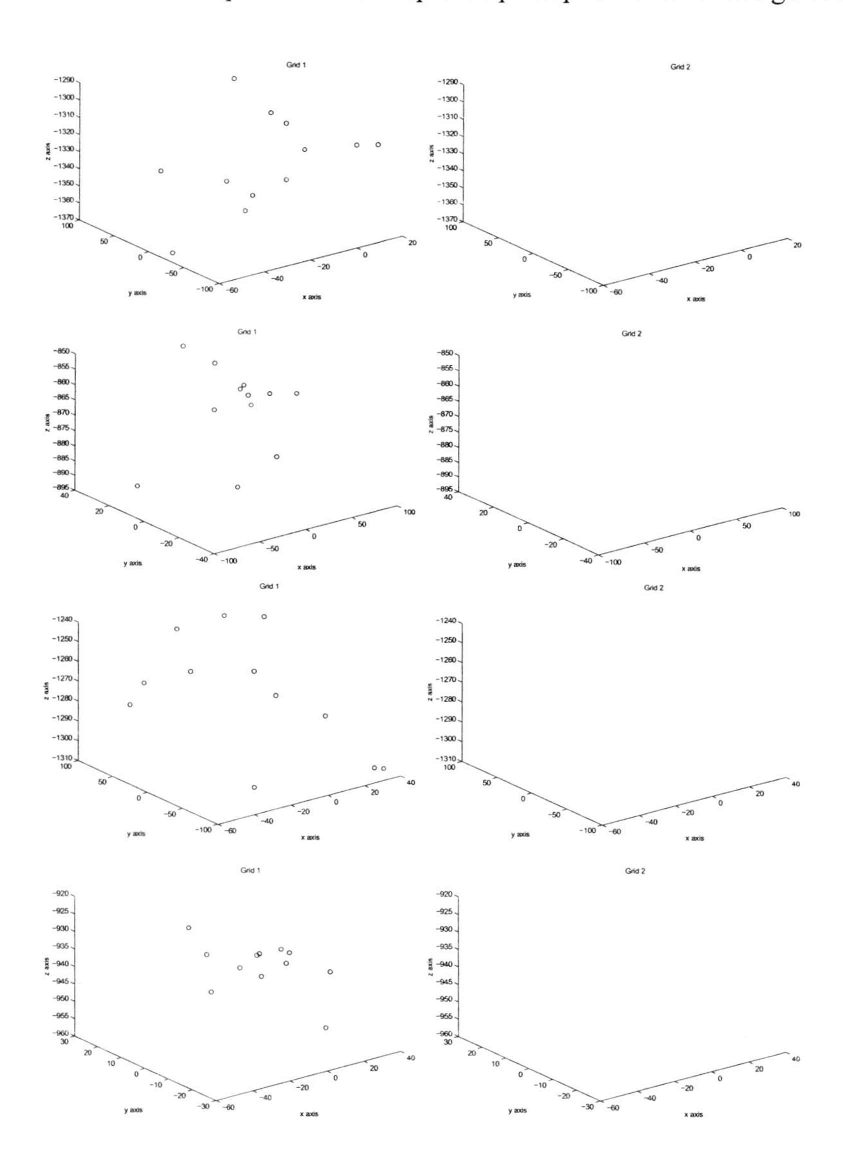

Figure 6. Duplicate the 3D shape represented as points (circles) in the left grid into the right grid.

2.1.3. Recognising 3D Shapes from Their Features

3D shapes are represented as shape distributions [16]. The shape distributions of six objects extracted from a total of 65 real range images downloaded from the same source above are depicted in the following figures. From the shape distribution, the original 3D shape may be recognised. The objective of this series of experiments is to identify what metric (Euclidean, χ^2, or KL) is used to choose the most similar lines. The experiments designed are presented in Figure 7. The task for this experiment is to identify the solid line from the candidate ones most similar to the dash line.

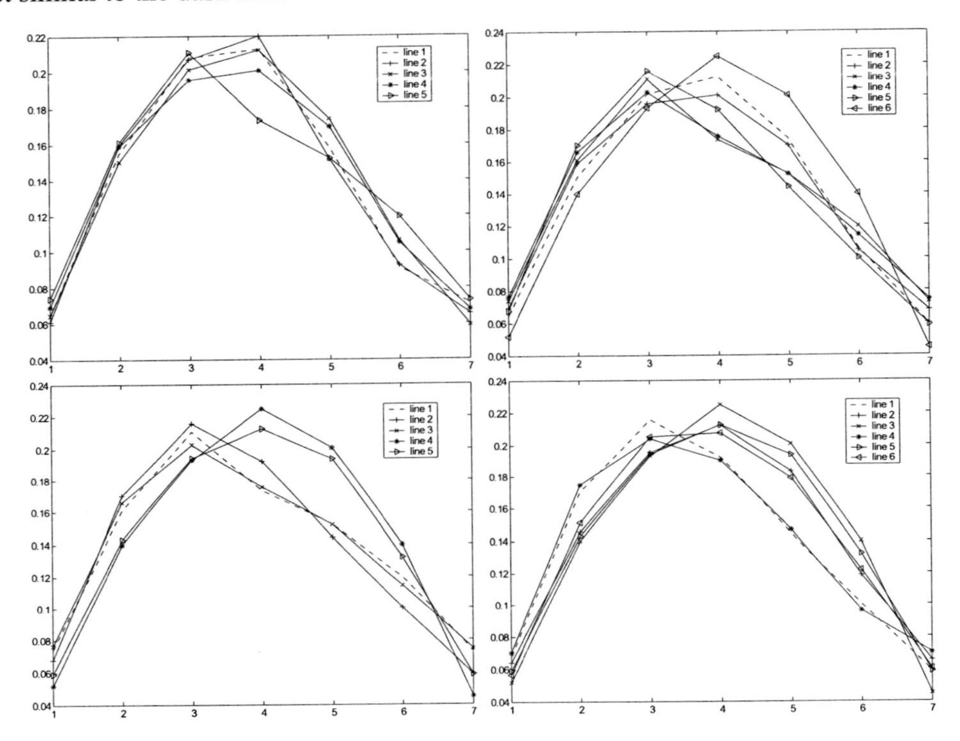

Figure 7. *Question*: Which solid line in the figure is most similar to dash line 1?

2.1.4. Comparing Features of Different 3D Shapes

3D shapes can be represented as features. The objective of this series of experiments is to compare different features from different 3D shapes. Through experiments, it can be revealed whether the feature comparison chooses most similar shapes in the sense of a certain metric and has to conserve the topological relationship between different features. The experiments in this section were designed as follows. Three pairs of point matches are randomly selected with uniform distribution from those established by the CPC between the two images to be registered. For the simulation of appearance and disappearance of points, a number of n points are randomly selected with uniform distribution from the first and second images respectively where n is randomly selected with uniform distribution from the interval [0, 3]. For the sake of avoiding an easy selection of features, we artificially added another line in the last three figures so that the figures created are a bit more confusing with regard to selecting similar features. The experiments designed are presented in Figure 8. The task for this experiment is to identify the dash line in one set most similar to the solid line 2 in another.

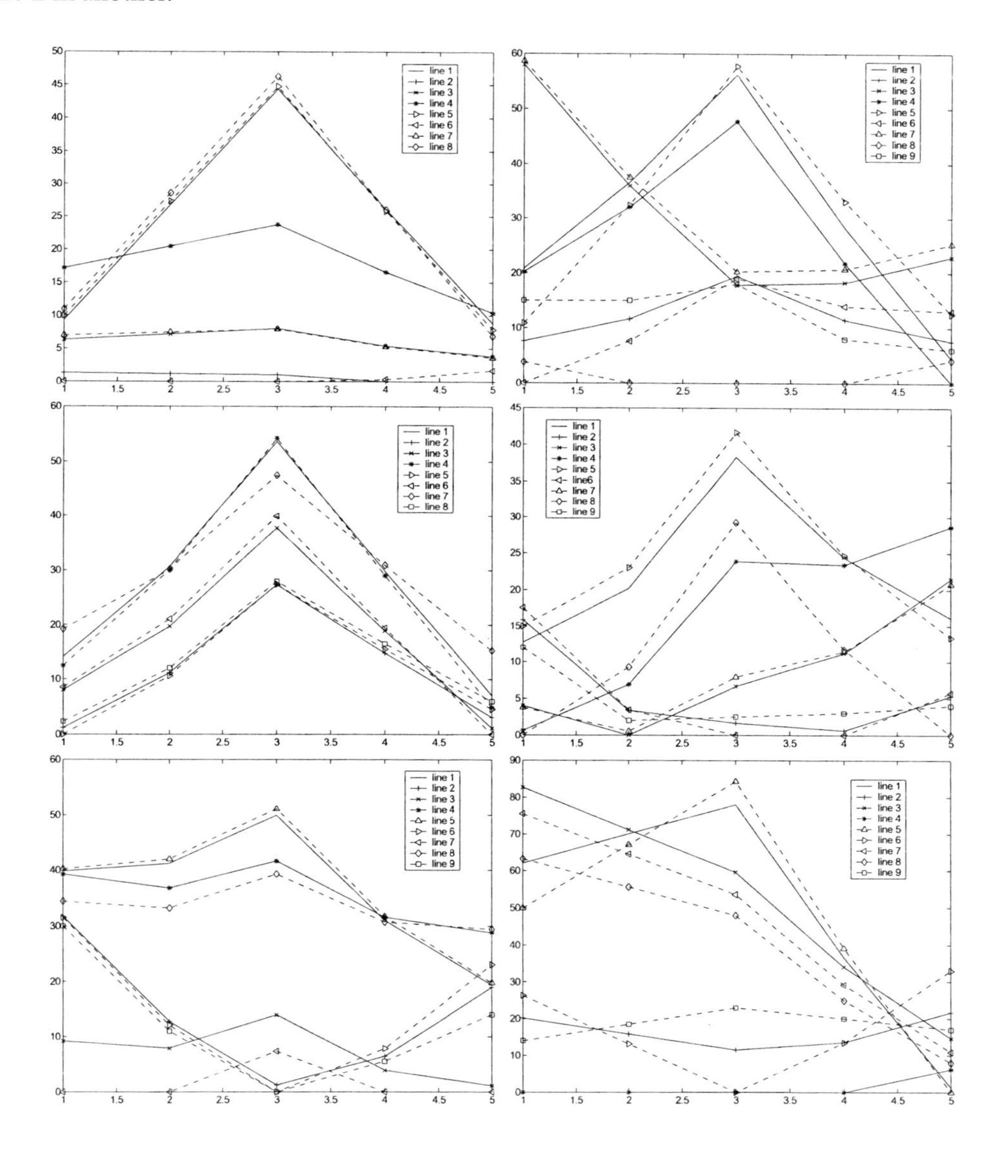

Figure 8. *Question*: given that solid line 1 in model shape is most similar to dash line 5 in data shape, which dash line in the data shape is most similar to solid line 2 in the model shape?

From the experiments designed above, it can be seen that the exact point correspondences from computer simulation, the point matches sampled from real images, the shape distribution from certain objects, and similar lines from the projections of 3D points onto a 2D plane are all known and thus, serve as reference for the performance evaluation of participants.

2.2. Participants

The first test was carried out on 17 November 2004, on 31 second year university students taking the Artificial Intelligence module. The test duration was 30 minutes. The second and third tests took place on 9 February, 2005 and 9 March, 2005, with durations of 45 minutes and 40 minutes respectively, at a comprehensive school. There were 25 year 7 students taking test 2 and 29 year 9 students taking test 3 (Figure 9).

We did not choose particular participants on purpose. All these participants were chosen due to their availability. The reason why some children were invited as experimental subjects is that we would like to obtain genuine insights into how the subjects respond to the given tasks they have never seen and thus their either conscious or subconscious decision making process in performing the given tasks can be revealed. On the other hand, both children and adults were invited as participants so that their behaviours in performing the given tasks can be compared and more objective findings can thus be made about how the participants in different age groups respond to the given point pattern and line matching tasks without any prior knowledge.

Figure 9. Participants in the psychological tests. Left: Year 2 undergraduates; Middle: Year 7 students; Right: Year 9 students.

2.3. Implementation Procedures

After the question papers were distributed to each participant, only necessary explanation was provided about what corresponding points mean. Then the participants worked independently on their own papers. No discussion and communication among the participants in the process of the tests was allowed. If one was not clear about the tasks to be performed, then s/he could ask the relative staff with minimum interference of others possible.

For the undergraduates, it took less than 5 minutes to explain what they were required to do in the experiments. Then the students worked independently on their own papers. But for the year 7 students, it was much harder to explain what they should do in the experiments. So we had to explain first, then drew figures on the whiteboard to illustrate, finally answered their questions and provided more explanations, if necessary. The only requirement for them to join points is that the points represent the same physical points in the three-dimensional space. For the final test, we learned our lessons and thus, we prepared a toy horse to facilitate the explanation of the problem in the test. With an overhead projector, we projected the toy horse onto a whiteboard and then drew 7 points on the joints of horse legs, tails, and heads, then moved the horse a little bit around and drew 10 points. Finally, we joined points before and after the small movement of the horse so that they represent the same leg joints, tails, and heads. Through such a projection based explanation, the participants quickly realized what they should do in the test.

3. Typical Explanations from Participants and Their Analysis

It is interesting and useful to list the explanations from the participants, since the explanations can reveal what knowledge was used to do the tests by the participants. In the following sections, the number of plus signs in the parentheses following an explanation represents the number of times that explanation was mentioned or given by different participants. The more plus signs following an explanation, the more frequently that explanation was given. In order to respect the explanations from the participants, we kept them intact without revising their relative spelling and expression errors.

3.1. Point Match Establishment

3.1.1. Undergraduates

- The closest points is most likely the right choice (+++++++++).
- Nearest if a lot of close circle and plus signs. Look for rotation patterns, some going left direction some right direction. Try to make lines of similar length–in theory every point has moved the same distance(ish). guesswork(+).
- So the lines went in the same general direction while joining as many as possible (+).
- They point the same way and move by roughly the same amount(+).
- Mostly by distance and alignment of points toward a certain location. Quite a bit is just intuitive guesswork(+).
- In this instance, the possibility of point close together, then point of the same structure and also their similar direction and length between the two point sets. Consistency also(+).

- Even if the points have been rotated in 3D they will still be about the same distance from each other(++).
- Just draw what I saw matched up(+).
- Not too sure, but it looks right to me(+).
- Where points exist that occupy a similar small region, the points were joined. The exception is where a shape in both data sets is clearly visible; a shape is then transposed to the 2nd data set(+).
- Try and keep a consistent angle (relative motion)(+).
- I try to visualise the points in a shape and then rotate and skew the points to match patterns(+).
- I had a look and tried to make it look it the similar shape in order to see if I could get the same point, but in some cases this failed so I visit for what sound right(+).
- I joined the points like this because this is how I interpret them using the given graph(+).
- Chose the most logical points to join(+).
- Play(+).
- no idea(+).

3.1.2. Year 7 Children

- Because the plus signs and circles should be joined is there is the opposite symbol is near(+).
- because they are close to each other and they are easier to match up(+).
- I joined the points like this because I joined the closest points and crosses together. It is complicated and strange(+).
- I didn't understand it properly but it was OK(+).
- I don't know(+).
- This is very complicated. I don't really now what I am doing and it is strange(++).
- It was generally easy because I just joined the closest crosses to the noughts together(+).
- I do not really no if my work is right or wrong(+).
- Because it seems like the right answer and because it seems line they are both the same point. I think this test was helpful to us when it comes to the computers(+).

- I found this very hard, but it was beginning to get easier as I got to the end and also think that it would be easier if it was explained better and more understandable(+).
- At first I found it really hard at first then I started to get what he was saying(+).
- This was hard(++).
- I found this difficult and found it hard to find some sort of object but as I went on it made it clearer what we will see(+).
- it's a clever way to use your imagination to think(+).
- it makes a imagination come to life and makes you think(+).
- I try to make pictures of a cartoon. I think it was very hard(+).
- Random. Very hard don't know what I'm doing(++).
- I joined the dots like that because they look like people walking(+).
- At the start I found it really confusing but when he said to try make face shapes and changes(+).

3.1.3. Year 9 Children

- Because they are closer and make more sense than half way across the page(+);
- Because they make sense because the cross and circle are close together(++++);
- Because it was the first thing I thought of. Its very hard and confusing(+);
- Because I think if a x and o are close together they should join up. This is fun but its very hard and I'm starting to get a headache from all the dots(+);
- because the object seemed to have moved that way(+);
- Because if it's a small moment they will probably all be small(+);
- because when the points come together I can tell which direction the object has been moved and when there are missing points I can tell that the position of the object has been changed slitely. I can imagine that almost all the object are animal figures. I think questions 7 and 8 are human faces(+);
- because they just look right. It is easy(+);
- because they are close together because they wouldn't be that far apart. That is what your brain sees and that's what you think that's why I have done the points like that and it just seems appropriate(+);
- because this is how I think that they are meant to be joined up. I think it was confusing(+);

- I think it's the outside of the shape the main points(+);
- I put them like that because the data shape points look like they have moved higher than the model shape points. I find it very difficult to imagine the object(+);
- It looks like something that I recognize but I can't name it. It's very confusing I don't know what I'am doing this for(+);
- all my patterns are facing upwards(+++);
- I joined the points like this because I was trying to imagine what 3D shape would be there. My comments are that it was quite hard because I was using my brain more than I usually(+);
- Because it looks like you have to do it like that because they are close together and most of it was just guessing because it didn't resemble anything(+);
- I joined the points where I thought they should go. There is mostly always a circle left(+);
- I joined the points like this because it seemed to me that they corresponded with each other(+);
- The one's without a partner are not joined up and the one's which are joined up I think have a partner(+);
- It is confusing and difficult probably not mean use knowing what they represent(+);
- I don't know(+);

Within 85 participants altogether, some did not explain why they joined points like that. But from the above lists of explanations, it can be clearly seen that the CPC has been consciously and most frequently used by 9 out of 31 undergraduate participants, 3 out of 25 year 7 children and 8 out of 29 year 9 children. If including those who used the CPC as one of the criteria, then 12 out of 31 undergraduate participants explicitly used the CPC for point match establishment. Note that within all possible criteria for point match establishment, only the CPC is the most commonly and consciously used. All other criteria were used by significantly fewer participants. This clearly shows that the CPC is really a good choice for human beings to interpret 3D shapes from different viewpoints.

At the beginning of the test, it is difficult to explain to the year 7 children what was required in the test. Then a human face had to be used as an example. Unfortunately, they then always tried to imagine the shape that the points may represent, even though we emphasized that join the points, as long as you think they represent the same physical points in 3D space. The points do not necessarily represent some specific shapes. Because the participants tried to use the shape information to help establish correct point matches, if the points do look like some specific shape, such as a cat and a bird, then they clearly have found more correct point matches. Otherwise, they may fail to find any correct point matches. This shows that children have a tendency to mimic what they are told in a passive way, involving little their own thinking.

From the comments listed above, it can be seen that while some year 7 children did apply the CPC to establish point matches, quite large a proportion of the participants complained that the test was very hard, especially when little explanation was given. This is expected, since the test is difficult not only for children, but for adults as well, sometimes, even for experts. This justifies that automatic approaches have to be developed for the point match establishment.

3.2. Duplicating a Figure

3.2.1. Undergraduates

- Because how it is in grid 1(+).
- It's what I thought you had to do(+).
- I was asked to draw the same pattern so I tried to place a cross in every circle's location according to the placement of other crosses and the axis (+++).
- Draw what I saw from the original model pattern using coordinate amounts(+).
- Try to use axes, but in the end just used good observation and judgement of angle/distance(+).
- Place the plus signs in the location opposite to the one in grid 1. I put the +'s in the same place although that wasn't asked for... it just needed the same pattern(+).
- I tried to mimic the relationship between the dots and the graph(+).
- Because it seems to match up(+).
- To look as similar to grid 1(+).
- The pattern was copied(++).
- Copy each point by comparison to its neighbours(+).
- Because I haven't got enough time to do it(+).

3.2.2. Year 7 Children

- I liked this section(+).
- Because we were asked to copy grid 1 pattern and it is a similar pattern. I think this part of the test is pretty hard (+).
- very hard(+++++).
- Because it said to copy it(+).
- I think it was quite easy but I did not copy the crosses in the exact place(+++).
- It a lot harder than joining the point matches(+).

- no clue(+).
- It was easier in one way but more difficult in another slightly stronger way(+).
- it is a clever way to use the brain and it makes you think(+).
- I think this is quite a hard test but in the end it is easy(+).
- I did not understand(+).

3.2.3. Year 9 Children

- because its found the same area(++);
- It seems appropriate(++++);
- Because it looks the same. It playing trick on my eyes all I can see is dots(+);
- Because I looked at them carefully and I tried to put them in the same place. This was quite hard but it was much easier than the other kind of patterns(++);
- because that seems close to the same thing(+);
- I drew a pattern like that because I thought it looked similar(++);
- because that's where I thought the + went(+);
- Its how I imaging the objects points and how it has been moved. I don't think when I write down the points I just do what my brain tells me(+);
- you would just have to mark the points where you think they should go(+);
- I do because I am trying to make it similar to the circle(+);
- Because they tell you to. It was a little harder(+);
- because that's how I think it's like(+);
- I drew the pattern like that because that is how it looked in my head. After doing a few it got hard and confusing so it gave me bit of headache(++);
- This is what is saw(+);

For this part of experiments, most participants can hardly explain why they duplicated the point patterns like that. But when they duplicated the point patterns, they subconsciously conserved the relationship between neighbouring points. This shows that when they perceive the real world, they subconsciously organise descriptive data in a certain order in spite of the fact that sometimes, they do not realise.

3.3. Recognising 3D Shapes from Their Features

3.3.1. Undergraduates

- I looked for a line of most similar shape and with many points it shares with the compared line or only skew and x or y translation, or change direction in similar places(+++++++).
- Comparing points and angles, then relative sizes of lines(+).
- because they were the closest(+++++++).
- Seemed logical(+).
- Either most like line or follows similar points(++).
- guessed(+).

3.3.2. Year 7 Children

- Because that is the closest line to it. Very easy(+).
- I guessed(++).
- don't know(+++)
- very hard(+++).
- because it follow the same pattern just with different angles. I found this easy(+).
- I think this was the easiest one out of them all(+).
- It was hard because some of the lines were too close together and its hard to see which line is which(+).
- This was a good section(+).
- I did not understand most of it(+).
- it looked right to me. This one confused me so I just guess what to do(+).

3.3.3. Year 9 Children

- I chose the ones closest together and are more similar(+);
- because they ran the longest along line 1. They are closest and it makes sense(++++++);
- I think it is(+);
- They have the same shape. Too easy(+++);

- because that's the lines I thought were write so I looked at it carefully and I saw that they were alike(++);
- because you look at the lines and look at which one looks the same. It wasn't as easy(+);
- Not as hard as the another ones(+);
- Because the lines looked most alike the line 1(++);
- Its near and bends the same similarly(+);
- Put down which ones started and ended from the same point. More straight forward than joining the points(+);
- That's what I have seen(+);
- It was quite confusing because there were so many lines(+);

For this part of experiments, different participants have quite different opinions with regard to what criteria should be used to recognise the query lines. Some prefer to find the one with the most similar shape without paying too much attention to the location of the line, while others tend to choose the closest line. Actually, both criteria are effective. If the line represents structural features, then the similar shape criterion is preferred. If the line represents the location of salient feature points, then the closest line criterion is preferred. All other criteria are not frequently used by the participants. But they do represent the intuitive knowledge of the participants.

3.4. Comparing Features of Different 3D Shapes

3.4.1. Undergraduates

- Trying to follow the line closest to the one chosen(++++).
- I looked for a line that had similar shape, or who's points might average out to a similar shape, if too close to call between two lines (+++++).
- Just tried to see why the lines were a good match in example by points + angles + distance, then find line that most matches, with some amount of intuition(+).
- I tried to make the line that fitted line 2 best(+).
- By simply look at the facts from the first set and apply to the second(+).
- Correlating translations and/or skews were found(+).
- look for similar patterns(+).
- seemed logical(+).
- look best(+).
- 'cos you told me to(+).

3.4.2. Year 7 Children

- Because that is the closest line to it. Very easy(+).
- It was hard(++).
- I don't know(+).
- I do not actually know I think I was guessing but I didn't understand the last bit. It was complicated and hard(+).
- I did not like this test(+).
- Because the lines are very close together and you have to see which line matches with which(+).

3.4.3. Year 9 Children

- because they are closest to the line(++++);
- It seemed appropriate(+);
- I done that it looks the same. That was soild(+);
- I do because its nearest and looks similar(+);
- Because that's what my brain tells me(+);
- Tried my best to see which ones go the same way. Quite difficult to find line 2(+);
- I followed the line to see wich line fitted best(+);
- because I looked at all the lines to see which ones look the same. This was quite easy(++);
- because you just have to look at the lines very carefully to see which one is more accurate and then you just choose the one that you think looks exactly it. But it is really really hard(+++);
- It was quite difficult to identify which line was which(+);
- It was quite confusing because there were so many lines(+);
- I don't know(+);

The analysis of the explanations for this part of experiments is similar to that in the last section. However, it is worth pointing out that since the questions are different, thus, for this section of experiments, the participants were expected to use the relative relationship between the given line pairs. Unfortunately, few participants made use of the relationship between the given line pairs. They ignore the fact that it is necessary to conserve the topology between all similar line pairs. Once this criterion was ignored, then they cannot guarantee that similar queries will find similar lines. This reflects in real life as a phenomenon: in a situation, we make a decision, but in a similar situation, we make quite different decision without convincing justification. If the given line pairs represent the past experience, then the selection of the similar line to the query one reflects whether we can learn from our past experience. This is the main subject of case based reasoning [11].

4. Results

After experiments, we carefully compared the answer book from each participant against the standard answer determined at the phase of data design. Analysis of results appears not easy. This is the nature of 3D free form shape matching whose evaluation is still arguable: average matching error, statistical test, success rate, or subjective judgement. However, since we have reference answers for all the questions designed, it is thus still feasible to evaluate the performance of each participant.

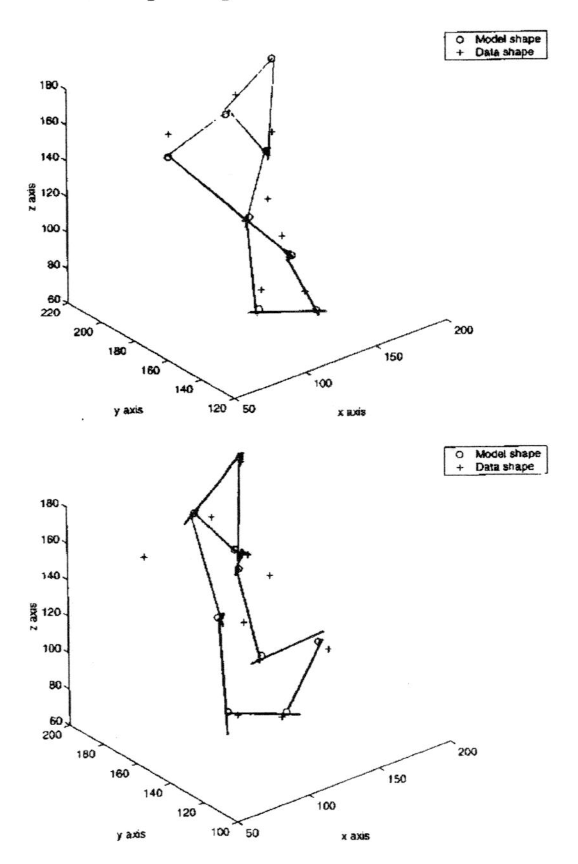

Figure 10. The given task was not correctly understood.

Although participation, especially among the younger students, was not always total (did not understand the requirements, did not take the tests seriously), as demonstrated in Figure 10, their response is still typical of any student sample, on account of different abilities and attitudes toward the given tasks. No matter how well the tasks are explained, some participants find it hard to understand the tasks, making it difficult to ensure that everyone can be fully engaged into the tests.

Our statistics are nevertheless based on all participants, reflecting the actual situation, since one aim of our psychological studies is to reveal genuine insight into how the subjects responded to the given unseen free form shape matching tasks. To this end, we need to analyse what the participants actually had done. Even though they may not complete the given tasks, their partial responses are still useful and may inspire us in the sense of deepening our understanding of the subjective criteria for overlapping free form shape matching and developing novel automatic overlapping free form shape matching algorithms.

The analysis will show whether there is a correlation between the scores of different questions and participants and whether the CPC is either consciously or subconsciously used for the interpretation of the point clouds perceived from different nearby viewpoints. For point pattern duplication, since the task was either not attempted or completed with a high standard. In this case, it is not meaningful to do performance analysis of the participants. Consequently, we left it out.

4.1. Relative Scores of Different Questions and Participants

The score surface of questions and participants for the undergraduates is presented in Figure 11 and its statistics are presented in Figure 12, those for the year 7 children are presented in Figures 13 and 14; and those for the year 9 children Figures 15 and 16. Summary statistics for the three groups are presented in Tables 1, 2 and 3.

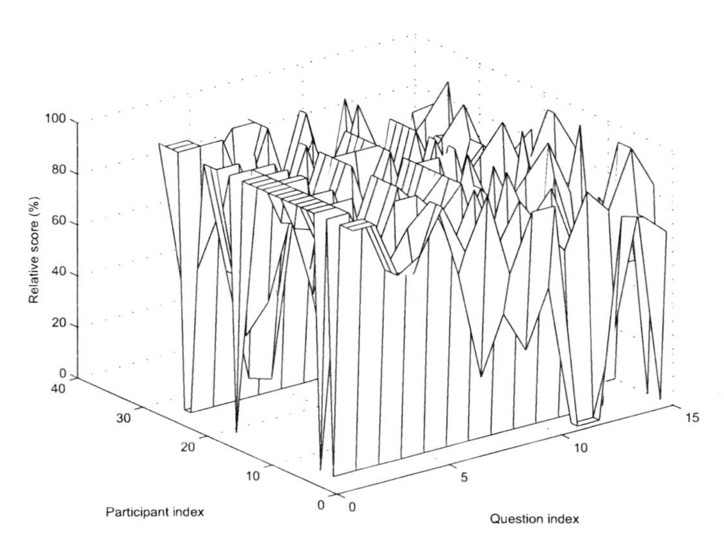

Figure 11. Score surface of different questions and participants for year 2 undergraduates.

From Table 1 and comparison of Figures 12, 14 and 16, it can be seen that the best performance for the point match establishment is achieved by the year 9 children. This is because the problem was more clearly explained and understood. The best performance for similar line recognition and comparison is achieved by the year 2 undergraduates. This is because they are most knowledgeable and competent at independent problem solving.

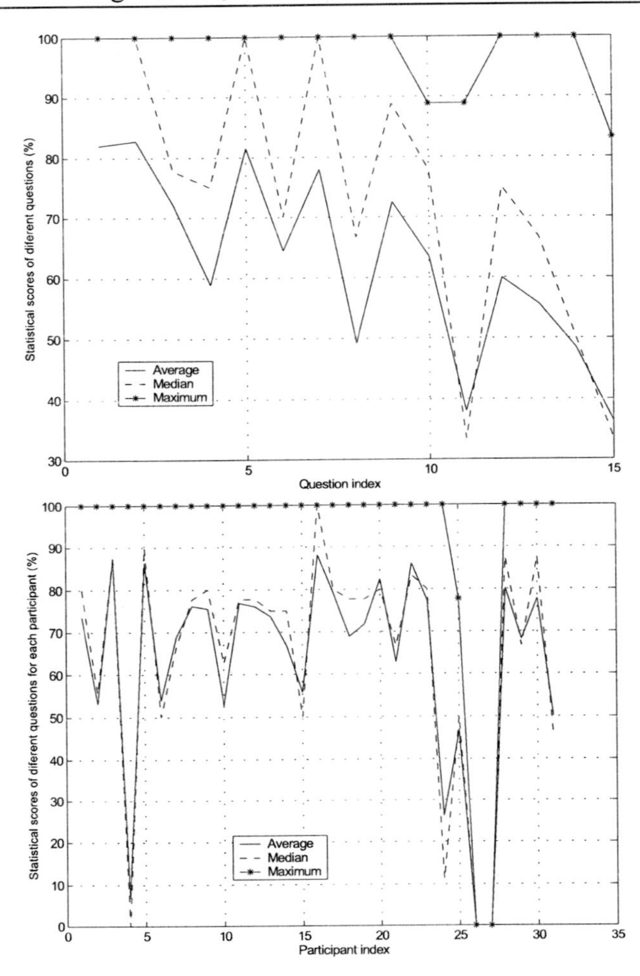

Figure 12. The statistical results of the scores of different questions and participants for year 2 undergraduates. Left: Different questions; Right: Different participants.

Table 1. The average scores of Questions 1 through 7, 8 through 13, Question 14, and Question 15 for different groups of participants.

Group	Group 1 (%)	Group 2 (%)	Group 3 (%)	Group 4 (%)
Year 2 undergraduates	74	56	48	36
Year 7	61	56	23	9
Year 9	78	65	44	30

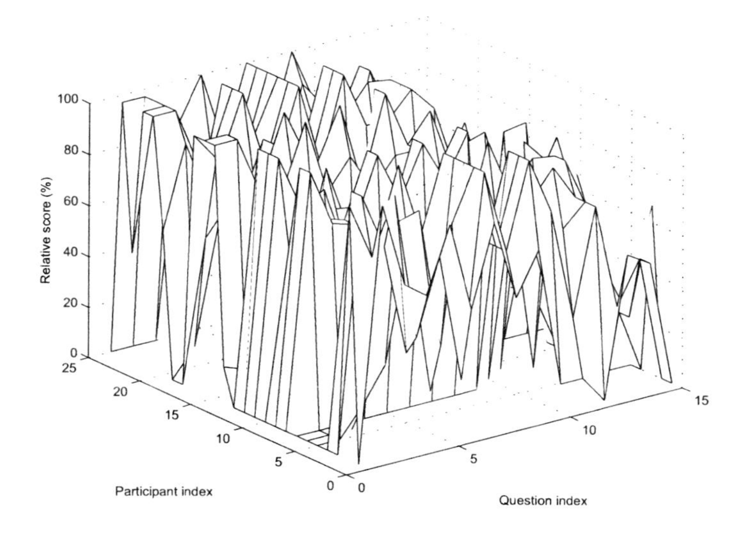

Figure 13. Score surface of different questions and participants for year 7 children.

Table 2. The correlation coefficient between the average scores of different questions relative to different groups of participants.

Group	year 2 undergraduates	year 7	year 9
Year 2 undergraduates	1	0.66	0.93
Year 7	0.66	1	0.67
Year 9	0.93	0.67	1

We did the test at different places with different groups of participants and different amounts of interaction between staff and participants. These artificial factors may affect the final performance of the participants. Thus, it is interesting to do some correlation analysis between the performance of different groups of the participants. Such correlation analysis does make sense since the participants in the same group completed the test under the same condition, correlation analysis can remove the effect of artificial factors on the behaviour of the participants and thus, reveal something they were subconsciously doing.

Firstly, we do some correlation analysis between the average scores of different questions relative to different groups of the participants. The result is presented in Table 2. From Table 2, it can be clearly seen that overall, the performance of different groups is highly correlated, especially, the correlation coefficient between the average scores of different questions answered by the year 9 children and the year 2 undergraduates is as high as 0.93. This shows that there is something consistent behind this high correlation coefficient, which is probably the fact that the competence of children in the sense of performing given tasks grows proportionally with ages.

Secondly, the 15 questions were grouped according to their data generation. Group 1 (Questions 1 through 7) were generated using computer simulation, group 2 (Questions 8

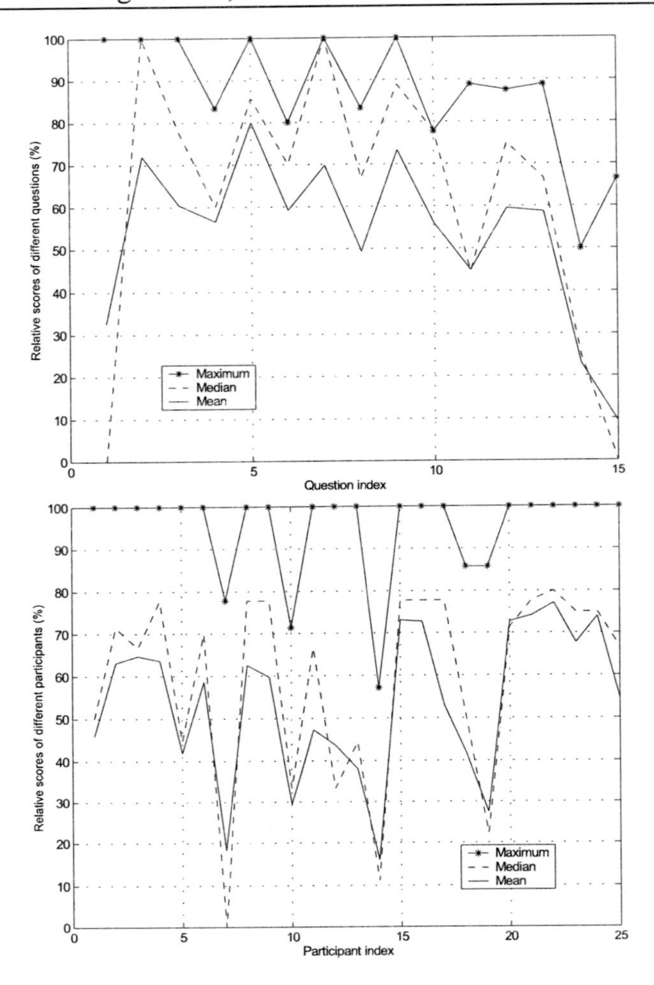

Figure 14. The statistical results of the scores of different questions and participants for year 7 children. Left: Different questions; Right: Different participants.

through 13) were generated using uniform sampling from real images, group 3 (Question 14) concerned the recognition of similar lines, and group 4 (Question 15) concerned the comparison of similar lines in the form of a 2D visualization of 3D points. The correlation analysis of the average scores of different groups of questions answered by different groups of the participants is presented in Table 3. It can be seen that the performance of different groups of the participants is even more highly correlated. For example, the lowest correlation coefficient between the year 7 children and the year 2 undergraduates is surprisingly as high as 0.91, while that between the year 9 children and the year 2 undergraduates is as high as 0.98.

The fundamental reason underlying the consistent performance of the participants in different tests is because they either consciously or subconsciously used the CPC to varying degrees for problem solving. This conclusion has been drawn from two facts: (1) what the participants drew and wrote on the answer books, as exemplified in Figures 17 through 20

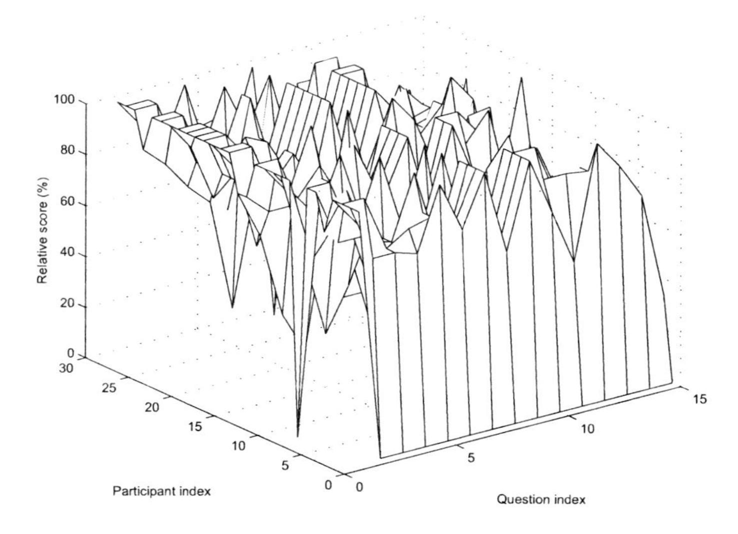

Figure 15. Score surface of different questions and participants for year 9 children.

below. Even though the participants may not realise, it can be clearly observed from the answer books that they indeed used the CPC in one way or another to perform the given tasks, and (2) what the participants explained how they performed the given tasks. Within those who explained why they performed the given tasks like that, 18 year 2 undergraduates out of 28 (64%), 13 year 9 children out of 26 (50%), and 5 year 7 children out of 17 (29%) *explicitly* stated that the CPC was used to establish possible point matches and identify similar lines. The more mature the children, the more likely it is for them to consciously apply the CPC for point pattern matching and similar line identification. On the other hand, maturing children become more and more knowledgeable and likely to learn from their past experience. Consequently, they are often more active in searching, evaluating and consistently utilizing relative knowledge (e.g., relative motion and relative distance) for point pattern matching and similar line identification. Even though they may be acting subconsciously and unable to articulate what knowledge they have used, this process does take place. Furthermore, the more explanation is given and the more interaction between teacher and students is provided, the easier it is for students to approach the problem.

Table 3. The correlation coefficient between the average scores of different groups of questions between different groups of participants.

Group	year 2 undergraduates	year 7	year 9
Year 2 undergraduates	1	0.91	0.97
Year 7	0.91	1	0.98
Year 9	0.97	0.98	1

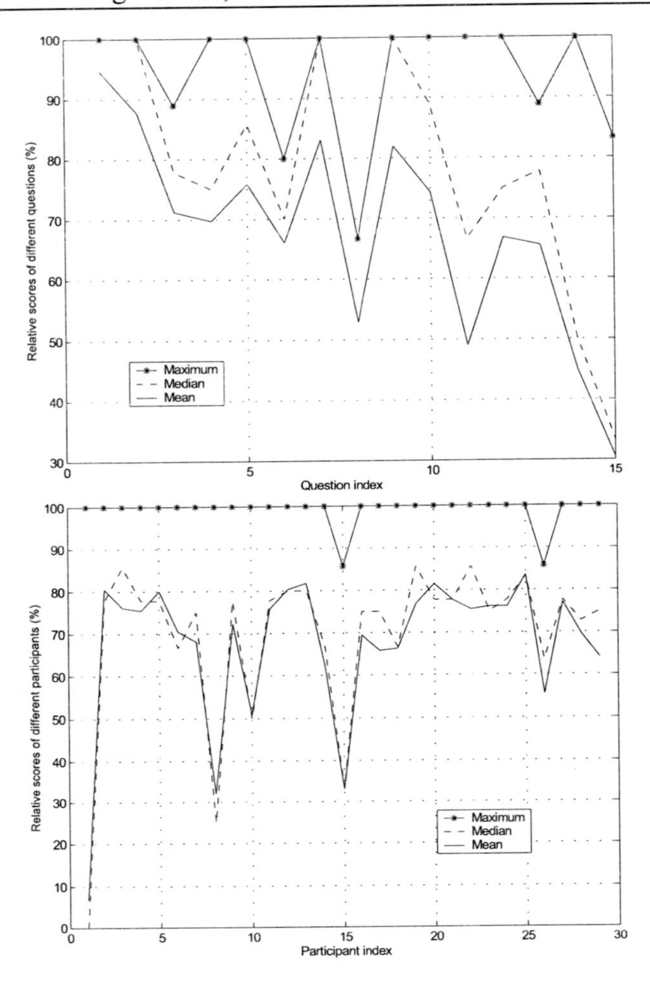

Figure 16. The statistical results of the scores of different questions and participants for year 9 children. Left: Different questions; Right: Different participants.

4.2. Matching Trials

In this section, we analyse the experimental results using the significance F test with regard to whether the average score δ over different questions of each participant in different tests is as expected: $\delta_1 \leq \delta_2$ is accepted if $F = \frac{s_1^2}{s_2^2} < F_p$ and $\int_0^{F_p} F(n_1 - 1, n_2 - 1)dv = 1 - p$ where δ_1, δ_2, s_1 and s_2 are the average and standard deviation of scores of a number n_1 and n_2 of the participants in different tests respectively. The results are: $[F = 0.56; F_p = 2.40; d.f. = 24, 30; p < 0.01]$ between the year 7 children and the year 2 undergraduates, $[F = 1.12; F_p = 2.70; d.f. = 28, 24; p < 0.01]$ between the year 7 and 9 children, and $[F = 1.99; F_p = 2.55; d.f. = 30, 28; p < 0.01]$ between the year 9 children and the year 2 undergraduates. From the significance tests, it can be clearly seen that the average performance over different questions of the participants in the different tests is as expected and age has no significant impact on the average performance of the participants. This analysis is consistent with the correlation analysis reported in the previous section.

We conjecture that there must exist some approaches invariably applied by different participants in different age groups. That is the criterion of closest point, which was applied in one way or another by different participants. This can be clearly observed from their answer books, as exemplified in Figures 17 through 20, no matter whether they realised or not.

4.3. Point Match Establishment

From the explanations listed in the last section, the following criteria can be extracted for point match establishment:

1. Closest points;
2. One-to-one mapping: a point in one point set can only correspond to a point in another;
3. Relative motion between point matches: different point matches have underwent similar motions;
4. Relative direction between point matches: different point matches have moved in similar directions;
5. Relative distance between point matches: different point matches have covered similar distances;
6. Overlapping area maximization: points were joined as many as possible;
7. The shape the point sets represent: two point sets represent a similar shape;

Accurate establishment of point matches has been determined by various factors, including point distribution, camera motion, the relative size of points in the overlapping area with regard to appearing and disappearing points, and the shape the points represent. Unfortunately, it is not always easy to establish correct point matches not only due to various factors above to be considered, but also the fact that 3D point sets have been projected onto a 2D space makes it difficult to imagine the shape the point sets represent and thus, it is difficult for the participants to make use of shape information to help establish real point matches. Moreover, the camera underwent a 3D motion, including both rotation and translation. After projection, the 3D point sets appear skewed and deformed. Finally, all these participants have no knowledge about 3D shape interpretation, image alignment, feature extraction and representation, or projective geometry. Thus, it is indeed challenging for the participants to find all correct point matches.

If the points well distribute, including a number of appearing or disappearing points, then it is much easier to establish point matches as demonstrated as Figures 3a, 3b, and 3e by different participants. In this case, hardly anybody made mistakes in establishing correct point matches, as illustrated in Figure 17. This phenomenon reveals that the 3D shape may be recognised by a limited number of salient feature points [21] like a human face. In this case, human beings can easily identify what and where the eyes, ears, mouth, nose,

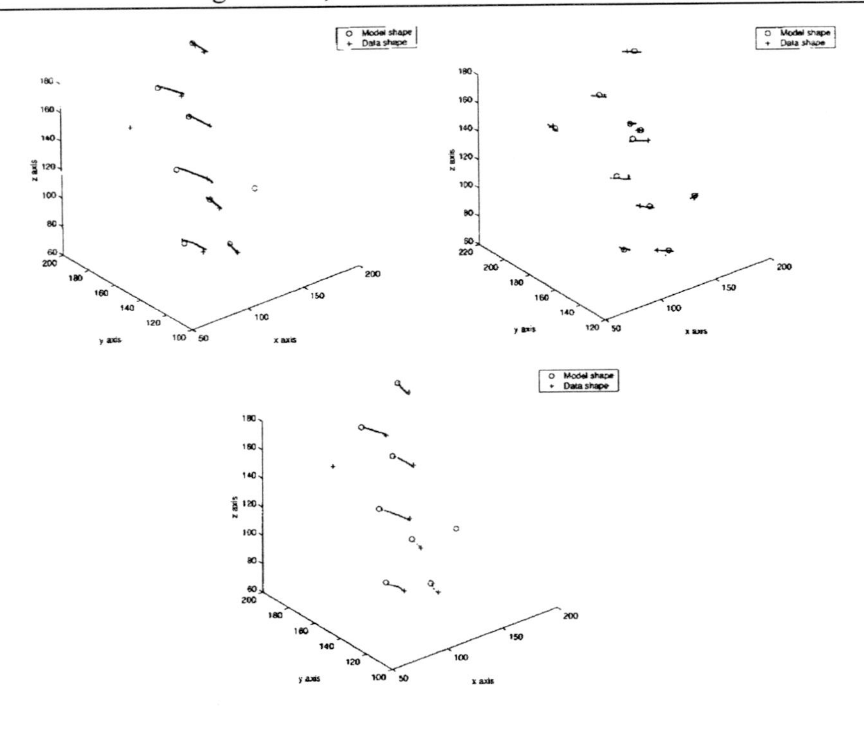

Figure 17. Correct point matches have been found.

or eye brows are from which a human face can be recognised efficiently. For Figures 3c and 3d, it is necessary to consider not only the closet points, but the one-to-one mapping and the relative motions of different point matches as well so that correct point matches can be found. Some participants do manage to find correct point matches as illustrated in Figure 18. Otherwise, some participants will get confused about how to find the correct point matches. Thus, the average scores for these two questions are relatively low, as illustrated in Figure 12. While appearing and disappearing points appear on the left and right hand sides (Figure 3e), it is relatively easy to find correct point matches, as long as the participants consider the relative motion of point matches. In Figure 3f, the appearing and disappearing points appear in the middle of point sets. In this case, the participants must consider not only the closest points, but also the one-to-one mapping and relative motion (direction, distance, and magnitude). Thus, it is more challenging for the participants to find correct point matches and the average score for this question is even lower. For questions 8 to 13, the point sets were generated using real images. As a result, their distribution is beyond expectation about how they distribute and thus they are even more challenging for the participants, even for those with image alignment knowledge, to establish correct point matches, as illustrated in Figure 19. Thus for the undergraduates, the average score for these questions based on the real images is 56%, while the average score for those based on the computer simulation is as high as 74%.

From Figure 14, it can be seen that the average score for the year 7 children is relatively low. This is because at the beginning of the test, some participants were struggling about

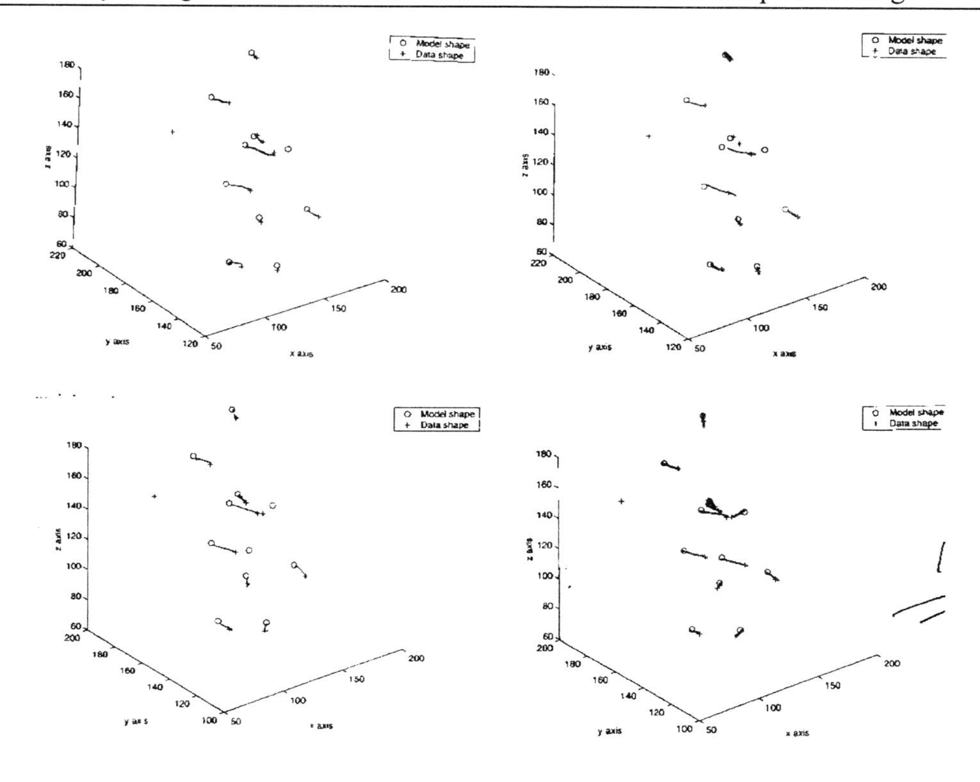

Figure 18. Correct point matches still have been found or partially found.

what to do. They even did not know the requirement of the test. So we had to explain the test again and again. Finally, they got what they should do. Thus, they achieved better scores for Questions 2, 3, and 4. For Question 5, the disappearing points appear in the middle of the data set. In order to find real point matches, they had to consider the relative motion of different points. Thus, it is a bit more challenging than the previous questions. Consequently, they got lower scores with an average of 56%. Surprisingly, even though Questions 8 through 13 are challenging, in this case, the year 7 children performed just as well as the undergraduates. This shows that once children made clear what they should do, they could do as well as adults and in this case, they reached the same level in recognising correct point matches in the test.

From Figure 16 and Tables 3 and 2, it can be seen that the year 9 students performed best. This is because a clearer explanation was provided so that from the beginning of the test, the participants had clear idea about what they should do in the test. Consequently, the average score for Questions 1 through 7 is as high as 78%. Since Questions 8 though 13 are more challenging, they performed a little worse with an average score of 65%, as expected, which is still better than either the year 7 participants or the year 2 undergraduates did.

While the undergraduates and the year 9 children often imposed the one-to-one mapping constraint on the point matches established, the year 7 children often had no idea with regard to conscious applications of some criteria like relative motion, relative direction,

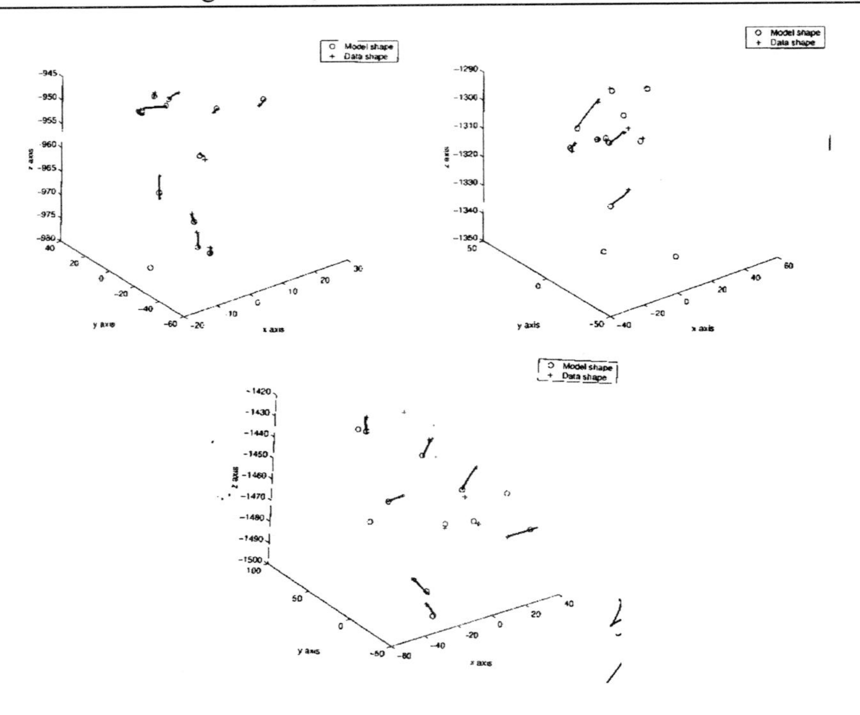

Figure 19. It is challenging to find correct point matches.

relative distance, and one-to-one mapping for the point match establishment. Consequently, the year 7 children often join more than one data point to a single model point and vice versa. It is interesting to note that the traditional CPC does imply three possible mappings: one-to-one, many-to-one, and one-to-many, as illustrated in Figure 20. For points in the overlapping area, then one-to-one mapping point matches are likely to be established. More than one disappearing point in the data shape often find a single point in the model shape as their correspondents. In this case, these point matches are called many-to-one mapping established point matches. If more than one point in the model shape are equidistant to a transformed point in the data shape, then this point match is called an one-to-many mapping established point match. This shows that the year 7 children are more likely to apply the brute force form of the CPC in joining point matches. Unfortunately, unless a careful manipulation is made, pure CPC often cannot exhibit stable performance in establishing real point matches. Thus, overall, the year 2 undergraduates perform better than the year 7 children in establishing real point matches in the psychological test.

While the year 2 undergraduates consciously applied some knowledge, such as relative motion, relative distance, and relative direction, to help establish real point matches, some year 7 children did not consider such knowledge, as illustrated in Figure 21, where while some point matches span relatively small distances, the others span relatively large distances, or two point matches cross each other. This shows that they did not think a lot yet about the surroundings and how to interpret them using consistent criteria. While few year 2 undergraduates inconsistently applied the CPC to establish point matches, some year 7 children inconsistently use this criterion as illustrated in Figure 21. Thus, the year 2 under-

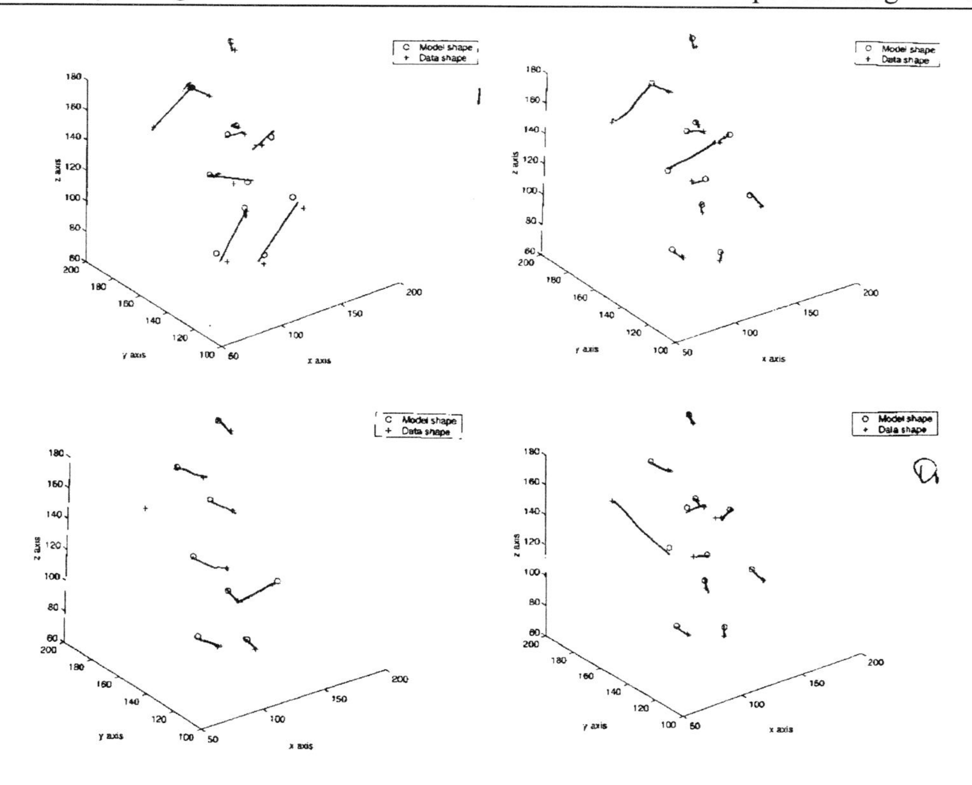

Figure 20. Different mappings implied by the traditional closest point criterion. Top two: many-to-one mapping; bottom two: one-to-many mapping.

graduates show more consistent performance in the test than the year 7 children. This shows that children have not yet exhibited consistent behaviour with regard to problem solving.

For the experiments based on computer-generated points, it is relatively easy for the participants to establish correct point matches. This is because the points were generated uniformly in the 3D space and thus, relatively well distribute. In contrast, the points from real images, even though they were uniformly sampled, however, they could only be sampled from the real object surfaces that are actually 2D spaces. Some points stay very close to each other in 2D space. In this case, the points do not well distribute and thus, render it difficult to make use of shape information to help establish correct point matches. In order to establish correct point matches, the participants have to not only imagine the object shape in 3D space, but also imagine the 3D motion the camera underwent and then use a number of criteria aforementioned to establish real point matches. The problem is that sometimes, the correct point matches may stay far away from each other. The nearest points do not mean that they represent real point matches. The participants have to consider the camera motion relative to particular points.

Some interesting findings are listed as follows:

1. Even though some participants claimed that they joined points that were nearest to each other, actually, they often did not use this criterion consistently, as illustrated in

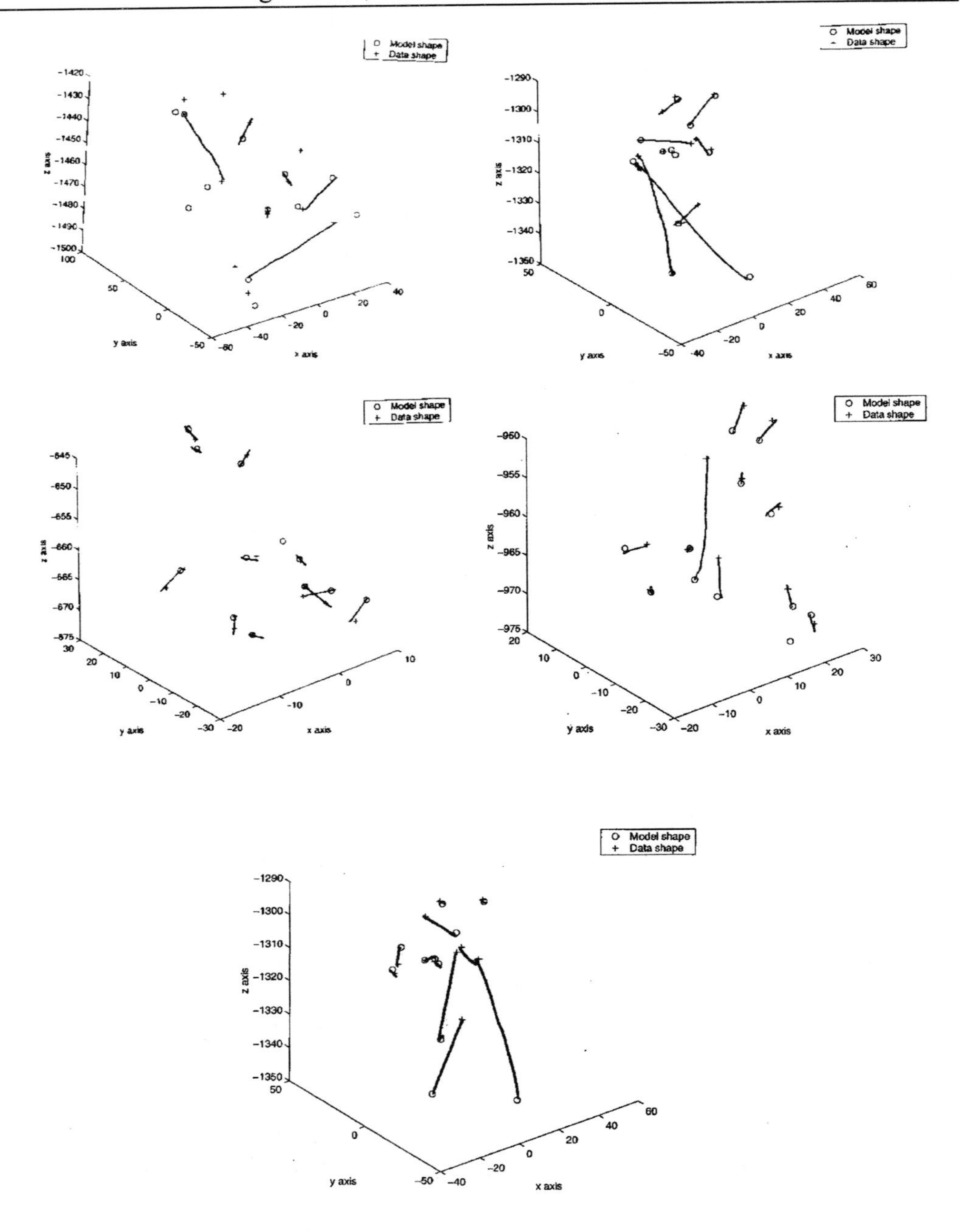

Figure 21. Closest point criterion has been inconsistently used for point match establishment.

Figure 21.

2. While the CPC may lead more than one point in one point set to select a single point in another as correspondents, almost few participants did like this. This shows that they thought that the one-to-one mapping is more important than the CPC. Unfortunately, such an idea does not guarantee that correct point matches can always be established. This may explain why some participants inconsistently use the CPC, as illustrated in Figure 21.

3. While the CPC was inconsistently used in one case, then it was often inconsistently used later on. This shows that different individuals have their own thinking and judging approaches. This means that different individuals prioritize different criteria for the establishment of point matches between two overlapping point sets from different viewpoints. This explains why different participants have relatively consistent performances in the psychological tests as demonstrated in Figures 11, 13, and 15.

4.4. Duplicating a Figure

While some participants found this part of the test is easy, others found it hard. Even so, those who attempted the questions often correctly duplicated the point patterns, conserving the topological relationship between points. This is the case especially for the year 7 participants. But once the participants made clear that they needed to use the plus signs to draw a pattern in Grid 2 that is as similar to the pattern in Grid 1 as possible, then they finished the duplication quickly. Some examples are illustrated in Figure 22 by different participants. The reason why the participants can easily complete this part of experiments is that the participants all have been trained in arts classes or clubs. So they know how to draw or sketch the scenes they are perceiving. This shows that the conservation of point topology plays an important role in interpreting 3D shapes from different viewpoints or human brains do store the perceived scene in a certain order. Such knowledge is subconsciously used in picture drawing or sketching. Even though the exact scale is not used, which is difficult due to 2D perception of 3D points, the relative relationship between points are retained in the process of duplication. This may encourage that the topology among points should be used to the free form shape matching in the machine vision literature.

4.5. Recognising 3D Shapes from Their Features

The following criteria have been most frequently used to recognise the most similar lines:

1. Similar shape, even not the same size; and

2. Closest to each other.

This part of the test is challenging to children, as they had no clear idea about how to measure the similarity of two lines. Since all lines deviate from each other, it is not easy to find the most similar ones. In order to do so, they first have to read different lines carefully, then decide the weight for the shape or path deviating from the query ones. Different opinions on the weights can lead different lines to be selected. The average score of the

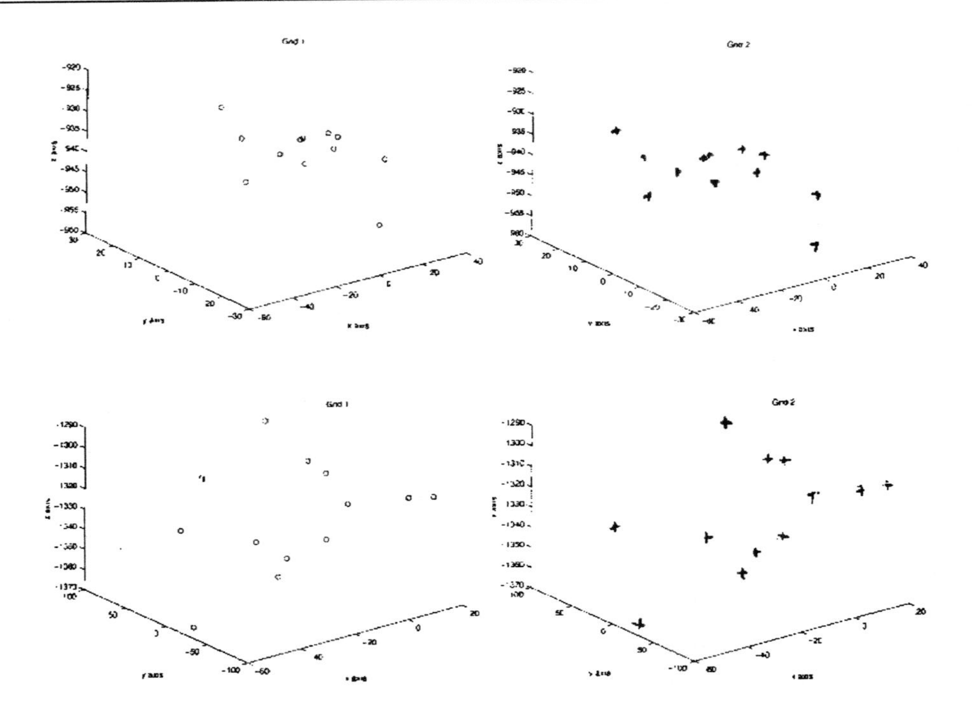

Figure 22. It is relatively easy to duplicate point patterns, conserving the relationship between points. Left and third columns: original; Second and right column: duplicated.

questions in this part of experiments is 48% for the undergraduates, 23% for the year 7 students, and 44% for the year 9 children. The low scores may be due to the following reasons:

- We did not explain the questions. As a result, all participants did the questions completely dependent on their own understanding and knowledge;
- Some participants misjudged the lines. For example, if some parts of a line deviate a lot from the query line but its overall shape is similar, then the line was not selected;
- Some participants did not read lines carefully. For example, they misread lines with plus signs as the ones with crosses;
- Some participants did not take it seriously; and
- Some participants did not attempt the questions.

4.6. Comparing Features of Different 3D Shapes

The following criteria have been frequently used to recognise the most similar lines:

1. Nearest one; and

2. Similar shape.

This part of the test is even more challenging. Some participants got confused about what to do. Some complained that the lines are too close together. However, this is the case for 3D shape alignment where for any point in the data shape, each point in the model shape could be a candidate for the correspondent of that point. In this case, of course, it is confusing to determine which point in the model shape is its real correspondent. Actually, given the line pairs and the query line, participants were expected to search for the most similar lines in a certain area, instead of everywhere. If they ignore the given line pairs, then they are prone to choose wrong lines. Most participants did not make use of information of the given line pairs. Some participants got confused why they were provided with the line pairs for choosing the most similar line to the query one. As a result, they considered the questions in this part of the test the same as those in the last section. Consequently, the average score for the questions in this part of the test is as low as 36% for the undergraduates, 9% for the year 7 children, and 30% for the year 9 children. The following reasons may partially explain these low scores:

- The questions were not explained at all. Thus, the participants had to answer the questions based on their own understanding and knowledge;
- Some participants did not consider the topology between the given line pairs;
- Some participants had no time to finish the questions, since this is the last part of the experiments within the given time;
- Some participants misread the lines to which the most similar lines are to be found.
- Some participants did not take it seriously; and
- Some participants did not attempt the questions.

For the last two questions, the year 7 children performed much worse than the undergraduates. 8 out of 25 participants did not attempt or completely failed the questions of recognising similar lines. Most participants did not know how to measure the similarity between two lines. The average score for this part of the test is just 23%. 17 out of 25 participants did not attempt, completely failed, or had no time to perform the given tasks about how to compare similar lines in the last part of the test. They either misread the lines or had no clear idea about how to use the knowledge about the given line pairs to find the most similar lines to the query ones. This suggests that children still have a lot to learn.

Since the last two questions were not explained at all, their average scores are more objective with regard to the reflection of the performance of the participants. The average scores of the last two questions for the year 9 children are 44% and 30% respectively, which are clearly lower than 48% and 36% for the undergraduates, but higher than 23% and 9% for the year 7 participants. These scores are obviously convincing, since they reflect a fact that the older the children, the more competent they are at solving problems by their own. This is expected.

From Figures 12, 14 and 16, it can be seen that with the given tasks becoming more challenging, the participants achieved worse results, especially when they have no background knowledge about 3D shape matching. Even so, their performance is satisfactory and reveals that the CPC is either consciously or subconsciously used for 3D shape matching. Through explicit education, children can quickly grasp the knowledge necessary for shape matching and become more interested in computer science and cognition research in the future.

5. Discussion

Based on the comments and what they performed in the tests, we can rank the criteria that were used by the participants to perform the given tasks in the tests:

1. The application of the CPC can be clearly observed from the answer books, as exemplified in Figures 17 through 20 and thus it is the most widely used, either consciously or subconsciously. This is especially the case when the camera motion is small and the participants have no idea about either what the point clouds represent or the size of the camera motion that the point clouds were subject to. In this case, the participants may even not be bothered by the camera motion and the object that the point clouds represent and simply drew lines between the closest points, which is actually a plausible choice. Unfortunately, this criterion is sometimes inconsistently used, joining points further away, due to the fact that they took into account other factors such as one-to-one mapping and relative distance and motion direction;

2. One-to-one mapping. A single data shape point can only be joined to a unique model shape point. This is true for accurate matching and is required for accurate interpretation of the data shape using the model shape. But this criterion is not true for the CPC. This implies that the CPC alone introduces false matches into the matching process and thus requires a careful evaluation of possible point matches established for accurate free form shape matching;

3. Overlapping area maximization. Usually, the participants joined up points as many as possible. This is often a hideous criterion that is rarely used and is difficult to formulate. But for accurate matching of 3D shapes, it seems that we have to take this criterion into consideration. There are two scenarios that necessitate the overlapping area maximization. One is that when the shapes intersect in 3D space, the points on the intersection line tend to have small matching errors and a rigid motion can be recovered from these points. However, in general, this motion is wrong. The other is that when the shape does not include rich structural information, the matching problem is often ill defined and the matching algorithm tends to have a sliding error. The overlapping area maximization has a potential to compensate these two scenarios for accurate free form shape matching;

4. Relative motions of different point matches. Since the data shape is subject to a small motion, the data shape points can only be joined to some model shape points with a small motion. However, in practice, it is not easy to recognise the small motion due to the fact that both the model and data shapes are described in 3D space and the motion is also in 3D space. After the data shape has been moved, the projection of some points in either model or data shape can appear far away from each other or too

close together. But when both the model and data shape points are well distributed, the small motion can be relatively easily recognised; and

5. Object structure. Once the components of an object are known in the two overlapping point clouds, it is relatively easy to find which is which. But the actual situation is that the shape that the point cloud represents is unknown, making it difficult to extract features from the point clouds even from dense structured points [10] due to noise in the imaging process caused by point sampling from object surfaces, mechanical and electronic errors, surface discontinuities, and different reflectance characteristics. But knowledge about the object structure does help to exclude infeasible point matches.

At the best of our knowledge, we have not yet seen any algorithm that applies all these five useful criteria for 3D free form shape matching. Thus, the psychological tests described in this chapter may reveal something useful for free form shape matching algorithm development and education.

6. Conclusion

In this chapter, we have employed psychological tests to reveal what criteria are either consciously or subconsciously used by the participants to interpret the perceived point patterns from different viewpoints. Even though the participants have no knowledge about overlapping free form shape matching, they performed satisfactorily. What the participants drew and wrote on the answer books and comments on why they performed the given tasks like that have clearly shown that they usually subconsciously use the CPC in conjunction with other criteria such as one-to-one mapping, relative motion, relative direction, relative distance, and overlapping area maximization to interpret the point sets from different viewpoints. Even though the point sets were subject to 3D rigid motions, composed of 3D rigid rotations and translations, as long as the points are well distributed, then the participants can correctly find real point matches.

Different participants show quite different performances in point match establishment and similar line identification. This not only reflects their different thinking approaches, but the knowledge they have as well. Some participants felt excited about the experiments. Through further guide, they may become more interested in vision research and computer science. The research and simulation of human vision and cognition is the main task of vision research and computer science.

Relative motion is very important for correct point match establishment. In [19], for many-to-one mapping established point matches $(\mathbf{p}_{ik}, \mathbf{p}'_{c(i)})(k = 1, 2, \cdots, r|r \geq 2)$, the ones with the smallest matching errors were kept and all other point matches were thrown away. But the experiments show that such an idea may not be effective. Instead, we can remove the ones with smallest matching errors from the closest point search $\mathbf{P}' \leftarrow \mathbf{P}' - \{\mathbf{p}'_{c(i)}\}$, then re-establish point matches.

Relative motion also implies that the direction of different point matches cannot be opposite. This criterion is also useful for the development of free form shape matching algorithms. First, we can calculate the consensus direction of point matches through calculating, for example, the average of the directions of point matches. Then any point match whose direction is opposite to the consensus direction is regarded as a false point match. Such a simple idea can help disambiguate some possible point matches and thus, accurately interpret 3D shape from different viewpoints.

Relative motion can also be used to help establish more accurate point matches. Based on the CPC, a set of possible point matches has been established from which the consensus relative motion can be calculated, for example, by median filtering the components of the relative motions associated with different possible point matches. Once this consensus relation has been estimated, the closest point should be selected relative to the sum of the original transformed point and this consensus motion. This simple calculation provides a new area for possible point match search.

In summary, the experiments do verify our conviction that human beings often subconsciously apply the CPC to interpret the point sets captured from different viewpoints. This leads to an average accuracy of as high as 78% for correct interpretation. In order to disambiguate the interpretation, human beings often turn to other means, such as changing viewpoints, colours, and the overall shape of objects. Machine vision research benefits from the experiments that new algorithms can be developed in the future through incorporating some criteria, such as one-to-one mapping, relative motion, and overlapping area maximization. For education, more and easy explanation and staff and student interaction should always be provided so that students can easily understand and approach the problem at hand. Thus, the experiments are fruitful, as planned.

Acknowledgements

Our sincere thanks go to the Joy Welsh Trust for their generous partial financial aid for the psychological tests. We also like to thank Mr Davey, head-teacher, Mrs Evans, the year 7 head-teacher, of Penglais Comprehensive school at Aberystywth to help organise the psychological tests. Thanks to Mr Lewis, geography teacher to the year 7, and Mr Taylor, mathematics teacher to the year 9, for allowing us to use their valuable teaching time for the tests, helping explain the requirement of the tests and maintaining good orders in the classrooms for the tests. Finally, we would like to express our gratitude to the year 2 undergraduates, selecting module CS26210 in academic year 2004-2005 at the University of Wales, Aberystwyth, the year 7 students, learning geography on 9 February 2005, and the year 9 students, learning math on 9 March 2005 at Penglais Comprehensive School, Aberystwyth, for their enthusiastic participation in the psychological tests.

References

[1] M. Andreetto, N. Brusco, G.M. Cortelazzo. Automatic 3D modelling of textured cultural heritage objects. *IEEE Trans. Image Processing*, 2004, 13, 354-369.

[2] P.J. Besl, N.D. McKay. A method for registration of 3D shapes. *IEEE Trans. PAMI*, 1992, 14, 239-256.

[3] I. Biederman, M. Bar. One-shot viewpoint invariance in matching novel objects. *Vision Research*, 1999, 39, 2885-2899.

[4] C. Bundesen, A. Larsen, S. Kyllingsbak, O.B. Paulson, I. Lay. Attentional effects in the visual pathways: a whole-brain PET study. *Exp Brain Research*, 2002, 147, 394-406.

[5] S. Carey and T. Williams. The role of object recognition in young Infants object segregation. *J. Exp. Child Psychology*, 2001, 78, 55-60.

[6] S. Edelman, and H.H. Bulthoff. Orientation depdence in the recognition of familiar and novel views of 3D objects. *Vision Research*, 1992, 32, 2385-2400.

[7] T. Funkhouser, P. Min, et al. A search engine for 3D models. *ACM Trans. Graphics*, 2003, 22, 83-105.

[8] S. Gold, A. Rangarajan, et al. New algorithms for 2-D and 3-D point matching: pose estimation and correspondence. *Pattern Recognition*, 1998, 31, 1019-1031.

[9] D. Hahnel, S. Thrun, and W. Burgard. An extension of the ICP algorithm for modelling nonrigid objects with mobile Robots. *Proceedings of International Joint Conference on Artificial Intelligence*, 2003, pp. 915-920.

[10] A. Johnson and M. Hebert. Using spin images for efficient object recognition in cluttered 3D scenes. *IEEE Trans. PAMI*, 1999, 21, 433-449.

[11] J.L. Kolodner. Case-based reasoning. Morgan Kaufmann Publishers, San Mateo, CA, 1993.

[12] R. Lawson. View sensitivity increases for same-shape matches if mismatches show pairs of more similar shapes. *Psychonomic Bulletin and Review*, 2004, 11, 896-902.

[13] R. Lawson, H.H. Bulthoff, S. Dumbell. Interactions between view changes and shape changes in picture-picture matching. *Perception*, 2003, vol. 32, 1465-1498.

[14] Y. Liu. Automatic registration of overlapping 3D point clouds using closest points. *Image and Vision Computing*, 2006, 24, 762-781.

[15] D. Marr, and T. Poggio. Cooperative computation of stereo dispartity. *Science*, 1976, 194, 283-287.

[16] R. Osada, T. Funkhouser, B. Chazelle, and D. Dobkin. Shape distributions. *ACM Trans. Graphics*, 2002, 21, 807-832.

[17] D. Page, A. Koschan, Y. Sun, and M. Abidi. Laser-based imaging for reverse engineering, *Sensor review*, 2003, 23, 223-229.

[18] J.J. Peissig, E.A. Wasserman, M.E. Young, I. Biederman. Learning an object from multiple views enhances its recognition in an orthogonal rotation axis in pigeons. *Vision Research*, 2002, 42, 2051-2062.

[19] M.A. Rodrigues and Y. Liu. On the representation of rigid body transformations for accurate registration of free form shapes. *Robotics and Autonomous Systems*, 2002, 39, 37-52.

[20] R.N. Shepard and J. Metzler. Mental Rotation of three dimensional objects. *Science*, 1971, 171, 701-703.

[21] J.T. Todd. The visual perception of 3D shape. *Trends in Cognitive Sciences*, 2004, 8, 115-121.

[22] R.Y. Tsai and T.S. Huang. Uniqueness and estimation of three-dimensional motion parameters of rigid objects with curved surfaces, *IEEE Trans. PAMI*, 1984, 6, 13-27.

[23] S. Ullman. Aligning pictorial descriptions: an approach to object recognition. *Cognition*, 1989, 32, 193-254.

In: Pattern Recognition in Biology
Editor: Marsha S. Corrigan, pp. 149-188
ISBN: 978-1-60021-716-6

Chapter 5

PROCEDURAL LEARNING DIFFERENCES IN OBSESSIVE-COMPULSIVE DISORDER: A BRIDGE BETWEEN COGNITIVE-BEHAVIORAL AND NEUROPSYCHOLOGICAL MODELS?

Heather M. Chik, Noelle K. Pontarelli and John E. Calamari*
Department of Psychology, Rosalind Franklin University of Medicine & Science

ABSTRACT

In this chapter we explore the nature of the information processing differences captured in studies of procedural learning and other types of implicit learning conducted with individuals with obsessive-compulsive disorder (OCD). We propose that findings from this line of research might function to help integrate several seemingly discrepant theories that have been proposed to explain the development and maintenance of obsessional disorders. We present an overview of this common psychiatric disorder and review cognitive, metacognitive, and neurobiological models of OCD. Following an extended review of procedural learning studies in OCD, we suggest that the implicit learning differences that characterize OCD are importantly related to a tendency to be hyperaware of thought processes. This propensity to focus on thought processes may be a possible cognitive risk factor for OCD. Exploration of the nature of this cognitive vulnerability and its relationship to procedural learning might help bridge theories and empirical findings, and help advance the treatment of this common psychiatric disorder.

WHAT IS OBSESSIVE-COMPULSIVE DISORDER?

Obsessive-compulsive disorder (OCD) is a complex but common anxiety disorder with lifetime prevalence estimates of 1.9% to 2.5% reported in a cross-national study (Weissman et al., 1994). Age of onset ranges from 21.9 years to 35.5 years (Weissman et al., 1994) with

* E-mail address: John.Calamari@rosalindfranklin.edu. Phone: 847-578-8747; Fax: 847-578-8765; Correspondent: John E. Calamari, Department of Psychology, Rosalind Franklin University of Medicine and Science, 3333 Green Bay Road, North Chicago, Illinois 60064

earlier onset reported in males (Rasmussen & Eisen, 1992). The *Diagnostic and Statistical Manual of Mental Disorders, Fourth Edition, Text Revision* [*DSM-IV-TR*; American Psychiatric Association (APA, 2000)] characterizes OCD as recurrent obsessions and/or compulsions that interfere substantially with daily functioning (APA, 2000). Obsessions are anxiety-provoking and persistent ideas, thoughts, impulses, or images. Compulsions, on the other hand, are repetitive behaviors (e.g., hand washing, ordering, checking) or mental acts (e.g., praying, counting, repeating words silently) that are performed to prevent or reduce anxiety or distress, or to avert some dreaded event. They can be performed in response to an obsession (e.g., repetitive hand-washing in response to obsessions about contamination), or in accordance to idiosyncratic rules (e.g., checking three times that the stove is switched off before leaving the house; APA, 2000). Most individuals with OCD recognize that their behavior is excessive or unreasonable. Development of the disorder is usually gradual, and the majority of patients either continue to meet full criteria for OCD or retain significant symptoms of the disorder over long periods of time (Eisen & Steketee, 1998). Moreover, OCD is frequently associated with impairments in general functioning, such as disruption of gainful employment and educational attainment, as well as disturbance in marital, familial, and interpersonal relationships (Karno, Golding, Sorenson, Burnam, & Audrey, 1988; Hollander et al., 1997). Many individuals with OCD suffer for years before seeking treatment. In one study, individuals first presented for psychiatric treatment over seven years after the onset of significant symptoms (Rasmussen & Tsuang, 1986).

Despite the bizarre and debilitating nature of OCD in its severest forms, intrusive thoughts and even anxiety neutralizing compulsions are quite common in the nonclinical population. Rachman and de Silva (1978), and later in a replication by Salkovskis and Harrison (1984), reported that 79% – 88% of normal individuals in their samples admitted to having intrusive thoughts. When comparing the intrusive thoughts of a nonclinical to a clinical sample with OCD, Rachman and de Silva (1978) identified similar content, form, and relationship with mood (i.e., occurred more frequently during periods of anxiety or depression). The "abnormal" intrusions differed from "normal" ones in that they were more frequent, more intense, longer lasting, more distressing, and more resistant to dismissal. Muris, Merckelbach, and Clavan (1997) reported similar findings in a study on compulsions in which 55% of their nonclinical sample admitted to performing ritualistic behaviors, suggesting that compulsive behaviors are also a common phenomenon among healthy individuals. These pivotal studies on the common occurrence of intrusive thoughts and ritualistic behaviors in the normal population have since permitted the study of normal intrusions and rituals as analogues of clinical OCD and also laid the foundation for cognitive models and treatments of OCD.

Cognitive Theories of OCD

Cognitive theorists have attempted to explain how commonly occurring intrusive thoughts evolve into an obsessional disorder and have often drawn from Beck's seminal work on depression (1976). Following the efforts of Beck, researchers began applying cognitive theories to OCD (Carr, 1974; McFall & Wollersheim, 1979; Salkovskis, 1985). These efforts led to the development of OCD specific cognitive therapies in hope of increasing the effectiveness of interventions and lessening the dropout rates associated with behavioral treatment for OCD (Salkovskis, 1998).

Salkovskis (1985, 1989) was the first to present a comprehensive cognitive model of OCD. He contended that the development and maintenance of OCD is influenced by dysfunctional beliefs that drive the negative appraisal of intrusive thoughts. In his detailed analysis, Salkovskis (1985, 1989) emphasized the role of inflated responsibility. He proposed that normal intrusions only become clinical obsessions when they have meaning and become salient to the individual experiencing them. When the occurrence and content of the intrusions are appraised as threatening or as having serious consequences for which the individual feels personally responsible, obsession (and anxiety) ensues. Attempts to suppress and neutralize intrusive thoughts, in the form of overt or covert compulsions, are made in an effort to remove the intrusions and prevent any perceived harmful consequences. Neutralizations are typically reinforced and strengthened because they are successful in temporarily reducing distress and removing the unwanted thoughts. In the long run, however, the use of neutralization is not adaptive because it helps maintain the individual's belief that the neutralizing act was responsible for preventing the feared event from occurring and that without neutralization, the discomfort caused by the obsessions would persist.

Rachman (1997, 1998) subsequently outlined a cognitive theory of obsessions, elaborating and extending Salkovskis' analyses. He proposed that obsessions are caused by catastrophic misinterpretations of the significance of one's intrusive thoughts, images, or impulses. He emphasized the relation between OCD and the cognitive bias of thought-action fusion (TAF), which involves the belief that having an unacceptable thought may actually influence the probability that the adverse event will occur, and the belief that having a repugnant unacceptable thought is morally equivalent to carrying out the related actions. It has been suggested that TAF increases the person's perceived responsibility for adverse events, and this in turn promotes feelings of guilt (Rachman, 1993). From Rachman's (1997) standpoint, an inflated sense of responsibility can both contribute to the occurrence of TAF, and also be the product of this and other cognitive biases.

Other types of dysfunctional beliefs may also be important in the etiology and maintenance of OCD, including beliefs about the importance of one's thoughts, the importance of controlling thoughts, and perfectionism (Freeston, Rhéaume, & Ladouceur, 1996). Drawing upon this outline, the Obsessive Compulsive Cognitions Working Group (OCCWG), an international group of investigators sharing a common interest in understanding the role of cognitive factors in OCD, began developing a consensus regarding the most important beliefs and appraisals relevant to OCD (OCCWG, 1997). As a result of the coordinated effort, two self-report measures that assess dysfunctional beliefs and appraisal of intrusions were developed. A 44-item Obsessional Beliefs Questionnaire (OBQ-44; OCCWG, 2005) measures three dysfunctional belief domains 1) inflated responsibility and overestimation of threat, 2) perfectionism and intolerance of uncertainty, 3) and over-importance and over-control of thoughts; while a 31-item Interpretation of Intrusions Inventory (III, OCCWG, 2005) measures negative interpretations of intrusive thoughts. These two measures have now become the gold standards for assessing dysfunctional beliefs and appraisals in OCD patients.

In summary, cognitive theories of OCD emphasize the ubiquitous nature of intrusive, negative thought experiences. Commonly occurring intrusive thoughts are posited to turn into obsessional problems through problematic appraisal processes (e.g., having a thought about dropping my baby on the floor means I am at risk of carrying out the act). The problematic appraisal processes are thought to be driven by several dysfunctional beliefs that are specific or related to OCD, as mentioned above. Cognitive theories have led to significant refinements in the cognitive-behavioral treatment of OCD (see Bouvard, 2002, for a review). Yet there has been limited empirical evaluations of the processes that underlie some patients' excessive focus on thought experiences and their difficulty in dismissing common negative intrusions, experiences that are easily disregarded by the majority of the population (Marker, Calamari, Woodard, & Riemann, 2006). Neuropsychological and metacognitive models of OCD, and recent findings that connect specific procedural learning problems to the disorder, may help to elucidate how intrusive thoughts become more salient and difficult to dismiss for some individuals.

Metacognitive Models of OCD

Theory and research in metacognition has emerged predominantly through cognitive developmental psychology (Flavell, 1979). During the past decade, there has been an increased interest in metacognitive processes, with several research groups positing that this focus will advance the understanding of OCD (Purdon & Clark, 1999; Wells & Matthews, 1994). Metacognition is broadly defined as any knowledge or cognitive process that is involved in the appraisal, monitoring, or control of cognition (Wells, 2000; Wells & Matthews 1997). In their metacognitive theory, Wells and Matthews (1994) proposed the Self-Regulatory Executive Function (S-REF) model of emotional disorders (Wells & Matthews, 1996). The model functions on three main levels: 1) an automatic and reflexive level of processing that functions outside of conscious awareness; 2) a level of voluntary processing that draws heavily on attentional resources in order to consciously appraise events and manage the control of thoughts and actions; 3) and a level of metacognitive beliefs that are stored in long-term memory. In OCD, the S-REF model serves the metacognitive function of determining the personal significance of thoughts and guiding subsequent attention allocation and cognitive processing (Wells & Matthews, 1996). Wells (2000) suggested that individuals with OCD monitor and pay excessive attention to their thoughts. As more attention is directed automatically or voluntarily towards thought and other internal processes, limited information about objective external events can be processed. Due to the finite nature of cognitive resources, the individual's coping capabilities are constrained as they deploy excessive attention to thought.

One of the important constructs that has emerged from this and other related efforts to comprehensively assess metacognition (Cartwright-Hatton & Wells, 1997) is cognitive self-consciousness (CSC), the tendency to focus attention on thought. Heightened CSC may function to enhance the salience of thought experiences broadly and increase opportunities for the detection of negative intrusive thoughts. Individuals' who attempt to force unwanted thoughts from their mind with various neutralizing strategies or rituals will likely fail. Their thought control failures may further function to increase their distress (e.g., Clark, 2004).

What is the Best Approach to Studying OCD Related Cognitive Processes?

McNally (2001) eloquently discussed the scientific adequacy of utilizing two distinct and potentially incompatible approaches to studying the cognitive model of anxiety disorders. While one method utilizes self-report to ascertain problematic beliefs or appraisals (i.e., cognitive content), the other utilizes overt behavioral indices, such as reaction time, to determine information-processing abnormalities (i.e., cognitive process). Unfortunately, self-report measures are insufficient in examining cognitive process issues that are hypothesized to be automatic or that might operate outside of individuals' awareness. On the other hand, information processing paradigms are often not useful in identifying beliefs and appraisals importantly related to many psychological disorders. As a result, McNally argued for methodological pluralism in order to investigate specific information processing differences, and differences in how specific experiences are perceived, recalled, and interpreted in individuals with anxiety disorders. A greater move towards the use of information-processing paradigms is essential for the study of cognition in OCD and this approach complements both self-report and neurobiological approaches and provides much needed additional knowledge on cognitive processing associated with the disorder.

Do Neuropsychological Differences Characterize OCD?

Current neuropsychological approaches to OCD suggest that neurobiological abnormalities play a crucial role in the etiology and course of the disorder (Greisberg & McKay, 2003; Kuelz, Hohagen, & Voderholzer, 2004). Many different investigators and theorists implicated the frontal-striatal network, or the cortico-striato-thalamo-cortical (CTSC) circuitry (Baxter, Schwartz, Guze, Bergman, & Szuba, 1990; Graybiel & Rauch, 2000; Insel, 1992; Modell, Mount, Curtis, & Greden, 1989; Rapoport & Wise, 1988; Rauch & Jenike, 1993; Rauch & Savage 2000; Saxena, Brody, Schwartz, & Baxter, 1998; Saxena & Rauch 2000), in the pathophysiology of OCD. A similar circuitry has also been implicated in the pathophysiology of other hyperkinetic disorders such as Parkinson's disease, Huntington's disease, and Tourette syndrome (Alegret et al., 2001; Anderson, Louis, Stern, & Marder, 2001; Cummings & Cunningham, 1992; Rauch et al., 2001). It is believed that the higher incidence of obsessive-compulsive symptoms in these disorders is due to the involvement of a shared circuitry, further strengthening the hypothesized involvement of the CSTC loop in the manifestation of OCD.

The Cortico-Striato-Thalamo-Cortical (CSTC) Circuitry and OCD

Much of what is currently known about the CSTC circuitry stems from the work of Alexander and colleagues (Alexander, 1994; Alexander, Crutcher, & DeLong, 1990; Alexander, DeLong, & Strick, 1986) and Cummings (Cummings, 1993). They identified five topographically-organized circuits (i.e., motor, oculomotor, dorsolateral prefrontal, lateral orbitofrontal, and limbic) that originate from the frontal lobes, projecting through the basal ganglia to the thalamus, and returning to the frontal lobes (see Figure 1). The basal ganglia, a collection of subcortical nuclei, are involved in motor control, cognition, emotions, and learning. They include the caudate, putamen, globus pallidus, nucleus accumbens, and substantia nigra. The caudate and putamen, often referred to as the striatum, receives most of the input from the cortex or frontal lobes. The globus pallidus, which is divided into globus pallidus externa (GPe) and globus pallidus interna (GPi), receives input from the striatum and

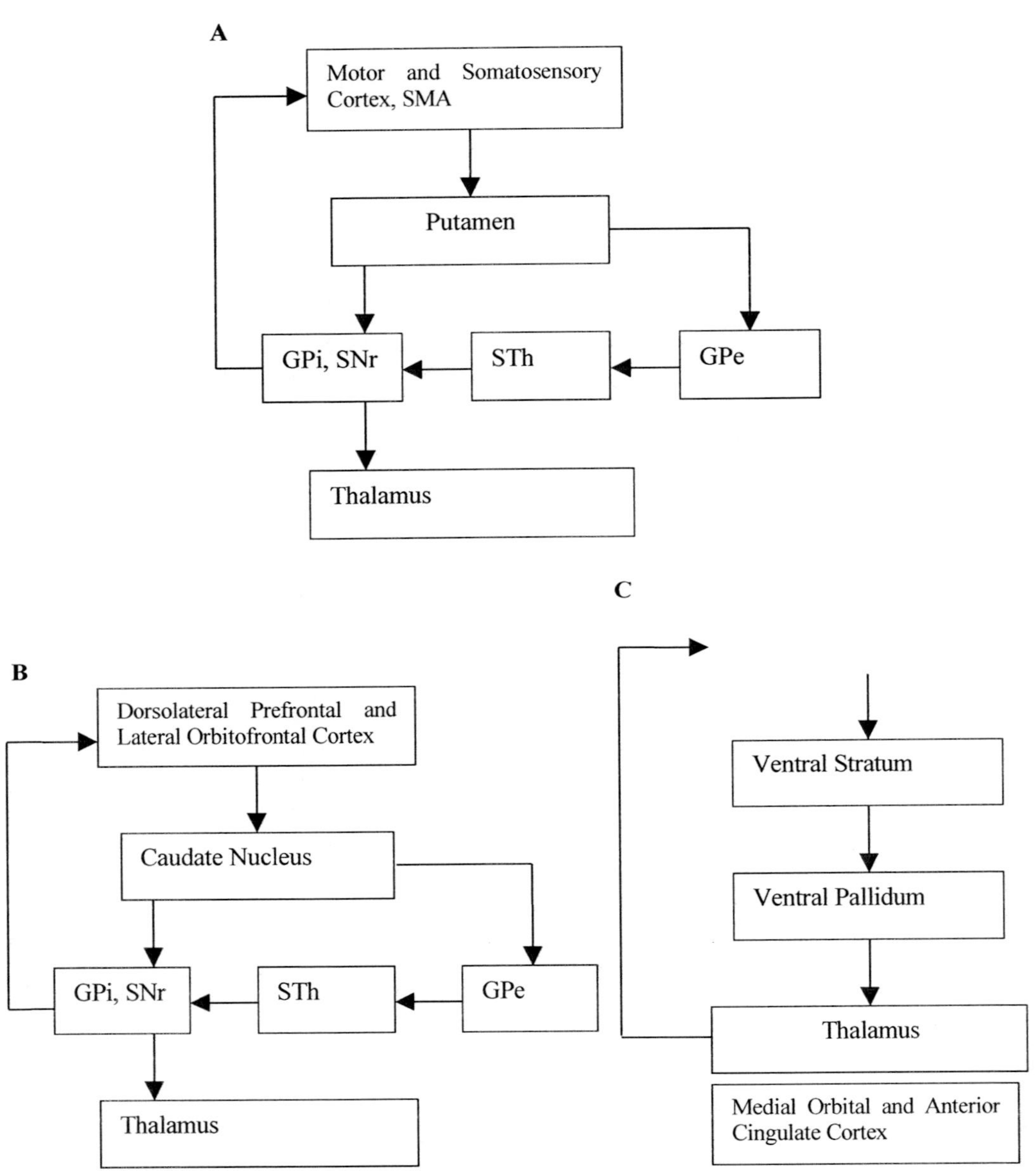

Note. A) motor loop; B) dorsolateral prefrontal loop; C) limbic loop; SMA = supplementary motor area; GPi = internal segmnt of globus pallidus; SNr = substantia nigra pars reticulate; STh = subthalamic nucleus; GPe = external segment of globus pallidus.

Figure 1. The Motor Loop, Dorsolateral Prefrontal Loop, and Limbic Loop of the CSTC Circuitry.

sends output to the thalamus (i.e., subthalamic nucleus). The substantia nigra is also divided into two components: substantia nigra pars compacta (SNc) and substantia nigra pars reticulata (SNr). The SNc sends dopamine projections to the striatum while the SNr sends gamma-aminobutyric acid (GABA) projections to the thalamus and brainstem nuclei. Degeneration of the SNc and its dopamine projections result in symptoms of Parkinson's disease.

For the purposes of this chapter, we will focus on the components of the relevant circuits and the functions that each is purported to mediate that might best explain the symptoms of OCD. The motor circuit entails projections from motor and somatosensory cortex as well as the supplementary motor area to the putamen, mediating motor movements, preparation for motor movements, and somatosensory capacities. The prefrontal circuit entails projections from the dorsolateral prefrontal and lateral orbitofrontal cortex to the head of the caudate, mediating behavioral and cognitive abilities including spatial memory, maintaining or switching sets, and behavioral inhibition. The limbic circuit entails projections from the medial orbitofrontal cortex and anterior cingulate cortex to the nucleus accumbens and related structures, together termed the ventral or limbic striatum, mediating emotional and motivational functions.

It is believed that the basal ganglia, and therefore the CSTC circuit, have two distinct pathways: one via the GPi (the direct pathway) and the other via the GPe (the indirect pathway). These pathways are analogous to the braking system of a car. In the direct pathway, cortical activity that excites the striatum leads to inhibition of the GPi and SNr. This inhibition, in turn, "releases the brakes" on the thalamus. In the indirect pathway, cortical activity that excites the striatum "puts the brakes on" or inhibits the thalamus via the GPe. These braking processes releases or inhibits the thalamus to relay signals to the cortex. Abnormalities to the "braking system" result in various symptoms of hyperkinetic disorders. For example, the presence of extraneous unwanted movements that are characteristic of Huntington's disease is believed to be a result of the over-activity of the direct pathway. By contrast, the poverty or difficulty with intended movements that are characteristic of Parkinson's disease is believed to be a result of the over-activity of the indirect pathway. Although this model of the CSTC loop is somewhat simplified, it illustrates some of the basic mechanisms of the circuitry and has implications for our understanding of OCD.

Neuroimaging Findings

Results from neuroimaging studies have been mixed, but provide indication of functional abnormalities of the CSTC circuitry in OCD (Saxena & Rauch, 2000). While the most consistent finding in the neuroimaging literature has been the presence of hyperfrontality, the basal ganglia and the thalamus have also been implicated in the pathophysiology of OCD (Cottraux & Gérard, 1998).

Structural Imaging

In several early case studies of individuals with OCD, lesions or atrophy in the basal ganglia, including the caudate nucleus and putamen, were reported based on results from magnetic resonance imaging (MRI; Laplane, et al., 1989; Weilburg et al, 1989; Trillet, Croisile,

Tourniaire, & Schott, 1990; Pénisson-Besnier, Le Gall, & Dubas, 1992). A small number of controlled studies using both MRI and computerized tomography (CT) have also been conducted. These investigations suggest abnormalities in the caudate nucleus (Luxenberg et al., 1988; Scarone et al., 1992; Calabrese, Colombo, Bonfanti, Scotti, & Scarone, 1993). However, the details regarding the abnormality in the caudate remain unclear. Other researchers found abnormalities in the overall volume of white matter in individuals with OCD. Garber, Ananth, Chiu, Griswold and Oldendorf (1989) reported abnormalities in frontal white matter, while Jenike et al. (1996) reported less white matter overall in OCD patients compared to controls. These differences in frontal white matter volume were not replicated by Szeszko et al. (2004) in a study examining pediatric patients, although a later investigation did find white matter abnormalities in the anterior cingulate (Szeszko et al., 2005).

More recently, Pujol et al. (2004) compared 72 individuals with OCD with 72 age and sex matched controls via MRI and found significant differences in gray matter volume. Specifically, those with OCD exhibited increased gray matter volume in the anterior cerebellum and the ventral region of the putamen, and reduced volume in the medial orbitofrontal cortex, the medial frontal gyrus, and the left insulo-opercular region. Szeszko et al. (2004) found that pediatric, drug-naïve patients had decreased globus pallidus volumes when compared to controls, but found no differences in the size of the caudate nucleus or the putamen between groups. Patients also exhibited increased total gray matter in the anterior cingulate gyrus when compared to controls. Despite these findings that implicate structures within the CSTC circuit, a number of investigations have not found structural differences between individuals with OCD and controls (Insel, Donnelly, Lalakea, Alterman, & Murphy, 1983; Kellner et al., 1991; Stein et al., 1993; Aylard et al., 1996; Stein, Coetzer, Lee, Davids, & Bouwer, 1997).

Functional Imaging

Functional imaging investigations have been conducted using a variety of methodologies. Researchers employed positron emission tomography (PET), single photon emission computed tomography (SPECT), and more recently, functional magnetic resonance imaging (fMRI) to examine abnormalities in brain activity in OCD patients. Several different types of experimental designs were utilized in order to elucidate these abnormalities. Initially, investigators compared the brain activity of OCD patients to controls during resting states. Lately, there has been a shift to making comparisons within groups during symptom provocation and/or pre and post treatment.

PET studies comparing individuals with OCD with healthy control participants in the resting state have implicated a wide array of neural structures. In general, increased glucose metabolism, or regional cerebral blood flow (rCBF), has been found in orbitofrontal regions of those with OCD (Baxter et al., 1987, 1988; Nordahl et al., 1989; Swedo et al., 1989; Sawle, Hymas, Lees, & Frackowiak, 1991). However, Perani et al. (1995) and Martinot et al. (1990) did not replicate these findings. There have also been some consistent findings with regard to increased rCBF in the basal ganglia (Baxter et al., 1987, 1988) and the thalamus (Swedo et al., 1989; Perani et al., 1995) in OCD studies.

Baseline SPECT studies comparing OCD patients with normal controls corroborate the findings of increased rCBF in the frontal cortex (Machlin et al., 1991; Rubin, Villanueva-Meyer, Ananth, Trajmar, & Mena, 1992; Harris, Hoehn-Saric, Lewis, Pearlson, & Streeter,

1994; Lacerda et al., 2003). More mixed findings are seen in the basal ganglia, however. Adams, Warneke, McEwan, and Fraser (1993) found asymmetric perfusion in the basal ganglia in OCD patients compared to normal controls, while Rubin et al. (1992) reported decreased rCBF in the head of the caudate in males with OCD. Furthermore, Lucey et al. (1995) evaluated a relatively large sample and found decreased rCBF in the frontal and basal ganglia regions in OCD patients.

It is believed that the examination of brain activation during resting states might not be completely valid because of the possible interference from obsessional intrusions during resting state testing (Cottraux & Gérard, 1998). As such, there has been a shift towards within subject designs comparing activation sites during "resting states" to metabolic activity during individualized symptom provocation. The OCD patient's specific obsessions are provoked using a variety of stimuli, including imaginal exposure, in vivo exposure, auditory presentation of obsessional content, and induction of the serotonin agonist meta-chlorophenylpiperazine (*m*CPP). Increased rCBF during symptom provocation was found in the frontal lobes (Hollander et al., 1991) and in global cortical regions (Hollander et al., 1995) after the introduction of *m*CPP. In vivo exposure to obsession related fears produced increased rCBF in the left anterior cingulate cortex, the right caudate nucleus, and bilateral orbitofrontal cortex (Rauch et al., 1994), as well as the posterior cingulate gyrus, the right inferior frontal gyrus, all of the basal ganglia, and the thalamus (McGuire et al, 1994). Interestingly, negative correlations were reported between symptoms and rCBF activation in the right superior frontal cortex, the temporo-parietal junction (McGuire et al., 1994), and the parieto-occipital regions (Zohar et al., 1989). Imaginal exposure has produced an increase in rCBF in the temporal region (Zohar et al., 1989). Breiter et al. (1996) and Cottraux et al. (1996) extended the within-group findings by comparing OCD patients to healthy controls during both resting states and symptom provocation, and found that symptom-related stimuli may also produce activation in the temporal regions.

The examination of brain activity in OCD patients pre- and post-treatment with monoamine oxidase inhibitors (MAOIs), tricyclic antidepressants, selective serotonin reuptake inhibitors (SSRIs), cognitive-behavioral therapy (CBT) and neurosurgery is an effective way of studying changes in brain activity and the relationship to symptom reduction. As discussed below, there are some data to suggest that brain activity changes in specific CSTC circuitry might mediate treatment response.

The majority of the treatment effect investigations employed ^{18}fluorodeoxyglucose (^{18}FDG) PET scans, a technique that measures cerebral glucose metabolism. Treatment with the MAOI Trazodone resulted in the increase of glucose metabolism bilaterally in the heads of caudates (Baxter et al., 1987). Responders to the tricyclic antidepressant clomipramine have exhibited decreased glucose metabolism in the orbitofrontal regions (Benkelfat et al, 1990; Swedo et al., 1992) and the left caudate (Benkelfat et al., 1990). Rubin, Ananth, Villanueva-Meyer, Trajmar, & Mena (1995) noted a decrease in rCBF in cortical areas after treatment with clomipramine. Treatment with the SSRIs fluoxetine and paroxetine produced decreased hyperfrontality (Hoehn-Saric, Pearlson, Harris, Machlin, & Camargo, 1991), and decreased glucose metabolism in the caudate (Baxter et al., 1992; Saxena et al., 1999; Ho Pian et al., 2005), the orbitofrontal cortex (Saxena et al., 1999; Ho Pian et al., 2005), the thalamus, and the putamen (Perani et al., 1995; Ho Pian et al., 2005). Schwartz, Stoessel, Baxter, Martin, and Phelps (1996) and Nakatani et al. (2003) found decreased metabolism in the caudate nucleus in response to behavioral therapy, although Nakatani et al. (2003) also observed increases in rCBF in other areas of the basal ganglia. Finally, Mindus, Nyman, Mogard, Meyerson, and Ericson (1991) reported decreased glucose metabolism in the orbital and caudate regions following a bilateral anterior capsulotomy in five patients, a surgical intervention for reducing OCD symptoms.

In summary, despite some discrepant findings, treatment response across all modalities has resulted in consistent decreases in glucose metabolism in the rCBF, in both regions of the orbitofrontal cortex and the basal ganglia, specifically the caudate nucleus.

The most recent development in neuroimaging design is to combine symptom provocation and treatment effect designs. Rauch et al., (2002) treated patients with fluvoxamine and found decreased rCBF in the orbitofrontal cortex and higher rCBF in posterior cingulate cortex during both symptom provocation and relaxed conditions. Hendler et al., (2003) found that patients who responded to sertraline exhibited lower perfusion in dorsal-caudal anterior cingulum and higher perfusion in right caudate compared to non-responders during symptom provocation. Furthermore, only responders demonstrated significant changes in perfusion overall from pre- to posttreatment.

Unfortunately, the great diversity of methodologies used in this line of research makes it difficult to reliably compare results among studies (Cottraux & Gérard, 1998). Despite these difficulties, Whiteside, Port, and Abramowitz (2004) conducted a meta-analysis of 13 studies that employed PET or SPECT that compared OCD patients to healthy controls. The authors concluded that consistent differences in brain activity between groups could be found in the orbital gyrus and the head of the caudate nucleus.

Neuropsychological Test Performance

Neuropsychological test performance provides information regarding the behavioral manifestations of neurological dysfunction. Various neuropsychological tests have suggested cognitive deficits in OCD patients, although findings have been mixed (Greisberg & McKay, 2003). In addition to providing information regarding cognitive functioning, the use of specific neuropsychological tests can also help to link neuroimaging findings to OCD symptomatology. For the purposes of this chapter, general trends in the neuropsychological test performance of OCD patients are summarized. Readers are directed to Greisberg and McKay (2003), Muller and Roberts (2003), and Kuelz et al. (2004) for a more comprehensive review.

Executive Functioning

As prefrontal region pathophysiology was posited to explain OCD symptomatology, a number of investigations have examined the performance of patients with OCD on tasks measuring executive functioning. Specifically, it has been hypothesized that individuals with OCD would demonstrate set shifting and planning deficits because of frontal lobe dysfunction. The Wisconsin Card Sorting Test (WCST; Grant & Berg, 1948; Heaton, Chelune, Talley, Kay, & Curtis, 1993), the Delayed Alternation Test (DAT; Freedman & Oscar-Berman, 1986), and the Object Alternation Test (OAT; Freedman, 1990) are commonly used to evaluate set shifting ability. Results in this area have been mixed. Many researchers have failed to find significant differences between OCD patients and healthy controls in set shifting performance on the WCST (e.g., Deckersbach, Otto, Savage, Baer, & Jenike, 2000; Moritz, Fricke, & Hand, 2001; Moritz et al., 2002; Moritz, Hübner, & Kluwe, 2004; Simpson et al., 2006). Others have reported deficits among OCD patients in their performance on the WCST (e.g., Lucey, Burness, & Costa, 1997; Okasha et al., 2000; Lacerda, et al., 2003; Lawrence et al., 2006). More substantial evidence in support of set shifting deficits has been found with the DAT and the OAT (Abbruzzese, Bellodi, Ferri, & Scarone, 1995; Abbruzzese, Ferri, & Scarone, 1997; Cavedini, Ferri, Scarone, & Bellodi, 1998; Gross-Isseroff et al., 1996; Moritz et al., 2001; Spitznagel & Sur, 2002). In sum, more research is needed in order to clarify whether set shifting deficits broadly characterize OCD.

Researchers have administered tasks such as the Tower of London (Shallice, 1982) and the Tower of Hanoi (Cohen, Eichenbaum, Deacedo, & Corkin, 1985) in order to examine the differential performance of OCD patients in the area of planning. The results in this area have more consistently pointed to performance deficits in OCD patients, but the interpretation of these findings is complex. In general, OCD patients do not make more errors than healthy controls, but they do take more time to complete the tasks (e.g., Veale, Sahakian, Owen, & Marks, 1996; Purcell, Maruff, Kyrios, & Pantelis, 1998). As a result, it is unclear whether overall performance deficits are due to impaired planning ability or reduced visual-spatial and visual-motor ability. Indeed, Boldrini et al. (2005) recently reported distinct visual-spatial deficits in OCD patients when compared to individuals with panic disorder with agoraphobia. van den Heuvel et al. (2005) took a novel approach to delineating the mechanisms responsible for planning deficits by utilizing fMRI during the Tower of London task. They reported decreased frontal-striatal activation in OCD patients during the task.

In summary, overall results of the assessment of executive functioning in OCD patients in comparison to healthy controls are equivocal. It may be that the use of heterogeneous clinical samples impedes researchers' ability to find clear evidence of executive functioning deficits (Simpson et al., 2006). That is, different subtypes of OCD may vary in executive functioning deficits. Further research in this area is necessary in order to elucidate the true nature of executive functioning in OCD.

Memory

The examination of possible memory deficits in OCD stems from the clinical observation that patients often doubt their own memory (McNally, 2000). Current research demonstrates that these memory difficulties may not be merely the subjective concerns of the patient (i.e., checking), but quantifiable deficits within specific domains of memory. Episodic memory (Tulving, 1972, 2002; the ability to remember the details of an event that has occurred) seems to be more relevant to OCD than semantic memory (the ability to remember facts; see Muller & Roberts, 2003). Tests of episodic memory assess the ability to recall or recognize information either verbally or non-verbally. There have been several reports of verbal memory deficits in patients with OCD (e.g., Deckersbach et al., 2000; Airaksinen, Larsson, & Forsell, 2005; Boldrini et al., 2005; Tuna, Tekcan, & Topçuoğlu, 2005), but other researchers have not found theses differences (e.g., Christensen, Kim, Dyksen, & Hoover, 1992; Dirson, Bouvard, Cottraux, & Martin, 1995; Radomsky & Rachman, 1999; Bohne et al., 2005).

Performance differences between OCD patients and healthy controls have been observed more consistently in non-verbal than verbal memory (e.g., Martinot et al, 1990; Savage et al., 1996, 1999; Deckersbach et al., 2000). Non-verbal memory performance is generally assessed with the Wechsler Memory Scale (WMS; Wechsler & Stone, 1974), the Rey-Osterrieth Complex Figure Test (RCFT; Osterrieth, 1944), or the Benton Visual Retention Test (BVRT; Benton, 1974; Sivan, 1992). Sher, Frost, Kushner, Crews, & Alexander (1989) found differences in WMS performance across OCD symptom subtype. That is, checkers scored lower on the WMS than non-checkers. Simpson et al. (2006) reported that pairwise comparisons between OCD patients and healthy controls on a large battery of neuropsychological tests revealed significant differences in only the BVRT. Studies that use the RCFT suggest that OCD patients' performance deficits in non-verbal memory may be the result of increased focus on visual details that were not relevant to the task (Savage et al., 1999, 2000; Deckersbach et al., 2000). OCD patients did not employ the more efficient strategy of copying the complex figure as a whole, but rather viewed the figure as fragmented parts, resulting in deficits in both immediate and delayed recall of the stimulus. Purcell et al. (1998) found that OCD patients displayed performance deficits on the Cambridge Neuropsychological Test Automated Battery (CANTAB; Robbins et al., 1994) similar to those exhibited by patients with frontal lobe lesions (i.e., used more inefficient strategies on tasks).

In sum, there is most consistent evidence that OCD patients suffer from impaired visual memory due to the use of poor encoding strategy. This circumscribed deficit is related to the aforementioned executive functioning difficulties, and further implicates the role of the frontal lobe and associated corticostriatal circuitry in the pathophysiology of OCD.

Attention

Individuals have a limited capacity for the amount of information that can be successfully processed at a given moment. Those who suffer from anxiety may exhibit attentional biases for threat-related information, thereby compromising the amount of task-relevant information that can be processed (McNally, 2000). Alternatively, individuals with anxiety may be unable to successfully inhibit task-irrelevant information overall (Enright & Beech, 1990, 1993a, 1993b). These hypotheses regarding attention deficits are congruent with the cognitive model of OCD in that one of the trademarks of the disorder is the inability to ignore unwanted intrusive thoughts.

Attention deficits are seen in patients with OCD only when the task includes distraction by irrelevant stimuli. Tasks that measure attention span and sustained attention without distracters, such as the Digit Span forward from the Wechsler Adult Intelligence Scale (WAIS-R; Wechsler, 1981), are not associated with impaired performance in patients with OCD (e.g., Savage et al., 1996; Okasha et al., 2000; Moritz et al., 2002). Impaired performance by OCD patients has been demonstrated with the Stroop task (Williams, Mathews, & MacLeod, 1996). In this task, participants are presented with a series of words and are asked to name the color that the word is presented in while ignoring the word itself. Response time for naming the color is delayed when the semantic content of the word differentially captures the participant's attention. Martinot et al. (1990) reported significantly slower scores on the Stroop task for OCD patients compared to healthy controls. However, these Stroop findings have not been consistently replicated (e.g., Schmidtke, Schorb, Winkelmann, & Hohagen, 1998; Mataix-Cols, Alonso, Pifarré, Menchón & Vallejo, 2002). Despite inconsistent findings, functional imaging studies have shown activation of the frontal-striatal and temporal regions during the Stroop task in OCD patients (van de Heuvel et al., 2005).

Modification of the Stroop task using threat-related words have also been used to examine attentional biases to threat-related information in OCD patients. It has been proposed that OCD patients would exhibit slower reaction times during the emotional Stroop tasks because their attention would be directed to the content of the threat-related word rather than the color (McNally, 2000). Foa, Illai, McCarthy, Shoyer, and Murdock (1993) found that OCD patients exhibited slower response times only to words that were relevant to the patient's symptomatology. These findings have sometimes been replicated (Lavy, van Oppen, & van den Hout, 1994; McNally et al., 1994), yet others have failed to find the same effect (Kyrios & Lob, 1998; Kampman et al., 2002). Besides symptom-specific words, OCD patients also exhibited attentional bias to words related to aversive somatic sensations (McNally, Riemann, Luro, Lukach, & Kim, 1992). In another variant of the Stroop task, Cohen, Lachenmeyer, and Springer (2003) examined the effects of symptom induction via the presentation of anxiety-producing scenarios on Stroop performance and saw significant deterioration of reaction time in OCD patients compared to healthy controls.

Another test for assessing selective attention to threatening stimuli is the visual dot probe task (MacLeod, Matthews, & Tata, 1986). In this task, two stimuli, one of which is neutral and one of which is threatening, appear randomly on either side of the screen. The stimuli are presented for a short duration (e.g., 500 ms) before a dot is presented in the location where one of the stimuli was presented. Participants are to respond as soon as the dot appeared. Faster reaction times were reported among anxious individuals when the probe follows the location of threat-related words (e.g., MacLeod & Matthews, 1988). For OCD patients with primary contamination obsessions, faster reaction times were demonstrated to probes following contamination-relevant words (Tata, Lebowitz, Prunty, Cameron, & Pickering, 1996). It is yet to be determined if these findings would generalize to other OCD subtypes.

OCD patients' difficulty in inhibiting negative automatic thoughts may translate into an overall inability to inhibit attention to irrelevant stimuli. The role that attentional inhibition plays in OCD symptomatology has been examined via a negative priming paradigm, in which an individual is asked to attend to a stimulus that was previously presented as an irrelevant distracter (Tipper, 1985). Healthy controls generally respond slower to this item because their first instinct is to inhibit processing of the stimulus. It was hypothesized that individuals with OCD would not exhibit this same delayed response because they are unable to inhibit processing of the distracter stimulus during the first presentation. Indeed, this lack of a negative priming effect in OCD patients has been demonstrated (Enright & Beech, 1990, 1993a, 1993b). In an extension of this original paradigm, McNally, Wilhelm, Buhlmann, and Shin (2001) examined the effect of negative priming on response time to valenced stimuli. They found evidence of deficits in negative priming after a short presentation time (100 ms) but not during the longer presentation interval (500 ms).

This brief review suggests that OCD patients may have difficulties with both selective attention biases and with the inhibition of attention to irrelevant stimuli. The presence of attention abnormalities may contribute to the development and maintenance of OCD. At this point, it is unclear whether attentional abnormalities are similar to or different from those found in other anxiety disorders that are characterized by attentional biases for threat information.

Contemporary Neurobiological Theories of OCD

Evidence from structural and functional neuroimaging studies has led to several neurobiological models of OCD. Like other hyperkinetic disorders, OCD is thought to result from hyperactivity of the direct pathway of the CSTC circuit (Baxter et al., 1992; Baxter et al., 1990; Rauch & Savage, 2000). Jackson and Houghton (1995) proposed that the indirect and direct pathways of the CSTC circuit are responsible for allocating attention to internal and external stimuli. Under normal circumstances, the direct pathway focuses attention toward salient stimuli, while the indirect pathway inhibits distraction from non-salient stimuli. Thus, familiar or innocuous inputs from the cortex (e.g., harmless thoughts) are recognized as not demanding conscious attention and are processed efficiently and nonconsciously by shifting the balance towards the indirect pathway. Novel or threatening stimuli, on the other hand, are recognized as demanding conscious attention and are processed consciously by shifting the balance towards the direct pathway.

In OCD however, Rauch and Savage (2000) proposed an impairment of nonconscious processing within the prefrontal CSTC circuit in which innocuous stimuli that are normally processed nonconsciously are instead processed consciously and inefficiently. They described this difference as a conscious process gating disturbance. As a result, individuals might be hyperaware of thought experiences with many cognitions occurring within awareness. Thoughts that might move through the mind of others with little or no conscious processing are likely to be processed consciously in individuals with these differences. This cognitive processing difference could make thoughts more difficult to dismiss and might set the stage for some thoughts becoming obsessional problems. The repetitive behaviors used by obsessional individuals might function to enhance the CSTC gating mechanism, in hopes of facilitating a shift back to appropriate nonconscious processing and reduced attentional allocation to thought (i.e., a shift from the direct to the indirect pathway). We discuss these ideas further below, connecting specific neuropsychological, metacognitive, and cognitive models to elucidate the nature of OCD.

Implicit and Procedural Learning and the CSTC

What are Implicit and Procedural Learning?

Implicit learning is the process whereby knowledge about the structural relations between objects or events is acquired incidentally (see Stadler & Frensch, 1998, for review). In other words, no attempts are made to discover rules or regularities in the sequences of objects or events. Moreover, implicit learning is assumed to be nonconscious and difficult to verbalize. Procedural learning, a subtype of implicit learning, is characterized by the acquisition of motor or perceptual skills without explicit awareness (Cohen & Squire, 1980; Squire, 1992; Squire & Zola-Morgan, 1988; Squire & Zola, 1996). Learning to ride a bike or play the piano all require procedural learning. In contrast, explicit or declarative learning or memory is associated with the conscious retrieval of information. Explicit learning or memory is believed to be mediated via prefrontal cortex and medial temporal structures, including the hippocampus, whereas procedural learning or memory is purportedly mediated via the CSTC circuits (Destrebecqz et al., 2005; Squire, 1992; Squire & Zola, 1996). Procedural learning results in faster and more accurate task performance by repeated exposure to a specific task. As a consequence, less attention is required and cognitive involvement is gradually reduced. The strongest evidence for this view comes from neuropsychological studies in which patients with global amnesia exhibit completely intact skill learning despite severely impaired declarative memory (Cohen & Squire, 1980; Nissen & Bullemer, 1987). Conversely, patients with injuries in motor systems that subserve performance exhibit impaired skill learning but not impaired explicit memory (Gabrieli, Stebbins, & Singh, 1997; Heindel, Butters, & Salmon, 1988; Willingham, Koroshetz, & Peterson, 1996).

Tasks that Measure Procedural Learning

A range of tasks have been developed to assess procedural learning that can be conceptualized along a motor to cognitive continuum (Saint-Cyr & Taylor, 1992). Tasks such as rotary pursuit (Heindel et al., 1988; Heindel, Salmon, Shults, Walicke, & Butters, 1989) are largely dependent on primary sensorimotor systems, whereas tasks such as mirror reading (Martone, Butters, Payne, Becker, & Sax, 1984), Tower of London (Shallice, 1982) and the Tower of Hanoi (Cohen et al., 1985) are largely dependent on visuospatial and cognitive capacities. The multiple CSTC circuits proposed by Alexander et al. (1990) would predict that the prefrontal circuits via the caudate would mediate cognitive procedural learning whereas the motor CSTC circuit via the putamen would mediate motor procedural learning. Therefore, the serial reaction time (SRT) task (Nissen & Bullemer, 1987), which depends on the combined motor and visuospatial systems, may be most suitable for the assessment of cognitive-motor procedural learning in OCD.

In the original version of the SRT task (Nissen & Bullemer, 1987), an asterisk appears at one of four positions horizontally arranged on a computer screen on each trial. Participants respond as quickly as possible to the position of the stimulus by pressing the corresponding response key directly below the stimulus. Response time decreases with practice when the sequence of positions follow a fixed pattern (e.g., when the same sequence of 10 positions are constantly repeating). However, when the repeating sequence is unexpectedly switched to a random sequence after prolonged training, response time increases markedly. If the SRT task is used to measure explicit learning, participants are told before the task begins that the stimuli are sequenced and that they are to learn the sequence over trials. Declarative knowledge of the sequence results in faster performance as subjects consciously anticipate the location of each successive target (Willingham, Nissen, & Bullemer, 1989). If the SRT task is used to measure implicit or procedural learning, participants are *not* told about a fixed sequence before the task begins. In the absence of explicit knowledge, participants should still perform faster with practice (although likely slower than if they knew explicitly about the sequence; Nissen & Bullemer, 1987). The difference in response times between a fixed sequence and a subsequent random-order test block would reveal procedural learning of the sequence structure. Typically, a generate task is administered at the end of the procedural learning trials to assess whether the participant gained awareness of the sequence spontaneously. This is done by telling the participants that the stimuli in the previous trials were sequenced and that they are now to predict this sequence (i.e., indicate where the next stimulus will appear). The SRT task offers an exceptionally well-controlled opportunity to compare procedural and declarative learning because it can be manipulated to be an implicit or an explicit task.

Variants of the original SRT task have been developed, mostly to help minimize the possible contribution of explicit learning to procedural learning. For example, in a dual-task paradigm (Grafton, Hazeltine, & Ivry, 1995; Hazeltine, Grafton, & Ivry, 1997), awareness was reduced by having subjects attend to a secondary task (e.g., tone counting). In the absence of the secondary task in the single-task blocks, subjects easily became aware of the presented sequence. Other researchers have employed a double SRT task in which simultaneous finger movements of the two hands were made in response to pairs of visual stimuli that were presented in a fixed or random order (van der Graaf, Maguire, Leender, & de Jong, 2006). The use of the SRT task in conjunction with functional imaging have greatly enhanced the understanding of the neuroanatomical basis of procedural learning, as described in the next section.

FUNCTIONAL IMAGING, PROCEDURAL LEARNING, AND THE CSTC CIRCUITRY

The neuroanatomical basis of procedural learning has been explored more extensively within the past two decades. Lesion studies have implicated the prefrontal regions, the cerebellum, and the basal ganglia as being critically involved in procedural learning (Beldarrain, Grafman, Pascual-Leone, & Garcia-Monco, 1999; Doyon et al., 1997; Exner, Koschack, & Irle, 2002; Vakil, Blachstein, & Soroker, 2004; Vakil, Kahan, Huberman, & Osimani, 2000). The use of functional neuroimaging has also implicated these and other areas as important for procedural learning in healthy adults. For example, using fMRI, Kassubek, Schmidtke, Kimmig, Lücking, and Greenlee (2001) found that non-motor procedural learning of the mirror reading paradigm was related to changes in the cortical blood oxygenation of the motor and premotor cortex, the parietal lobe, and the occipital lobe.

Most neuroimaging studies, however, exclusively used the SRT task as a measure of procedural learning. With PET, Grafton et al. (1995) showed, using dual-task conditions, that rCBF increases were related to the procedural learning of structured sequences in the contralateral motor cortex, supplementary motor area and putamen, a result replicated with color-coded rather than spatially-coded stimuli (Hazeltine et al., 1997). Under single-task conditions, procedural learning was found to be associated with a variety of brain regions. Rauch et al. (1995), using traditional sequenced asterisks, showed that procedural sequence learning was associated to rCBF activation in the right hemisphere in ventral premotor cortex, caudate nucleus and thalamus, and bilaterally in visual association areas in one study (Rauch et al., 1995), a result that was not replicated with numeral-coded stimuli (Honda et al., 1998). Instead, Honda et al. (1998) found rCBF activation in the contralateral primary sensorimotor cortex. Other PET studies pointed to the specific involvement of the cerebellum and striatum in the more automated late stage of the implicit learning process (Doyon, Owen, Petrides, Sziklas, & Evans, 1996), and the involvement of the striatum in the acquisition of complex higher-order sequences (Peigneux et al., 2000).

The importance of the striatum in procedural learning was further supported by fMRI studies (Rauch, Whalen, et al., 1997, 1998; Reiss et al., 2005). Rauch, Whalen, et al. (1997)

reported that the thalamus was deactivated while the striatum was recruited in the early phase of learning (Rauch et al., 1997), suggesting a thalamic gating mediated by the indirect CSTC pathway (Rauch et al., 1998). Willingham, Salidis, and Gabrieli (2002) demonstrated that procedural learning activated left prefrontal cortex, left inferior parietal cortex, and right putamen. Most recently, using a double SRT task, van der Graaf et al. (2006) showed specific activation related to procedural learning in the right ventrolateral prefrontal cortex, and less so in the medial prefrontal and right ventral premotor cortex.

The diversity of the results reported in these neuroimaging studies may be partly explained by several factors: the experimental design, the stimulus type (e.g., spatial locations, colors, Arabic numerals, letters), the structure of the material (e.g., fixed versus probabilistic sequences), or the relative contribution of explicit knowledge to performance as well as the way in which this contribution was assessed (e.g., verbal reports, structured questionnaires, generation or recognition tasks) and controlled for (e.g., dual-task conditions, double SRT tasks). Specific limitations of each neuroimaging technique and analysis methods (e.g., whole-brain or regional-or-interest investigation, a priori or post-hoc hypotheses, categorical or parametric analyses) also influenced the results of these experiments. As a consequence, the variety of brain areas activated in the SRT task may reflect various components of functional networks responsible for implicit sequential learning.

The Relationship between Implicit and Explicit Learning

One question arising from this line of research is whether implicit and explicit learning occur in parallel, or possibly interact with each other, during the acquisition of knowledge or skills. Based on the functional imaging studies described in the previous section, explicit sequence learning seems to activate completely different brain regions than implicit sequence learning. Hazeltine et al. (1997) reported that explicit sequence awareness during single task performance in the SRT task was associated with activation of the prefrontal, premotor, and temporal cortexes. Explicit learning of a sequence also activated the primary visual cortex, perisylvian cortex, and cerebellar vermis (Rauch et al., 1995) and the frontoparietal network (Honda et al., 1998). The activation of different brain regions suggest that implicit and explicit learning occur quite independently from each other. However, Willingham and his colleagues argued that this assumption may be faulty due to several experimental confounds in these studies (Willingham et al., 2002). First, the conclusion that neural areas are unique to procedural learning relies on null result. The lack of activation of these brain areas during explicit sequence learning does not preclude their involvement in explicit learning. Second, measuring implicit and explicit activation sequentially as participants become aware of the sequence (e.g., Honda et al., 1998; Rauch et al., 1995) confounds order, amount of practice, and performance levels with the change of awareness. There is more practice in the explicit condition because it always occurs after the implicit condition has terminated. Furthermore, performance is always superior in explicit learning conditions for such tasks. Grafton et al. (1995) and Hazeltine et al.'s (1997) attempted to control the amount of practice of each sequence by using two different sequences procedures to help distinguish between implicit and explicit learning. Trials were conducted with the presence or absence of a secondary task. This additional behavioral dimension might have affected the neural activity in the implicit condition in these studies. Once these confounds were addressed, the evidence pointed to the parallel acquisition of implicit and explicit knowledge.

Willingham and Goedert-Eschmann (1999) demonstrated the overlap of implicit and explicit learning by administering an implicit test of a sequence after the participants learned them explicitly in an SRT task. Participants showed procedural learning equivalent to those without the explicit training, implying that implicit knowledge had been acquired independent of explicit knowledge. Functional imaging studies corroborated this finding. Willingham et al. (2002) manipulated the awareness of the sequence by changing the color of the stimuli to match with or differ from the color used for random sequences. This allowed direct comparison of brain activation associated with implicit and explicit learning for an identical sequence. This cleverly designed study demonstrated that participants performed fastest under the "explicit-overt" condition in which they were aware of the repeating sequence appearing in red. Participants were slower in the "explicit-covert" condition in which the same sequence used for the "explicit-overt" condition appeared disguised as black circles, and in the 'implicit" condition in which a different sequence, also presented as black circles, was learned without awareness. Finally, participants performed slowest in the "random" condition in which black circles were presented in random locations. A common neural network (i.e., ventrolateral prefrontal cortex, parietal cortex, and putamen) was activated during the "explicit-covert," "implicit," and "explicit-overt" conditions. Moreover, additional widespread activation was found to be associated with explicit learning in the "explicit-overt" condition. Therefore, not only do these findings suggest that there are a specific set of brain regions involved in implicit learning with or without explicit awareness, they also suggest that the presence of explicit learning does not impair implicit learning.

Under some circumstances, however, explicit learning can impede implicit learning. Howard and Howard (2001) proposed that attempting to discover a pattern (e.g., sequence learning in a SRT task) can drain mental capacity from a covariation-detecting mechanism that underlies all forms of implicit learning. If the demand on cognitive resources is increased (e.g., if there is noise in the sequence) or if cognitive capacity is reduced (e.g., in the elderly), then implicit learning becomes vulnerable. Accordingly, the reason why most studies with the SRT task have not demonstrated the negative effects of explicit awareness in implicit learning is likely due to the minimal demands placed by the explicit search processes on cognitive resources. As such, Howard and Howard (2001) demonstrated the negative effects of explicit awareness in implicit learning by using a more demanding SRT task in which sequence elements alternated with random ones in a visual/spatial display, and by using an elderly population more likely to have deficits in processing capacity. Older participants, but not younger ones, showed impaired implicit learning after being told explicitly to learn the sequence (Howard & Howard, 2001). An fMRI study further supported Howard and Howard's (2001) results. In this study, Fletcher et al. (2005) placed participants in one of four conditions that varied in subjective intention to learn (explicit versus implicit) and sequence difficulty (a standard SRT task versus the alternating SRT task used by Howard & Howard, 2001). They showed that explicit attempts to learn the difficult sequence hindered implicit learning. A follow-up behavioral experiment showed that this failure represents suppression of learning itself rather than of the expression of learning. This suppression is associated with sustained right frontal activation and activation of learning-related changes in the medial temporal lobe and thalamus. Furthermore, this condition is characterized by a reversal of the fronto-thalamic connectivity observed with unimpaired implicit learning. These findings, therefore, emphasized the conscious search and monitoring processes necessary for explicit learning through a reduced emphasis on response speed. They also provided a neural basis for the deleterious impact of an explicit search upon implicit learning.

PROCEDURAL LEARNING IN PARKINSON'S, HUNTINGTON'S, AND TOURETTE SYNDROME

The potential importance of procedural learning to OCD can be evaluated by reviewing studies with related neurobiological disorders, although the association between striatal dysfunction and procedural learning in these patients have proved inconclusive to date. Impaired procedural learning in patients with Parkinson's and Huntington's disease was observed in the performance of cognitive tasks such as the Tower of Hanoi (Schmidtke, Manner, Kaufmann, & Schmolck, 2002; Vakil, Herishanu-Naaman, 1998) and its variant, Tower of London (Saint-Cyr, Taylor, & Lang, 1988). Performance on the rotary pursuit task, which measures motor learning, was impaired with patients with Huntington's disease (Heindel et al., 1989; Schmidtke, 2002). However, results for Parkinson's disease patients were more inconsistent. While one study reported rotary pursuit learning deficits (Harrington, Haaland, Yeo, & Marder, 1990), other studies did not (Bondi & Kaszniak, 1991; Heindel et al., 1989). Haaland, Harrington, O'Brien, and Hermanowica (1997) later observed and demonstrated that impaired rotary pursuit learning only occurred in these patients when rotation speeds were presented randomly, and not when they were presented in blocks.

A number of studies have shown impairment in SRT sequence learning in patients with Parkinson's disease (Brown et al., 2003; Jackson, Jackson, Harrison, Henderson, & Kennard 1995; Smith & McDowall, 2004; Stefanova, Kostic, Ziropadja, Markovic, & Ocic, 2000; Westwater, McDowall, Siegert, Mossman, & Abernethy, 1998) as well as Huntington's disease (Kim et al., 2004; Willingham & Koroshetz, 1993). Yet in other studies of patients with basal ganglia dysfunction, only minor sequence-specific learning deficits (Ferraro, Balota, & Connor, 1993; Knopman & Nissen, 1991; Pascual-Leone et al., 1993) were found. In some investigations, preserved sequence specific learning and only nonspecific impairments (e.g., reduced reaction time improvements over sequential trials) were found (Helmuth, Mayr, & Daum, 2000; Shin & Ivry, 2003; Sommer, Grafman, Clark, & Hallett, 1999; Werheid, Ziessler, Nattkemper, & Yves von Cramon, 2003; Werheid, Zysset, Müller, Reuter, & Von Cramon, 2003). Furthermore, normal SRT performance has previously been reported in both patients with Parkinson's disease (Kelly, Jahashahi, & Dirnberger, 2004; Smith, Siegert, McDowall, & Abernethy, 2001) and patients with Huntington's disease (Brown, Redondo-Verge, Chacon, Lucas, & Channon, 2001). These mixed results are very likely attributable to differences between the patient samples and to methodological variations of the SRT tasks employed. A recent meta-analysis conducted by Siegert, Taylor, Weatherall, and Abernethy (2006) on six of the more recent studies (i.e., Brown et al., 2003; Kelly et al., 2004; Shin & Ivry, 2003; Smith & McDowall, 2004; Sommer et al., 1999; Werheld et al., 2003) suggested reliable impairment in implicit sequence learning in Parkinson's disease patients..

Tourette syndrome, which also involves the basal ganglia and related CSTC circuitry, is often considered an OCD-spectrum disorder (Leckman, 2002; Rauch et al., 2001). To date,

Marsh and colleagues have conducted the only known studies on procedural learning in Tourette syndrome (Marsh et al., 2004; Marsh, Alexander, Packard, Zhu, & Peterson, 2005). Although they did not find deficits in procedural learning using the pursuit rotary task or the mirror-tracing task (Marsh et al., 2005), they did find significant impairment on a task of probabilistic classification learning, a form of habit learning that circumvents the use of declarative memory by probabilistically associating cues with specific task outcomes (Marsh et al., 2004; Marsh et al., 2005). The specific deficits in the probabilistic classification form of habit learning are likely a consequence of disturbances in specific CSTC circuits that are different from the ones that subserve the perceptual-motor form of habit learning of the rotary pursuit and mirror tracing tasks (Marsh et al., 2005).

Finally, it is important to point out that impaired implicit learning was not observed in specific animal phobia (Martis, Wright, McMullin, Shin, & Rauch, 2004), a type of anxiety disorder. This is likely due to the lack of striatal dysfunction in specific phobia patients, and suggests that the underlying neurocircuitry disturbance is different from Parkinson's disease, Huntington's disease, and OCD. This finding calls into question the construct validity and the categorical boundaries seen in the current psychiatric classification system and suggest the need for revision.

Procedural Learning and OCD

Within the last decade, several groups of investigators have hypothesized that OCD patients would exhibit performance deficits on procedural learning tasks because of the purported involvement of cortico-striatal circuitry in this type of learning. Evidence for performance deficits and or the activation of specific regions within the CSTC circuitry during tasks within the procedural learning paradigm would support specific theories regarding the pathophysiology of OCD (Rauch & Savage, 2000). Further, OCD patients may compensate for deficits in procedural learning by relying on explicit learning strategies when confronted with an implicit learning task (Rauch, Savage, et al., 1997). That is, individuals with OCD might actively and consciously encode information as it is presented, regardless of task instructions, instead of processing the information nonconsciously. These information processing differences may have important implications for understanding OCD symptoms as discussed below.

While the SRT task has proven to be a reliable and valid measure of procedural learning, performance results can be contaminated by the use of explicit learning strategies during the task. In one of the first studies of its kind, Rauch, Savage, et al. (1997) examined regions of brain activation in OCD patients during the SRT task using PET, and directly measured contamination by explicit learning strategies by asking participants to independently generate the embedded pattern at the end of the SRT task trials. Although the hypothesized performance differences were not found on the implicit learning task, possibly as a result of small sample size, Rauch, Savage, et al. (1997) made several notable findings. Unlike healthy controls, OCD patients did not display significant inferior striatal activity during the SRT task. OCD patients exhibited bilateral activation in the parahippocampal and hippocampal regions. The hippocampus is usually implicated in tasks that require explicit learning, leading the authors to suggest that individuals with OCD compensate for their difficulties with procedural learning by employing explicit learning strategies.

Another attempt to show the interference of explicit learning in implicit learning was made by Deckersbach et al. (2002) in which a dual task paradigm was introduced. Participants engaged in an explicit learning task (i.e., actively memorizing a seven-letter sequence) while performing the SRT task. Deckersbach et al. (2002) were able to demonstrate poorer performance on the SRT task in OCD patients compared to healthy controls under these dual task conditions and no differences were found between control and OCD participants in recognition of the embedded pattern following the SRT task. Thus, Deckersbach et al. (2002) successfully demonstrated that OCD patients display deficits in procedural but not explicit learning.

As an extension of the previous two studies, Kathmann, Rupertseder, Hauke, and Zaudig (2005) compared performance on the SRT task for OCD patients before and after intensive cognitive-behavioral treatment. They also examined correlations between procedural learning performance and symptom severity. While they replicated Deckersbach et al.'s (2002) results with a single-task paradigm, they did not find significant differences in task performance before and after treatment, nor did they find significant correlations between task performance and their measures of symptom severity (i.e., the Yale-Brown Obsessive-Compulsive Scale, Goodman et al., 1989; and the Hamilton Rating Scale for Depression, Hamilton, 1960). The lack of improvement in the SRT task from pre to posttreatment despite the reduced symptom levels of the OCD patients suggest that implicit sequence learning may be independent of symptom states.

Rather than follow convention with the SRT task, Joel et al. (2005) examined procedural learning deficits by utilizing a card betting task in which explicit learning strategies have been demonstrated to hinder rather than facilitate the acquisition of the task. In addition to comparing task performance between OCD patients and healthy controls, the investigators added a group of Parkinson's disease patients and patients with major depressive disorder to their analyses. They hypothesized that because Parkinson's disease patients also display procedural learning deficits, similarities in task performance between them and OCD patients would indicate that the task did indeed measure this specific type of deficit. Because both OCD and Parkinson's disease patients may also suffer from depression, a group of patients with major depressive disorder was included, in order to test the possibility that impaired learning in the card betting task may be a result of concurrent depression. Patients' performance on the card betting task supported the main hypotheses: the majority of OCD and Parkinson's disease patients did not acquire the task (i.e., performance did not improve over trials), while depressed patients' performance was comparable to that of healthy controls. Therefore, despite the use of a different task, these results were congruent with previous results and demonstrate that OCD patients are impaired on a procedural learning task.

Rauch et al. (2007) attempted to replicate their previous findings of abnormal striatal and hippocampal functioning during procedural learning tasks. They also tested the relationship between established factor-analyzed OCD symptom dimensions and rCBF activation during implicit learning with the SRT task. The factor structure, as derived by Leckman et al. (1997), included four dimensions: Factor 1 involved aggressive, sexual, religious, and somatic obsessions and checking compulsions; Factor 2 involved symmetry obsessions and ordering, arranging, counting, and repeating compulsions; Factor 3 involved contamination obsessions and cleaning and washing compulsions; and Factor 4 involved hoarding obsessions and compulsions. Overall, Rauch et al. (2007) found greater rCBF activation during implicit learning for OCD participants in the orbitofrontal cortex and the hippocampus than for healthy controls. No such differences in rCBF activation were found in the striatum or thalamus. Moreover, overall OCD symptom severity was not related to rCBF activation of the different brain regions during implicit learning. Yet upon further examination, a significant inverse relationship was found between activation of the right inferior caudal-ventral striatum and OCD symptom severity during implicit learning in both Factor 2 and Factor 3 while a significant direct relationship was found between activation of the left orbitofrontal cortex and OCD symptom severity during implicit learning for Factor 1 scores. Although their results replicated previous findings of hippocampal activation in OCD during implicit learning, they failed to replicate findings of striatal activation. Furthermore, activation of the orbitofrontal cortex was not previously reported. Rauch et al. (2007) suggested that this inconsistency may be attributable to differences across OCD symptom dimensions, although future replications that are sensitive to the heterogeneous nature of OCD are warranted. Nevertheless, this study adds to the growing literature on the need to recognize OCD symptom subtypes and the underlying mechanism or factors that may contribute to the symptom variations.

Procedural Learning, CSC, and OCD

While researchers have often highlighted the connections between procedural learning deficits and the proposed pathophysiology of OCD, few have focused on the broader implications of these deficits for understanding the core phenomenology of OCD. We propose that these findings might have implications for other theoretical models of the disorder. We have posited that the procedural learning differences, now known to be importantly connected to OCD, might function to integrate findings from neuropsychological models of the disorder with cognitive and metacognitive theories of OCD (e.g., Janeck, Calamari, Riemann, & Heffelfinger, 2003; Marker et al., 2006).

In Rauch and Savage's (2000) neuropsychological model of OCD, emphasis was placed on a breakdown of conscious processing gating, functions that they argue are governed by the frontal cortico-striatal circuits. Rauch and Savage posited that this system plays a critical role in maintaining a balance between what is processed consciously and what is dismissed without significant conscious awareness or the deployment of attentional resources. OCD is understood to result from impaired nonconscious processing, such that innocuous stimuli are processed consciously and inefficiently due to a misallocation of attention (Rauch & Savage, 2000).

Janeck et al. (2003) argued that the conscious processing gating disturbance that Rauch and Savage proposed characterizes OCD is conceptually similar to the metacognitive construct, cognitive self-consciousness (CSC), defined as the valence-neutral propensity to focus attention on one's own thought processes.

Specifically, Janeck et al. (2003) suggested that the apparent striatal under activation and procedural learning deficits seen in patients with OCD may be related to CSC. As discussed earlier, CSC was first introduced by Cartwright-Hatton and Wells (1997) as a subscale on their measure of metacognitive processes, processes thought importantly related to anxiety disorders. Janeck et al. (2003) proposed that elevations in CSC might explain the increased salience of negative intrusive thought experiences among individuals with OCD, and that

CSC might function as a cognitive risk factor for OCD. Significant elevations in CSC have been consistently found in OCD patients (Cartwright-Hatton & Wells, 1997; Janeck et al., 2003; Marker et al., 2006). Additionally, CSC has predicted OCD symptoms in a nonclinical sample (Cohen & Calamari, 2004) suggesting that CSC does not simply result from the development of an obsessional problem, thus lending support to the contention that CSC could be a cognitive risk factor for OCD.

Marker et al. (2006) suggested that the CSC construct might function to address a limitation of cognitive theory. The negative appraisal of intrusive thought experiences is an important mechanism in all cognitive models of OCD and specific types of social learning experiences (e.g., inadvertent action leading to detrimental outcomes) have been posited to precipitate OCD-related dysfunctional beliefs (Salkovskis, Shafran, Rachman, & Freeston, 1999). There has been limited empirical evaluation of this hypothesis, and more generally, of the processes responsible for some peoples' focus on thought experiences and related difficulties in dismissing common negative intrusive thoughts. With these issues in mind, Marker et al. (2006) evaluated the relationships between implicit learning, CSC, and OCD symptoms in OCD patients and healthy controls.

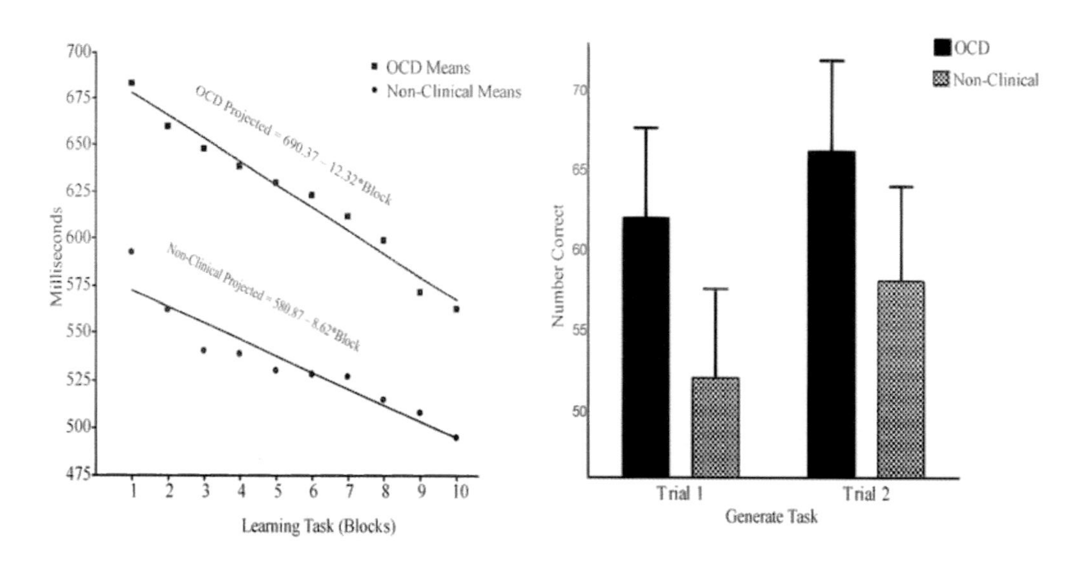

Figure 2. Mean reaction times for the serial reaction time (SRT) task and number of correct responses for a test of pattern recognition (generate task) for obsessive-compulsive (OCD) and nonclinical groups[1].

Marker et al. (2006) found that deficits in procedural learning, as measured by performance on the SRT task, were in fact correlated with CSC, as measured by the Cognitive Self-Consciousness Scale – Expanded (CSC-E; Janeck et al., 2003). Congruent with previous findings, Marker et al. (2006) found that OCD patients performed more poorly on the procedural learning task than healthy controls (i.e., displayed longer reaction times across

---

[1] From "Cognitive Self-Consciousness, Implicit Learning, and Obsessive-Compulsive Disorder" by C. D. Marker, J. E. Calamari, J. L. Woodard, and B. C. Riemann, 2006, *Anxiety Disorders, 20,* p. 400. Copyright 2005 by Elsevier.

trials). Most notable was the demonstration of *better* explicit learning performance in OCD patients compared to control participants as measured by recognition of the embedded pattern after the procedural learning trials. Results of the Marker et al. (2006) study are shown in Figure 2. Using multilevel modeling techniques, the authors found that CSC better predicted performance deficits on the SRT task than either depression or OCD symptom severity. The results of the Marker et al. study were congruent with both Rauch and Savage's (2000) model of impaired information processing in OCD and with cognitive and metacognitive models of the disorder. Both CSC and the implicit and explicit learning abnormalities may result from impaired inhibition of conscious thought processes. Individuals with OCD or individuals vulnerable to developing the disorder may experience thoughts as highly salient, providing increased opportunities for the negative appraisal of thoughts, a process posited to be central to the development of OCD in cognitive models of the disorder (e.g., Salkovskis, 1985).

FUTURE DIRECTIONS

OCD is a complex disorder that affects the lives of many individuals. Although significant advances have been made in etiologic theories and in the treatment of the disorder, much work remains to be done. As reviewed here, recent investigations on procedural learning differences in OCD have the potential to advance understanding of important issues and may function to integrate cognitive and neurobiological theories.

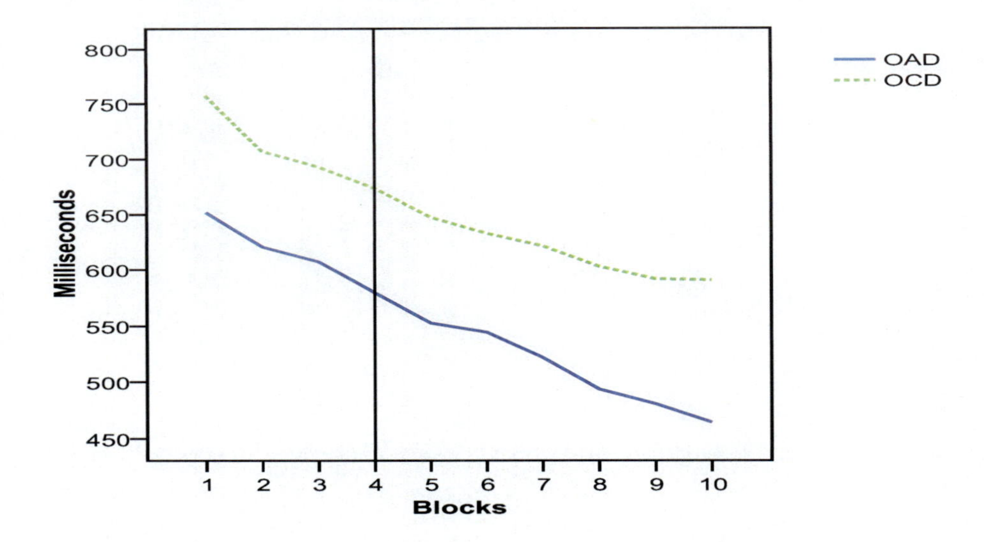

Figure 3. Mean reaction times for the serial reaction time (SRT) task across three practice blocks and seven learning trial blocks for obsessive-compulsive (OCD) and other anxiety disorder (OAD) groups.

In our laboratory, a replication of Marker et al.'s (2006) study is underway and preliminary results are encouraging (see Figure 3). Goldman et al. (2007) are attempting to evaluate SRT differences in OCD patients and anxious controls. Variability in SRT test performance is being tested across empirically defined OCD symptom subgroups in an attempt to address arguably the greatest problem in OCD research, the heterogeneous nature of the condition. Significant variability in implicit learning was found by Goldman et al.

(2007). Additional investigations are needed to determine whether procedural learning differences and related CSC differences are state or trait characteristics of OCD, and whether CSC is a cognitive risk factor for the disorder. Do current treatments for OCD normalize information processing or does the tendency to focus attention on thought remain elevated, leaving successfully treated OCD patients at greater risk to redevelop the disorder? Only future investigations on the nature of procedural learning and CSC in individuals with OCD will tell.

REFERENCES

Abbruzzese, M., Bellodi, L., Ferri, S., & Scarone, S. (1995). Frontal lobe dysfunction in schizophrenia and obsessive-compulsive disorder: A neuropsychological study. *Brain and Cognition,* **27**, 202-212.

Abbruzzese, M., Ferri, S., & Scarone, S. (1997). The selective breakdown of frontal functions in patients with obsessive-compulsive disorder and in patients with schizophrenia: A double dissociation experimental finding. *Neuropsychologia,* **35**, 907-912.

Adams, B. L, Warneke, L. B., McEwan, A. J. B., & Fraser, B. A. (1993). Single photon emission computerized tomography in obsessive-compulsive disorder: A preliminary study. *Journal of Psychiatry and Neuroscience,* **18**, 109-112.

Airaksinen, E., Larsson, M., & Forsell, Y. (2005). Neuropsychological functions in anxiety disorders in population-based samples: Evidence of episodic memory dysfunction. *Journal of Psychiatric Research,* **39**, 207-214.

Alegret, M., Junqué, C., Valldeoriola, F.; Vendrell, P., Matrí, M. J., & Tolosa, E. (2001). Obsessive-compulsive symptoms in Parkinson's disease. *Journal of Neurology, Neurosurgery & Psychiatry*, 70, 394-396.

Alexander, G. E. (1994). Basal ganglia-thalamocortical circuits: Their role in control of movements. *Journal of Clinical Neurophysiology*, 11, 420-431.

Alexander, G. E., Crutcher, M. D., & Delong, M. R. (1990). Basal ganglia-thalamocortical circuits: Parallel substrates for motor, oculomotor, "prefrontal" and "limbic" functions. *Progress in Brain Research*, 85, 119-146.

Alexander, G. E., DeLong, M. R., & Strick, P. L. (1986). Parallel organization of functionally segregated circuits linking basal ganglia and cortex. *Annual Review of Neuroscience*, 9, 357-381.

American Psychiatric Association. (2000). *Diagnostic and statistical manual of mental disorders* (4th ed., text revision). Washington, DC: Author.

Anderson, K. E., Louis, E. D., Stern, Y., & Marder, K. S. (2001). Cognitive correlates of obsessive and compulsive symptoms in Huntington's disease. *American Journal of Psychiatry,* 158, 799-801.

Baxter, L. R., Schwartz, J. M., Bergman, K., Szuba, M. P., Guze, B. H., Mazziota, J. C., et al. (1992). Caudate glucose metabolic rate changes with both drug and behavior therapy for obsessive-compulsive disorder. *Archives of General Psychiatry*, 49, 681-689.

Baxter, L. R., Schwartz, J. M., Guze, B. H., Bergman, K., & Szuba, M. P. (1990). Neuroimaging in obsessive-compulsive disorder: Seeking the mediating neuroanatomy. In M. A. Jenike, L. Baer, & W. E. Minichiello (Eds.), *Obsessive compulsive disorder: Theory and management* (2nd ed., pp. 167-188).Chicago: Year Book Medical.

Baxter, L. R., Schwartz, J. M., Mazziotta, J. C., Phelps, M. E., Pahl, J. J., Guze, B. H., & Fairbanks, L. (1988). Cerebral glucose metabolic rates in non-depressed obsessive-compulsives. *American Journal of Psychiatry,* **145**, 1560-1563.

Baxter, L. R., Thompson, J. M., Schwartz, J. M., Guze, B. H., Phelps, M. E., Mazziotta, J. C., et al. (1987). Trazodone treatment response in obsessive-compulsive disorder – correlated with shifts in glucose metabolism in the caudate nuclei. *Psychopathology,* **20**, 114-122.

Beck, A.T. (1976). *Cognitive therapy and the emotional disorders.* New York: International University Press.

Beldarrain, M. G., Grafman, J., Pascual-Leone, A., & Garcia-Monco, J. C. (1999). Procedural learning is impaired in patients with prefrontal lesions. *American Academy of Neurlogy,* **52**, 1853-1860.

Benkelfat, C., Nordahl, T.E., Semple, W.E., King, A.C., Murphy, D. L., & Cohen, R. M. (1990). Local cerebral glucose metabolic rates in obsessive-compulsive disorder: Patients treated with clomipramine. *Archives of General Psychiatry,* **47**, 840-848.

Benton, A. L. (1974). *Revised visual retention test* (4th ed). San Antonio, TX: The Psychological Corporation.

Bohne, A., Savage, C.R., Deckersbach, T., Keuthen, N.J., Jenike, M.A., Tuschen-Caffier, B., & Wilhelm. (2005). Visuospatial abilities, memory, and executive functioning in trichotillomania and obsessive-compulsive disorder. *Journal of Clinical and Experimental Neuropsychology,* **27**, 385-399.

Boldrini, M., Del Pace, L., Placidi, G. P. A., Keilp, J., Ellis, S.P., Signori, S., et al. (2005). Selective cognitive deficits in obsessive-compulsive disorder compared to panic disorder with agoraphobia. *Acta Psychiatrica Scandiavica,* **111**, 150-158.

Bondi, M. W., & Kaszniak, A. W. (1991). Implicit and explicit memory in Alzheimer's disease and Parkinson's disease. *Journal of Clinical and Experimental Neuropsychology,* **13**, 339-358.

Bouvard, M. (2001). Cognitive effects of cognitive-behavior therapy for obsessive compulsive disorder. In R. O. Frost & G. Steketee (Eds.), *Cognitive approaches to obsessions and compulsions: Theory, assessment, and treatment* (pp. 403-416). Kidlington, Oxford: Elsevier Science.

Breiter, H.C., Rauch, S.L., Kwong, K.K., Baker, J.R., Weisskoff, R. M. Kennedy, D. N., et al. (1996). Functional magnetic resonance imaging of symptom provocation in obsessive-compulsive disorder. *Archives of General Psychiatry,* **49**, 595-606.

Brown, R. G., Jahanshahi, M., Limousin-Dowsey, P., Thomas, D., Quinn, N. P., & Rothwell, J. C., (2003). Palidotomy and incidental sequence learning in Parkinson's disease. *Cognitive Neuroscience,* **14**, 21-24.

Brown, R. G., Redondo-Verge, L., Chacon, J. R., Lucas, M. L., & Channon, S. (2001). Dissociation between intentional and incidental sequence learning in Huntington's disease. *Brain,* **124**, 2188-2202.

Calabrese, G., Colombo, C., Bonfanti, A., Scotti, G., & Scarone, S. (1993). Caudate nucleus abnormalities in obsessive-compulsive disorder: Measurements of MRI signal intensity. *Psychiatry Research: Neuroimaging,* **50**, 89-92.

Carr, A. T. (1974). Compulsive neurosis: A review of the literature. *Psychological Bulletin,* **81**, 311-318.

Cartwright-Hatton, S., & Wells, A. (1997). Beliefs about worry and intrusions: The meta-cognitions questionnaire and its correlates. *Journal of Anxiety Disorders,* **11**, 279-296.

Cavedini, P., Ferri, S., Scarone, S., & Bellodi, L. (1998). Frontal lobe dysfunction in obsessive-compulsive disorder and major depression: A clinical-neuropsychological study. *Psychiatry Research,* **78**, 21-28.

Christensen, K. J., Kim, S. W., Dysken, M. W., & Hoover, K. M. (1992). Neuropsychological performance in obsessive-compulsive disorder. *Biological Psychiatry,* **31**, 4-18.

Clark, D. A. (2004). *Cognitive behavior therapy of OCD.* New York: Guilford Press.

Cohen, N.J., Eichenbaum, H., Deacedo, B.S., & Corkin, S. (1985). Memory dysfunctions: An integration of animal and human research from preclinical and clinical perspectives. In D.S. Olton, E. Gamzu, & S. Corkin (Eds.), *Different Memory Systems Underlying Acquisition of Procedural and Declarative Knowledge* (pp. 54-71). New York: New York Academy of Science.

Cohen, R. J., & Calamari, J. E. (2004). Thought-focused attention and obsessive compulsive symptoms: An evaluation of cognitive self-consciousness in a nonclinical sample. *Cognitive Therapy and Research,* **28**, 457-471.

Cohen, Y., Lachenmeyer, J.R., & Springer, C. (2003). Anxiety and selective attention in obsessive-compulsive disorder. *Behaviour Research and Therapy,* **41**, 1311-1323.

Cohen, N. J., & Squire, L. R. (1980). Preserved learning and retention of pattern-analyzing skill in amnesia: Dissociation of knowing how and knowing that. *Science,* **210**, 207-210.

Cottraux, J., & Gérard, D. (1998). Neuroimaging and neuroanatomical issues in obsessive-compulsive disorder: Toward and integrative model – perceived impulsivity. In Swinson, R.P, Antony, M.M, Rachman, S., & Richter, M.A. (Eds.), *Obsessive-compulsive disorder: Theory, research, and treatment* (pp. 154-180). New York, NY: Guilford Press.

Cottraux, J., Gerard, D., Cinotti, L., Froment, J. C., Deiber, M. P., Le Bars, D., et al. (1996). A controlled positron emission tomography study of obsessive and neutral auditory stimulation in obsessive-compulsive disorder with checking rituals. *Psychiatry Research,* **60**, 101-112.

Cummings, J. L. (1993). Frontal-subcortical circuits and human behavior. *Archives of Neurology,* **50**, 873-880.

Cummings, J. L., & Cunningham, K. (1992). Obsessive-compulsive disorder in Huntington's disease. *Biological Psychiatry,* **31**, 263-270.

Deckersbach, T., Otto, M. W., Savage, C. R., Baer, L. & Jenike, M. A. (2000). The relationship between semantic organization and memory in obsessive–compulsive disorder. *Psychotherapy and Psychosomatics,* **69**, 101–107.

Deckersbach, T., Savage, C. R., Curran, T., Bohne, A., Wilhelm, S., Baer, L., et al. (2002). A study of parallel implicit and explicit information processing in patients with obsessive-compulsive disorder. *American Journal of Psychiatry,* **159**, 1780-1782.

Destrebecqz, A., Peigneux, P., Laureys, S., Degueldre, C., Del Fiore, G., Aerts, J., et al. (2005). The neural correlates of implicit and explicit sequence learning: Interacting networks revealed by the process dissociation procedure. *Learning & Memory*, 12, 480-490.

Dirson, S., Bouvard, M., Cottraux, J., & Martin, R. (1995). Visual memory impairment in patients with obsessive-compulsive disorder: A controlled study. *Psychotherapy and Psychosomatics,* **63**, 22-31.

Doyon, J., Gaudreau, D., Laforce, Jr., R., Castonguay, M., Bédard, P. J., Bédard, F., & Bouchard, J-P. (1997). Role of the striatum, cerebellum, and frontal lobes in the learning of a visuomotor sequence. *Brain and Cognition,* **34**, 218-245.

Doyon, J., Owen, A.M., Petrides, M., Sziklas, V., & Evans, A.C. (1996). Functional anatomy of visuomotor skill learning in human subjects examined with positron emission tomography. *European Journal of Neuroscience,* **8**, 637–648.

Eisen, J. L., & Steketee, G. (1998). Course of illness in obsessive-compulsive disorder. In L. J. Dickstein, M. B. Riba, & J. M. Oldham (Eds.), *Review of psychiatry* (Vol. 16). Washington, D. C.: American Psychiatric Press.

Enright, S. J. & Beech, A. R. (1990). Obsessional states: anxiety disorders or schizotypes? An information processing and personality assessment. *Psychological Medicine,* **20**,621-627.

Enright, S. J. & Beech, A. R. (1993a). Further evidence of reduced inhibition in obsessive compulsive disorder. *Personality and Individual Differences,* **14**, 387-395.

Enright, S. J. & Beech, A. R. (1993b). Reduced cognitive inhibition in obsessive compulsive disorder. *British Journal of Clinical Psychology,* **32**, 67-74.

Exner, C., Koschack, J., & Irle, E. (2002). The differential role of premotor frontal cortex and basal ganglia in motor sequence learning: Evidence from focal basal ganglia lesions. *Learning & Memory,* **9**, 376-386.

Ferraro, F. R., Balota, D. A., & Connor, L. T. (1993). Implicit memory and the formation of new associations in nondemented Parkinson's disease individuals and individuals with senile dementia of the Alzheimer type: A serial reaction time (SRT) investigation. *Brain and Cognition,* **21**, 163-180.

Flavel, J. H. (1979). Metacognition and metacognitive monitoring: A new area of cognitive-developmental inquiry. *American Psychologist,* **34**, 906-911.

Fletcher, P. C., Zafiris, O., Frith, C. D., Honey, R. A., Corlett, P. R., Killes, K., & Fink, G. R. (2005). On the benefits of not trying: Bain activity and connectivity reflecting the interactions of explicit and implicit sequence learning. *Cerebral Cortex,* **15**, 1002-1015.

Foa, E. B., Ilai, D., McCarthy, P. R., Shoyer, B., & Murdock, T. (1993). Information processing in obsessive-compulsive disorder. *Cognitive Therapy and Research,* **17**, 173-189.

Freedman, M. (1990). Object alternation and orbitofrontal system dysfunction in Alzheimer's and Parkinson's disease. *Brain and Cognition.* **14**, 134–143.

Freedman, M., & Oscar-Berman, M. (1986). Comparative neuropsychology of cortical and subcortical dementia. *Canadian Journal of Neurlogical Science,* **13**, 410–414.

Freeston, M. H., Rhéaume, J., & Ladouceur, R. (1996). Correcting faulty appraisals of obsessional thoughts. *Behaviour Research and Therapy,* **34**, 433-446.

Gabrieli, J. D., Stebbins, G. T., & Singh, J. (1997). Intact mirror-tracing and impaired rotary-pursuit skill learning in patients with Huntington's disease: Evidence for dissociable memory systems in skill learning. *Neuropsychology,* **11**, 272-281.

Garber, H. J., Ananth, J. V., Chiu, L. C., Griswold, V. J., & Oldendorf, W. H. (1989). Nuclear magnetic resonance study of obsessive-compulsive disorder. *American Journal of Psychiatry,* **146**, 1001-1005.

Goldman, B. L., Martin, E., Calamari, J. E., Woodard, J. W., Messina, M, Chik, H. M., & Pontarelli, N. K. (2007). *Procedural learning and thought focused attention in obsessive-compulsive disorder subgroups*. Manuscript in preparation.

Goodman, W. K., Price, L. H., Rasmussen, S. A., Mazure, C., Fleischmann, R. L., Hill,

C. L., et al. (1989). The Yale-Brown obsessive-compulsive scale, I: Development, use and reliability. *Archives of General Psychiatry,* **40**, 1006-1011.

Grafton, S.T., Hazeltine, E., & Ivry, R. (1995). Functional mapping of sequence learning in normal humans. *Journal of Cognitive Neuroscience,* **7**, 497–510.

Grant, D. A., & Berg, E. A. (1948). A behavioral analysis of degree of reinforcement and ease of shifting to new responses in a Weigl-type card sorting problem. *Journal of Experimental Psychology,* **34**, 404-411.

Graybiel, A. M., & Rauch, S. L. (2000). Toward a neurobiology of obsessive-compulsive disorder. *Neuron,* **28**, 343-347.

Greisberg, S., & McKay, D. (2003). Neuropsychology of obsessive-compulsive disorder: A review and treatment implications. *Clinical Psychology Review,* **23**, 95-117.

Gross-Isseroff, R., Sasson, Y., Voet, H., Hendler, T., Luca-Haimovici, K., Kandel Sussman, H., & Zohar, J. (1996). Alternation learning in obsessive-compulsive disorder. *Biological Psychiatry,* **39**, 733-738.

Haaland, K. Y., Harrington, D. L., O'Brien, S., & Hermanowicz, N. (1997). Cognitive-motor learning in Parkinson's disease. *Neuropsychology,* **11**, 180-186.

Hamilton, M. (1960). A rating scale for depression. *Journal of Neurology, Neurosurgery, Psychiatry,* 23, 56–62.

Harrington, D. L., Haaland, K. Y., Yeo, R. A., & Marder, E.. (1990). Procedural memory in Parkinson's disease: Impaired motor but not visuoperceptual learning. *Journal of Clinical and Experimental Neuropsychology,* **12**, 323-339.

Harris, G. J., Hoehn-Saric, R. H., Lewis, R., Pearlson, G. D., Streeter, G. (1994). Mapping of SPECT regional cerebral perfusion abnormalities in obsessive-compulsive disorder. *Human Brain Mapping,* **1**, 237-248.

Hazeltine, E., Grafton, S. T., & Ivry, R. (1997). Attention and stimulus characteristics determine the locus of motor-sequence encoding. *Brain,* **120**, 123-140.

Heaton, R. K., Chelune, G. J., Talley, J. L., Kay, G. G., & Curtis, G. (1993). *Wisconsin Card Sorting Test Manual, Revised and Expended.* Odessa: Psychological Assessment Resources.

Heindel, W. C., Butters, N., & Salmon, D. P. (1988). Impaired learning of a motor skill in patients with Huntington's disease. *Behavioral Neuroscience,* **102**, 141-147.

Heindel, W. C., Salmon, D. P., Shults, C. W., Walicke, P. A., & Butters, N. (1989). Neuropsychological evidence for multiple implicit memory systems: A comparison of Alzheimer's Huntington's, and Parkinson's disease patients. *Journal of Neuroscience,* **9**, 582-587.

Hendler, T., Goshen, E., Zwas, S. T., Sasson, Y., Gal, G., & Zohar, J. (2003). Brain reactivity to specific symptom provocation indicates prospective therapeutic outcome in OCD. *Psychiatry Research: Neuroimaging,* **124**, 87-103.

Hoehn-Saric, R., Pearlson, G.D., Harris, G., Machlin, S., & Camargo, E. (1991). Effects of fluoxetine on regional cerebral blood flow in obsessive-compulsive patients. *American Journal of Psychiatry,* **148**, 1243-1245.

Hollander, E., DeCaria, C., Gully, R., Nitescu, A., Suckow, R. F., Gorman, J. M., et al. (1991). Effects of chronic fluoxetine treatment on behavioral and neuroendocrine responses to meta-chloro-phenylpiperazine in obsessive-compulsive disorder. *Psychiatry Research,* **36**, 1-17.

Hollander, E., Prohovnik, I., & Stein, D.J. (1995). Increase cerebral blood flow during m-CPP exacerbation of obsessive-compulsive disorder. *Journal of Neuropsychiatry and Clinical Neuroscience,* **5**, 104-107.

Hollander, E., Stein, D., Kwon, J. H., Rowland C, Wong, C. M., Broatch, J., & Himelein, C. (1997). Psychosocial function and economic costs of obsessive-compulsive disorder. *CNS Spectrums,* **3**, 48-58.

Honda, M., Deiber, M. P., Ibáñez, V., Pascual-Leone, A., Zhuang, P., & Hallett, M. (1998). Dynamic cortical involvement in implicit and explicit motor sequence learning: A PET study. *Brain,* **121**, 2159-2173.

Ho Pian, K. L., van Megen, H. J. G. M, Ramsey, N. F., Mandl, R., van Rijk, P. P., Wynne, H. J., & Westenberg, H. G. M. (2005). Decreased thalamic blood flow in obsessive compulsive disorder patients responding to fluvoxamine. *Psychiatry Research: Neuroimaging,* **138**, 89-97.

Howard, D. V., & Howard, J. H., Jr. (2001). When it does hurt to try: Adult age differences in the effects of instructions on implicit pattern learning. *Psychonomic Bulletin & Review,* **8**, 798-805.

Insel, T. R. (1992). Toward a neuroanatomy of obsessive-compulsive disorder. *Archives of General Psychiatry,* **49**, 739-744.

Insel, T. R., Donnelly, E. F., Lalakea, M. L., Alterman, I. S., & Murphy, D. L. (1983). Neurological and neuropsychological studies of patients with obsessive-compulsive disorder. *Biological Psychiatry,* 18, 741-751.

Jackson, S., & Houghton, G. (1995). Sensorimotor selection and the basal ganglia. In J. C. Houk, J. L. Davis, & D. G. Beiser (Eds.), *Models of information processing in the basal ganglia* (pp. 337-367). Cambridge, MA: MIT Press.

Jackson, G. M., Jackson, S. R., Harrison, J., Henderson, L., & Kennard, C. (1995). Serial reaction time learning and Parkinson's disease: Evidence for a procedural learning deficit. *Neuropsychologia,* **33**, 577-593.

Janeck, A. S., Calamari, J. E., Riemann, B. C., & Heffelfinger, S. K. (2003). Too much thinking about thinking?: Metacognitive differences in obsessive-compulsive disorder. *Journal of Anxiety Disorders,* **17**, 181-195.

Jenike, M. A., Breiter, H. C., Baer, L., Kennedy, D. N., Savage, C. R., Olivares, M. J., et al. (1996). Cerebral structural abnormalities in obsessive-compulsive disorder: A quantitative morphometric magnetic resonance imaging study. *Archives of General Psychiatry,* **53**, 625-632.

Joel, D., Zohar, O., Afek, M., Hermesh, H., Lerner, L., Kuperman, R., et al. (2005). Impaired procedural learning in obsessive-compulsive disorder and Parkinson's disease, but not in major depressive disorder. *Behavioural Brain Research,* **157**, 253-263.

Kampman, M., Keijsers, G. P. J., Verbraak, M. J. P. M., Naring, G., & Hoogduin, C. A. L. (2002). The emotional Stroop: A comparison of panic disorder patients, obsessive-compulsive patients, and normal controls, in two experiments. *Journal of Anxiety Disorders,* **16**, 425-441.

Karno, M., Golding, J. M., Sorenson, S. B., Burnam, M., & Audrey, M. (1998). The epidemiology of obsessive–compulsive disorder in five US communities. *Archives of General Psychiatry,* **45**, 1094–1099.

Kassubek, J., Schmidtke, K., Kimmig, H., Lücking, C. H., & Greenlee, M. W. (2001). Changes in cortical activation during mirror reading before and after training: An fMRI study of procedural learning. *Brain Research. Cognitive Brain Research,* **10**, 207-217.

Kathmann, N., Rupertseder, C., Hauke, W., & Zaudig, M. (2005). Implicit sequence learning in obsessive-compulsive disorder: Further support for the fronto-striatal dysfunction model. *Biological Psychiatry,* **58**, 239-244.

Kellner, C. H., Jolley, R. R., Holgate, R. C., Austin, L., Lydiard, R. B., Laraia, M., & Ballenger, J. C. (1991). Brain MRI in obsessive-compulsive disorder. *Psychiatry Research,* **36**, 45-49.

Kelly, S. W., Jahanshahi, M., & Dirnberger, G. (2004). Learning of ambiguous versus hybrid sequences by patients with Parkinson's disease. *Neuropsychologia,* **42**, 1350–1357.

Kim, J. S., Reading, S. A., Brashers-Krug, T., Calhoun, V. D., Ross, C. A., & Pearlson, G. D. (2004). Functional MRI study of a serial reaction task in Huntington's disease. *Psychiatry Research: Neuroimaging,* 23-30.

Knopman, D., & Nissen, M. J. (1991). Procedural learning is impaired in Huntington's disease, Evidence from the serial reaction time task. *Neuropschologia,* **29**, 245-254.

Kuelz, A. K., Hohagen, F., & Voderholzer, U. (2004). Neuropsychological performance in obsessive-compulsive disorder: A critical review. *Biological Psychology,* **65**, 185-236.

Kyrios, M. & Lob, M. A. (1998). Automatic and strategic processing in obsessive compulsive disorder: attentional bias, cognitive avoidance or more complex phenomena. *Journal of Anxiety Disorders,* **12**, 271-292.

Lacerda, A. L. T., Dalgalarrondo, P., Caetano, G. L, Camargo, E. E., Etchebehere, E. C. S. C., & Soares, J. C. (2003). Elevated thalamic and prefrontal regional cerebral blood flow in obsessive-compulsive disorder: A SPECT study. *Psychiatry Research: Neuroimaging,* **123**, 125-134.

Lacerda, A. L. T., Dalgalarrondo, P., Caetano, D., Haas, G. L., Camargo, E. E., & Keshavan, M. S. (2003). Neuropsychological performance and regional cerebral blood flow in obsessive compulsive disorder. *Progress in Neuro-Psychopharmacology & Biological Psychiatry,* **27**, 657-665.

Laplane, D., Levasseur, M., Pillon B., Dubois, B., Baulac, M., Mazoyer, B. M., et al. (1989). Obsessive-compulsive and other behavioural changes with bilateral basal ganglia lesions: A neuropsychological, magnetic resonance imaging and positron tomography study. *Brain,* **112**, 699-725.

Lavy, E., van Oppen, P., & van den Hout, M. N. (1994). Selective processing of emotional information in obsessive compulsive disorder. *Behaviour Research and Therapy,* **32**, 243-246.

Lawrence, N. S., Wooderson, S., Mataix-Cols, D., David, R., Speckens, A., & Phillips, M. L. (2006). Decision making and set shifting impairments are associated with distinct symptom dimensions in obsessive-compulsive disorder. *Neuropsychology,* **20**, 409-419.

Leckman, J. F. (2002). Tourette's syndrome. *Lancet,* **360**, 1577-1586.

Leckman, J. F., Grice, D. E., Boardman, J., Zhang, H., Vitale, A., Bondi, C., Alsobrook, J., Peterson, B. S., Cohen, D. J., Rasmussen, S. A., Goodman, W. K., McDougle, C. J., & Pauls, D. L. (1997). Symptoms of obsessive-compulsive disorder. *American Journal of Psychiatry,* **154**, 911-917.

Lucey, J. V., Burness, C. E., & Costa, D. C. (1997). Wisconsin card sorting task (errors) and cerebral blood flow in obsessive-compulsive disorder. *British Journal of Medical Psychology,* **70**, 403-411.

Lucey, J. V., Costa, D. C., Blanes, T., Busatto, G. F., Pilowsky, L. S., Takei, N., Marks, I. M., Ell, P. J., & Kerwin, R. W. (1995). Regional cerebral blood flow in obsessive-compulsive disordered patients at rest—Differential correlates with obsessive-compulsive and anxious-avoidant dimensions. *British Journal of Psychiatry,* **167**, 629–634.

Luxenberg, J. S., Swedo, S. E., Flament, M. F., Friedland, R. P., Rapoport, J., & Rapoport, S. I. (1988). Neuroanatomical abnormalities in obsessive-compulsive disorder detected with quantitative X-ray computed tomography. *American Journal of Psychiatry,* **145**, 1089-1093.

Machlin, S. R., Harris, G. J., Pearlson, G. D., Hoehn-Saric, R., Jeffery, P., & Camargo, E. E. (1991). Elevated medial-frontal cerebral blood flow in obsessive-compulsive patients: A SPECT study. *American Journal of Psychiatry,* **154**, 1240-1242.

MacLeod, C. & Mathews, A. (1988). Anxiety and the allocation of attention to threat. *Quarterly Journal of Experimental Psychology: Human Experimental Psychology,* **40**, 653-670.

MacLeod, C., Matthews, A., & Tata, P. (1986). Attentional bias in emotional disorders, *Journal of Abnormal Psychology,* **95**, 15-20.

Marker, C. D., Calamari, J. E., Woodard, J. L., & Riemann, B. C. (2006). Cognitive self-consciousness, implicit learning and obsessive-compulsive disorder. *Journal of Anxiety Disorders,* **20**, 389-407.

Marsh, R., Alexander, G. M., Packard, M. G., Zhu, H., & Peterson, B. S. (2005). Perceptual-motor skill learning in Gilles de la Tourette syndrome. Evidence for multiple procedural learning and memory systems. *Neuropsychologia,* **43**, 1456-1465.

Marsh, R., Alexander, G. M., Packard, M. G., Zhu, H., Wingard, J. C., & Peterson, B. S. (2004). Habit learning in Tourette syndrome. *Archives of General Psychiatry,* 61, 1259-1268.

Martinot, J. L., Allilaire, J. F., Mazoyer, B. M., Hantouche, E., Huret, J. D., Legaut-Demare, F., et al. (1990). Obsessive-compulsive disorder: A clinical, neuropsychological and positron emission tomography study. *Acta Psychiatrica Scandinavica* **82**, 233–242.

Martis, B., Wright, C. I., McMullin, K. G., Shin, L. M., & Rauch, S. L. (2004). Functional magnetic resonance imaging evidence for a lack of striatal dysfunction during implicit sequence learning in individuals with animal phobia. *American Journal of Psychiatry,* **161**, 67-71.

Martone, M., Butters, N., Payne, M., Becker, J. T., & Sax, D. S. (1984). Dissociations between skill learning and verbal recognition in amnesia and dementia, *Archives of Neurology,* **41**, 965-970.

Mataix-Cols, D., Alonso, P., Pifarré, J., Menchón, J. M., & Vallejo, J. (2002). Neuropsychological performance in medicated vs unmedicated patients with obsessive compulsive disorder. *Psychiatry Research,* **109**, 255-264.

McFall, M. E., & Wollersheim, J. P. (1979). Obsessive-compulsive neurosis: A cognitive behavioral formulation and approach to treatment. *Cognitive Therapy and Research,* **3**, 333-348.

McGuire, P. K., Bench, C. J., Frith, C. D., Marks, I. M., Frackowiak, R. S., & Dolan, R. J. (1994). Functional anatomy of obsessive-compulsive phenomena. *British Journal of Psychiatry,* **164**, 459-468.

McNally, R.J. (2000). Information-processing abnormalities in obsessive-compulsive disorder. In W.K. Goodman, M.V. Rudorfer, & J.D. Maser (Eds.), *Obsessive-compulsive disorder: Contemporary issues in treatment* (pp. 106-116). Mahwah, NJ: Lawrence Erlbaum Associates Publishers.

McNally, R. J. (2001). On the scientific status of cognitive appraisal models of anxiety disorder. *Behaviour Research and Therapy,* **39**, 513-521.

McNally, R. J., Amir, N., Louro, C. E., Lukach, B. M., Riemann, B. C., & Calamari, J. E. (1994). Cognitive processing of idiographic emotional information in panic disorder. *Behaviour Research and Therapy,* **32**, 119-122.

McNally, R. J., Riemann, B. C., Luro, C. E., Lukach, B. M., & Kim, E. (1992). Cognitive processing of emotional information in panic disorder. *Behaviour Research and Therapy,* **30**, 143-149.

McNally, R. J., Wilhelm, S., Buhlmann, U., & Shin, L. M. (2001). Cognitive inhibition in obsessive-compulsive disorder: Application of a valence-based negative priming paradigm. *Behavioural and Cognitive Psychotherapy,* **29**, 103-106.

Mindus, P., Nyman, H., Mogard, J., Meyerson, B. A., & Ericson, K. (1991). Orbital and caudate glucose metabolism studied by positron emission tomography (PET) in patients undergoing capsulotomy for obsessive-compulsive disorder. In M. A. Jenike, & M. Asberg (Eds.), *Understanding obsessive compulsive disorder (OCD)* (pp. 52-57). Toronto, ON: Hogrefe and Huber Publishers.

Modell, J. G., Mountz, J. M., Curtis, G. C., & Greden, J. F. (1989). Neurophysiologic dysfunction in basal ganglia/limbic striatal and thalamocortical circuits as a pathogenetic mechanism of obsessive-compulsive disorder. *Journal of Neuropsychiatry & Clinical Neurosciences,* **1**, 27-36.

Moritz, S., Birkner, C., Kloss, M., Jahn, H., Hand, I., Haasen, C., & Krausz, M. (2002). Executive functioning in obsessive-compulsive disorder, unipolar depression and schizophrenia. *Archives of Clinical Neuropsychology,* **17**, 477-483.

Mortiz, S., Fricke, S., & Hand, I. (2001). Further evidence for delayed alternation deficits in obsessive-compulsive disorder. *Journal of Nervous and Mental Disorders,* **189**, 562-564.

Moritz, S., Hübner, M., & Kluwe, R. (2004). Task switching and backward inhibition in obsessive-compulsive disorder. *Journal of Clinical and Experimental Neuropsychology,* **26**, 677-683.

Muller, J., & Roberts, J. E. (2005). Memory and attention in obsessive-compulsive disorder: A review. *Anxiety Disorders,* **19**, 1-28.

Muris, P., Merckelbach, H., & Clavan, M. (1997). Abnormal and normal compulsions. *Behaviour Research and Therapy,* **35**, 249-252.

Nakatani, E., Nakgawa, A., Ohara, Y., Goto, S., Uozumi, N., Iwakiri, M., et al. (2003). Effects of behavior therapy on regional cerebral blood flow in obsessive-compulsive disorder. *Psychiatry Research: Neuroimaging,* **124**, 113-120.

Nordahl, T. E., Benkelfat, C., Semple, W. E., Gross, M., King, A. C., & Cohen, R. M. (1989). Cerebral glucose metabolic rates in obsessive-compulsive disorder. *Neuropsychopharmacology,* **2**, 23-28.

Nissen, M. J., & Bullemer, P. (1987). Attentional requirements of learning: Evidence from performance measures. *Cognitive Psychology,* **19**, 1-32.

Obsessive Compulsive Cognitions Work Group. (1997). Cognitive assessment of obsessive-compulsive disorder. *Behaviour Research and Therapy,* **35**, 667-681.

Obsessive Compulsive Cognitions Working Group. (2005). Psychometric validation of the Obsessive Beliefs Questionnaire and the Interpretation of Intrusions Inventory – Part 2: Factor analyses and testing of a brief version. *Behaviour Research and Therapy,* **43**, 1527-1542.

Okasha, A., Rafaat, M., Mahallawy, N., El Nahas, G., Seif El Dawla, A., Sayed, M., & El Kholi, S. (2000). Cognitive dysfunction in obsessive-compulsive disorder. *Acta Psychiatria Scandinavica,* **101**, 281-285.

Osterrieth, P. (1944). Test of copying a complex figure: Contribution to the study of perception and memory. *Archives de Psychologie,* **30**, 206–356.

Pascual-Leone, A., Grafman, J., Clark, K., Stewart, M., Massaquoi, S., Lou, J. S., & Hallett, M. (1993). Procedural learning in Parkinson's disease and cerebellar degeneration. *Annals of Neuroloy,* **34**, 594-602.

Peigneux P, Maquet P, Meulemans T, Destrebecqz A, Laureys S, Degueldre C, et al. (2000). Striatum forever, despite sequence learning variability: a random effect analysis of PET data. *Human Brain Mapping,* **10**, 179-194.

Penisson-Besnier, I., Le Gall, D., & Dubas, F. (1992) Obsessive-compulsive behavior (arithmomania). Atrophy of the caudate nuclei. *Revista de Neurologia,* **148**, 262–267.

Perani, D., Colombo, C., Bressi, S., Bonfanti, A., Grassi, F., Scarone, S., Bellodi, L., Smeraldi, E., & Fazio, F. (1995). [18F] FDG PET Study in obsessive compulsive disorder: A clinical / metabolic correlation study after treatment. *British Journal of Psychiatry,* **166**, 244-250.

Pujol, J., Soriano-Mas, C., Alonso, P., Cardoner, N., Menchón, J. M., Deus, J., & Vallejo, J. (2004). Mapping structural brain alterations in obsessive-compulsive disorder. *Archives of General Psychiatry,* **61**, 720-730.

Purcell, R., Maruff, P., Kyrios, M., & Pantelis, C. (1998). Neuropsychological deficits in obsessive–compulsive disorder: A comparison with unipolar depression, panic disorder, and normal controls. *Archives of General Psychiatry,* **55** 415–423

Purdon, C., & Clark, D. A. (1999). Metacognition and obsessions. *Clinical Psychology and Psychotherapy,* **6**, 102-110.

Rachman, S. (1993). Obsessions, responsibility and guilt. *Behaviour Research and Therapy,* **31**, 149-154.

Rachman, S. (1997). A cognitive theory of obsession. *Behaviour Research and Therapy,* **35**, 793-802.

Rachman, S. (1998). A cognitive theory of obsessions: Elaborations. *Behaviour Research and Therapy,* **36**, 385-401.

Rachman, S., & de Silva, P. (1978). Abnormal and normal obsessions. *Behaviour Research and Therapy,* **16**, 233-248.

Radomsky, A. S. & Rachman, S. (1999). Memory bias in obsessive-compulsive disorder (OCD). *Behaviour Research and Therapy,* **39**, 813-822.

Rapoport, J. L., & Wise, S. P. (1988). Obsessive-compulsive disorder: Evidence for basal ganglia dysfunction, *Psychopharmacology Bulletin*, **24**, 380–384.

Rasmussen, S. A., & Eisen, J. L. (1992). The epidemiology and clinical features of obsessive-compulsive disorder. *Psychiatric Clinics of North America,* **15**, 743-758.

Rasmussen, S. A., & Tsuang, M. T. (1986). Clinical characteristics and family history in DSM-III obsessive-compulsive disorder. *American Journal of Psychiatry,* **143**, 317-382.

Rauch, S. L., & Jenike, M. A. (1993). Neurobiological models of obsessive-compulsive disorder. *Psychosomatics: Journal of Consultation Liaison Psychiatry,* **34**, 20-32.

Rauch, S. L., Jenike, M. A., Alpert, N. M., Baer, L., Breiter, H. C., Savage, C. R., et al. (1994). Regional cerebral blood flow measured during symptom provocation in obsessive-compulsive disorder using oxygen 15-labeled carbon dioxide and positron emission tomography. *Archives of General Psychiatry,* **51**, 62-70.

Rauch, S. L., & Savage, C. R. (2000). Investigating cortico-striatal pathophysiology in obsessive-compulsive disorders: Procedural learning and imaging probes. In W. K. Goodman, M. V. Rudorfer, & J. D. Maser (Eds.), *Obsessive-compulsive disorder: Contemporary issues in treatment* (pp. 133-154). Mahwah, NJ: Lawrence Erlbaum Associates Publishers.

Rauch, S. L., Savage, C. R., Alpert, N. M., Dougherty, D., Kendrick, A., Curran, T., et al. (1997). Probing striatal function in obsessive-compulsive disorder: A PET study of implicit sequence learning. *Journal of Neuropsychiatry and Clinical Neuroscience,* **9**, 568-573.

Rauch, S. L., Savage, C. R., Brown, H. D., Curran, T., Alpert, N. M., Kendrick, A., et al. (1995). A PET investigation of implicit and explicit sequence learning. *Human Brain Mapping,* **3**, 271–286.

Rauch, S. L., Shin, L. M., Dougherty, D. D., Alpert, N. M., Fischman, A. J., & Jenike, M. A. (2002). Predictors of fluvoxamine response in contamination-related obsessive compulsive disorder: A PET symptom provocation study. *Neuropsychopharmacology,* **27**, 782-791.

Rauch, S. L., Wedig, M. M., Wright, C. I., Martis, B., McMullin, K. G., Shin, L. M., et al. (2007). Functional magnetic resonance imaging study of regional brain activation during implicit sequence learning in obsessive-compulsive disorder. *Biological Psychiatry,* **61**, 330-336.

Rauch, S. L., Whalen, P. J., Curran, T., McInerney, S., Heckers, S., & Savage, C. R. (1998). Thalamic deactivation during early implicit sequence learning: A functional MRI study. *NeuroReport,* **9**, 865-870.

Rauch, S. L., Whalen, P. J., Curran, T., Shin, L. M., Coffey, B. J., Savage, C. R., et al. (2001). Probing striato-thalamic function in obsessive-compulsive disorder and Tourette syndrome using neuroimaging methods. *Advances in Neurology,* **85**, 207-224.

Rauch, S. L., Whalen, P. J., Savage, C. R., Curran, T., Kendrick, A., Brown, H. D., et al. (1997). Striatal recruitment during an implicit sequence learning task as measured by functional magnetic resonance imaging. *Human Brain Mapping,* **5**, 124–132.

Reiss, J. P., Campbell, D. W., Leslie, W. D., Paulus, M. P., Stroman, P. W., Polimeni, J. O., et al. (2005). The role of the striatum in implicit learning: A functional magnetic resonance imaging study. *NeuroReport: For Rapid Communication of Neuroscience Research,* **16**, 1291-1295.

Robbins, T. W., James, M., Owen, A. M., Sahakian, B. J., McInnes, L., & Rabbitt, P. (1994). Cambridge Neuropsychological Test Automated Battery (CANTAB): A factor analytic study of a large sample of normal elderly volunteers. *Dementia,* **5**, 266–281.

Rubin, R. T., Villanueva-Meyer, J., Ananth, J., Trajmar, P. G. & Mena, I. (1992). Regional ^{133}Xe cerebral blood flow and cerebral 99m-HMPAO uptake in unmedicated obsessive-compulsive disorder patients and matched normal control subjects: Determination by high-resolution single-photon emission computed tomography. *Archives of General Psychiatry,* **49**, 695-702.

Rubin, R. T., Ananth, J., Villanueva-Meyer, J., Trajmar, P. G., & Mena, I. (1995). Regional ^{133}Xenon cerebral blood flow and cerebral Tc-HMPAO: Uptake in patients with obsessive-compulsive disorder before and during treatment. *Biological Psychiatry,* **38**, 429-437.

Saint-Cyr, J. A., Taylor, A. E., & Lang, A. E. (1988). Procedural learning and neostraital dysfunction in man. *Brain,* **111**, 941-959.

Salkovskis, P. M. (1985). Obsessive-compulsive problems: A cognitive-behavioural analysis. *Behaviour Research and Therapy,* **23**, 571-583.

Salkovskis, P. M. (1989). Cognitive-behavioral factors and the persistence of intrusive thoughts in obsessional problems. *Behaviour Research and Therapy,* **27**, 677-682.

Salkovskis, P. M. (1998). Psychological approaches to the understanding of obsessional problems. In R. P. Swinson, M. M. Antony (Eds.), *Obsessive-compulsive disorder: Theory, research, and treatment* (pp. 33–50). New York: Guilford.

Salkovskis, P. M., & Harrison J. (1984). Abnormal and normal obsessions: A replication *Behaviour Research and Therapy,* **22**, 549-552.

Salkovskis, P. M., Shafran, R., Rachman, S., & Freeston, M. H. (1999). Multiple pathways to inflated responsibility beliefs in obsessional problems: possible origins and implications for therapy and research. *Behavioural Research and Therapy,* **37**, 1055-1072.

Savage, C. R., Keuthen, N. J., Jenike, M. A., Brown, H. D. Baer, L., Kendrick, A. D., et al. (1996). Recall and recognition memory in obsessive-compulsive disorder. *Journal of Neuropsychiatry,* **8**, 99-103.

Savage, C. R., Baer, L., Keuthen, N. J., Brown, H. D., Rauch, S. L., & Jenike, M. A. (1999). Organizational strategies mediate nonverbal memory impairment in obsessive–compulsive disorder. *Biological Psychiatry,* **45**, 905–916.

Savage, C. R., Deckersbach, T., Wilhelm, S., Rauch, S. L., Baer, L., Reid, T., & Jenike, M. A. (2000). Strategic processing and episodic memory impairment in obsessive compulsive disorder. *Neuropsychology,* **14**, 141–151.

Sawle, G. V., Hymas, N. F., Lees, A. J., & Frackowiak, R. S. J. (1991). Obsessional slowness: Functional studies with positron emission tomography. *Brain,* **114**, 2191-2202.

Saxena, S., Brody, A. L, Maidment, K. M., Dunkin, J. J., Colgan, M., Alborzian, S., et al. (1999). Localized orbitofrontal and subcortical metabolic changes and predictors of response to paroxetine treatment in obsessive-compulsive disorder. *Neuropsychopharmacology,* **21**, 683-693.

Saxena, S., Brody, A. L., Schwartz, J. M., & Baxter, L. R. (1998). Neuroimaging and frontal-subcortical circuitry in obsessive-compulsive disorder. *British Journal of Psychiatry, Suppl.* **35**, 26-37.

Saxena, S., & Rauch, S. L. (2000). Functional neuroimaging and the neuroanatomy of obsessive-compulsive disorder. *The Psychiatric Clinics of North America,* **23**, 563-86.

Scarone, S., Colombo, C., Livian, S., Abbruzzese, M., Ronchi, P., Locatelli, M., et al. (1992). Increased right caudate nucleus size in obsessive-compulsive disorder: Detection with magnetic resonance imaging. *Psychiatry Research: Neuroimaging,* **45**, 115-121.

Schmidtke, K., Hendrik, M., Kaufmann, R., & Schmolck, H. (2002). Cognitive procedural learning in patients with fronto-striatal lesions. *Learning & Memory,* **9**, 419-429.

Schmidtke, K., Schorb, A., Winkelmann, G., & Hohagen, F. (1998). Cognitive frontal dysfunction in obsessive-compulsive disorder. *Biological Psychiatry,* **43**, 666-673.

Schwartz, J. M, Stoessel, P. W., Baxter, L. R., Martin, K. M., & Phelps, M.E. (1996). Systematic changes in cerebral glucose metabolic rate after successful behavior modification treatment of obsessive-compulsive disorder. *Archives of General Psychiatry,* **53,** 109-113.

Shallice, T. (1982). Specific impairments of planning. *Philosophical Transactions of the Royal Society of London,* **298**, 199-209.

Sher, K. J., Frost, R. O., Kushner, M., Crews, T. M., & Alexander, J. E. (1989). Memory deficits in compulsive checkers: Replication and extension in a clinical sample. *Behaviour Research and Therapy,* **27**, 65-69.

Shin, J. C., & Ivry, R. B. (2003). Spatial and temporal sequence learning in patients with Parkinson's disease or cerebellar lesions. *Journal of Cognitive Neuroscience,* **15**, 1232-1243.

Siegert, R. J., Taylor, K. D., Weatherall, M., & Abernethy, D. A. (2006). Is implicit sequence learning impaired in Parkinson's disease? A meta-analysis. *Neuropsychology,* **20**, 490-495.

Simpson, H. B., Rosen, W., Huppert, J. D., Lin, S. H., Foa, E. B., & Liebowitz, M. R. (2006). Are there reliable neuropsychological deficits in obsessive-compulsive disorder? *Journal of Psychiatry Research,* **40**, 247-57.

Sivan, A.B. (1992). *Benton visual retention test manual* (5th ed). San Antonio, TX: The Psychological Corporation.

Smith, J. G., & McDowall, J. (2004). Impaired higher order implicit sequence learning on the verbal version of the serial reaction time task in patients with Parkinson's disease. *Neuropsychology,* **18**, 679-691.

Smith, J., Siegert, R. J., McDowall, J., & Abernethy D. (2001). Preserved implicit learning on both the serial reaction time task and artificial grammar in patients with Parkinson's disease. *Brain and Cognition,* **45**, 378-391.

Sommer, M., Grafman, J., Clark, K., & Hallett, M. (1999). Learning in Parkinson's disease: Eyeblink conditioning, declarative learning, and procedural learning. *Journal of Neurology, Neurosurgery, and Psychiatry,* **67**, 27-34

Spitznagel, M. B. & Suhr, J. A. (2002). Executive function deficits associated with symptoms of schizotypy and obsessive-compulsive disorder. *Psychiatry Research,* **110**, 151-163.

Squire, L. R. (1992). Declarative and nondeclarative memory: Multiple brain systems supporting learning and memory. *Journal of Cognitive Neuroscience,* **4**, 232-243.

Squire, L. R., & Zola, S. M. (1996). Structure and function of declarative and nondeclarative memory systems. *Proceedings of the National Academy of Sciences of the United States of America,* **93**, 13515-13522.

Squire, L. R., Zola-Morgan, S. (1988). Memory: Brain systems and behavior. *Trends in Neurosciences,* 11, 170-175.

Stadler, M. A., & Frensch, P. A. (Eds.). (1998). *Handbook of Implicit Learning.* Thousand Oaks: Sage Publications.

Stefanova, E. D., Kostic, V. S., Ziropadja, L., Markovic, M., & Ocic, G. G. (2000). Visuomotor skill learning on serial reaction time task in patients with early Parkinson's disease. *Movement Disorder,* **15**, 1095-1103.

Stein, D. J., Coetzer, R., Lee, M., Davids, B., & Bouwer, C. (1997). Magnetic resonance brain imaging in women with obsessive-compulsive disorder and trichotillomania. *Psychiatry Research: Neuroimaging,* **74**, 177-182.

Stein, D. J., Hollander, E., Chan, S., DeCaria, C. M., Hilal, S., Liebowitz, M. R., & Klein, D. F. (1993). Computed tomography and neurological soft signs in obsessive-compulsive disorder. *Psychiatry Research: Neuroimaging,* **50**, 143-150.

Swedo, S. E., Pietrini, P., Leonard, H., Schapiro, M. B., Rettew, D. C., Goldberger, E. L., et al. (1992). Cerebral glucose metabolism in childhood-onset obsessive-compulsive disorder: revisualization during pharmacotherapy. *Archives of General Psychiatry,* **49**, 690-694.

Swedo, S. E., Schapiro, M. G., Grady, C. L., Cheslow, D. L., Leonard, H. L., Kumar, A., et al. (1989). Cerebral glucose metabolism in childhood onset obsessive-compulsive disorder. *Archives of General Psychiatry,* **46**, 518-523.

Szeszko, P. R., MacMillan, S., McMeniman, M., Chen, S., Baribault, K., Lim, K. O., Ivey, J., Rose, M., Banerjee, S. P., Bhandari, R., Moore, G. J., & Rosenberg, D. R. (2004). Brain abnormalities in psychotropic drug-naïve pediatric patients with obsessive-compulsive disorder. *American Journal of Psychiatry,* **161**, 1049-1056.

Szesko, P. R., Ardekani, B. A., Ashtari, M., Malhotra, A. K., Robinson, D. G., Bilder, R. M., & Lim, K.O. (2005). White matter abnormalities in obsessive-compulsive disorder: A diffusion tensor imaging study. *Archives of General Psychiatry,* **62**, 782-790.

Tata, P. R., Leibowitz, J. A., Prunty, M. J., Cameron, M., & Pickering, A. D. (1996). Attentional bias in obsessional compulsive disorder. *Behaviour Research and Therapy,* **34**, 53-60.

Tipper, S. P. (1985). The negative priming effect: Inhibitory priming by ignored objects. *Quarterly Journal of Experimental Psychology: Human Experimental Psychology,* **37**, 571-590.

Trillet, M., Croisile, B., Tourniaire, D., & Schott, B. (1990) Disorders of voluntary motor activity and lesions of caudate nuclei. *Revista de Neurologia,* **146**, 338–344.

Tulving, E. (1972). Episodic and semantic memory. In E. Tulving & W. Donaldson (Eds.), *Organization of Memory* (pp. 381–403). New York, N. Y.: Academic.

Tulving, E. (2002). Episodic memory: From mind to brain. *Annual Review of Psychology,* **53**, 1-25.

Tuna, S., Tekcan, A. I., & Topçuoğlu, V. (2005). Memory and metamemory in obsessive compulsive disorder. *Behaviour Research and Therapy,* **43**, 15-27.

Vakil, E., Blachstein, H., & Soroker, N. (2004). Differential effect of right and left basal ganglionic infarctions on procedural learning. *Cognitive Behavioral Neurology: Official Journal of the Society for Behavioral and Cognitive Neurology,* **17**, 62-73.

Vakil, E., & Herishanu-Naaman, S. (1998). Declarative and procedural learning in parkinson's disease patients having tremor or bradykinesia as the predominant symptom. *Cortex,* **34**, 611-620.

Vakil, E., Kahan, S., Huberman, M., & Osimani, A. (2000). Motor and non-motor sequence learning in patients with basal ganglia lesions: The case of serial reaction time (SRT). *Neuropsychologia.* **38**, 1-10.

van der Graaf, F. H., Maguir, R. P., Leenders, K. L., & de Jong, B. M. (2006). Cerebral activation related to implicit sequence learning in a Double Serial Reaction Time task. *Brain Research,* **1081**, 179-190.

van den Heuvel, O. A., Veltman, D. J., Groenewegen, H. J., Cath, D. C., van Balkom, A. J. L. M., van Hartskamp, J., et al. (2005). Frontal-striatal dysfunction during planning in obsessive-compulsive disorder. *Archives of General Psychiatry,* **62**, 301-310.

Veale, D. M., Sahakian, B. J., Owen, A. M., & Marks, I. M. (1996). Specific cognitive deficits in tests sensitive to frontal lobe dysfunction in obsessive-compulsive disorder. *Psychological Medicine,* **26**, 1261-1269.

Wechsler, D. (1981). *WAIS-R: Manual: Wechsler Adult Intelligence Scale—Revised.* San Antonio, TX: The Psychological Corporation.

Wechsler, D., & Stone, P. S. C. (1974). *Wechsler Memory Scale Manual.* New York, N. Y.: The Psychological Corporation.

Weilburg, J. B., Mesulam, M. M., Weintraub, S., Buonnano, F., Jenike, M. A. & Stakes, J. W. (1989). Focal striatal abnormalities in a patient with obsessive-compulsive disorder. *Archives of Neurology,* **46**, 233-235.

Weissman, M. M., Bland, R. C., Canino, G. J., Greenwald, S., Hwu, H., Lee, C. K., et al. (1994). The cross national epidemiology of obsessive compulsive disorder: The Cross National Collaborative Group. *Journal of Clinical Psychiatry,* **55**(Suppl. 3), 5-10.

Wells, A. (2000). *Emotional disorders and metacognition: Innovative cognitive therapy.* West Sussex, England: John Wiley & Sons.

Wells, A., & Matthews, G. (1994). *Attention and emotion: A clinical perspective.* Hove: Erlbaum.

Wells, A., & Matthews, G. (1996). Modeling cognition in emotional disorder: The S-REF model. *Behaviour Research and Therapy,* **34**, 881-888.

Wells, A., & Matthews, G. (1997). *Cognitive therapy of anxiety disorder: A practice manual and conceptual guide.* Chichester: Wiley.

Werheid, K., Ziessler, M., Nattkemper, D., & von Cramon, D. Y. (2003).Sequence learning in Parkinson's disease: The effect of spatial stimulus-response compatibility. *Brain and Cognition,* **52**, 239-49.

Werheid, K., Zysset, S., Muller, A., Reuter, M., & von Cramon, D. Y. (2003). Rule learning in a serial reaction time task: An fMRI study on patients with early Parkinson's disease. *Brain Research. Cognitive Brain Research,* **16**, 273-84.

Westwater, H., McDowall, J., Siegert, R., Mossman, S., & Abernethy, D. (1998). Implicit learning in Parkinson's disease: Evidence from a verbal version of the serial reaction time task. *Journal of Clinical and Experimental Neuropsychology,* **20**, 413-418.

Whiteside, S. P., Port, J. D., & Abramowitz, J.S. (2004). A meta-analysis of functional neuroimaging in obsessive-compulsive disorder. *Psychiatry Research: Neuroimaging,* **132**, 69-79.

Williams, J. M. G., Mathews, A., & MacLeod, C. (1996). The emotional Stroop task and psychopathology. *Psychological Bulletin,* **120**, 3-24.

Willingham, D. B. (1998). A neuropsychological theory of motor skill learning. *Psychological Review,* **105**, 558-584.

Willingham, D. B., & Goedert-Eschmann, K. (1999). The relation between implicit and explicit learning: Evidence for parallel development. *Psychological Science,* **10**, 531-534.

Willingham, D. B., & Koroshetz, W. J. (1993). Evidence for dissociable motor skills in Huntington's disease patients. *Psychobiology,* **2**, 173-182.

Willingham, D. B., Koroshetz, W. J., & Peterson, E. W. (1996). Motor skills have diverse neural bases: Spared and impaired skill acquisition in Huntington's disease. *Neuropsychology,* **10**, 315-321.

Willingham, D. B., Nissen, M. J., & Bullemer, P. (1989). On the development of procedural knowledge. *Journal of Experimental Psychology. Learning, Memory and Cognition,* **15**, 1047-1060.

Willingham, D. B., Salidis, J., & Gabrieli, J. D. (2002). Direct comparison of neural systems mediating conscious and unconscious skill learning. *Journal of Neurophysiology,* **88**, 1451-1460.

Zohar, J., Insel, T. R., Berman, K. F., Foa, E. B., Hill, J. L., & Weinberger, D. R. (1989). Anxiety and cerebral blood flow during behavioral challenge: Dissociation of central from peripheral and subjective measures. *Archives of General Psychiatry,* **46**, 505-510.

In: Pattern Recognition in Biology
Editor: Marsha S. Corrigan, pp. 189-204
ISBN: 978-1-60021-716-6

Chapter 6

MODELING OF AMBULATORY HEART RATE USING LINEAR AND NEURAL NETWORK APPROACHES

***Vitaliy Kolodyazhniy*[a]*, Monique C. Pfaltz*[b] *and Frank H. Wilhelm*[c]**
Clinical Psychophysiology Laboratory,
Department of Clinical Psychology and Psychotherapy,
Institute for Psychology, University of Basel, CH-4055 Basel, Switzerland

ABSTRACT

This chapter presents new results on modeling 24 hour (circadian) human heart rate data collected with the LifeShirt system using a variety of linear regression and neural network models. Such modeling is important in biopsychology, chronobiology, and chronomedicine where signals collected continuously from human subjects for one or several days need to be interpreted. Ambulatory heart rate is influenced by a variety of factors, including physical activity, posture, and respiration, and our models try to predict heart rate based on these factors. The analyses described in the chapter indicate that neural and especially neuro-fuzzy techniques provide better results in the modelling of human heart rate at the circadian scale than conventional linear regression. The advantages of the neuro-fuzzy approaches consist in their computational efficiency, better interpretability, and the possibility of incorporation of prior knowledge for easier model construction.

1. INTRODUCTION

Biological systems are very complex and their behavior is typically influenced by a large number of factors. Consider the human heart rate (HR): at a shorter time scale (seconds to minutes) it is influenced by respiration, speech, physical activity, and body posture. At a longer scale (minutes to hours), HR is influenced by emotions, food intake, and thermoregulation (environmental and body temperature). At even longer scale, HR is

[a] E-mail address: v.kolodyazhniy@unibas.ch
[b] E-mail address: monique.pfaltz@unibas.ch
[c] E-mail address: frank.wilhelm@unibas.ch

influenced by the circadian (near 24 hour long) patterns of everyday life: during sleep our hearts beat slower than when we are awake.

The character of these dependencies is often assumed to be linear for the ease of modeling. Linear techniques have been successfully applied to modeling the human HR at shorter time scales, e.g. to discriminate between physical and emotional activation causing an increase in HR [23].

However, a general linear model of the human HR is hardly feasible because of the complexity of nonlinear static and dynamic behavior of the regulation systems of our organisms, including the internal clock, the thermoregulation system, the cardiovascular system, and the nonlinear interactions of these systems.

Circadian variations of the human HR are of interest for biopsychology, chronobiology, and chronomedicine. In order to detect such variations, and especially to reliably estimate the underlying parameters of the circadian clock like its phase, amplitude, and free-running period (which is usually slightly different from 24 hours), nonlinear modeling approaches are required. This is a new and challenging field, and our goal here is to present a comparison of different modeling approaches that can be used for the investigation of the human circadian clock based on ambulatory measurements, when test subject keep their everyday life schedule while their physiological parameters are continuously recorded with a wearable device throughout several days or weeks.

Artificial neural networks (ANNs) [8] have been widely used in recent years to solve a wide range of problems such as data mining and processing of signals of different nature under uncertainty as to the structure and parameters of the underlying model. The ANNs possess universal approximation properties and learning capabilities, so they can be trained to identify unknown and very complex nonlinear input-output relationships. However, the ANNs represent the black-box approach to systems modelling since they are nontransparent models, and the interpretation of the knowledge stored in an ANN can be difficult. We will consider three popular types of neural networks for nonlinear regression: the multilayer perceptron (MLP), the radial basis function network, and the generalized regression neural network (GRNN).

Hybrid neuro-fuzzy approaches [5] emerged as a synergism of the neural nets and fuzzy systems [11], the two major directions in computational intelligence [10]. The neuro-fuzzy systems possess the learning capabilities similar to those of neural networks, and provide the interpretability and "transparency" of results inherent to the fuzzy approach. We will consider here two neuro-fuzzy models, the neo-fuzzy neuron (NFN) and the novel neuro-fuzzy Kolmogorov's network (NFKN) that were designed to overcome such disadvantages of most neuro-fuzzy systems such as the slow convergence of the gradient-based learning procedures or high computational load of the genetic algorithm-based optimization of their parameters.

2. Regression Models

In this section, we consider a number of regression models that will be further used for modeling the ambulatory HR. The most general form of such models is

$$\hat{y} = \hat{f}(x), \ x = [x_1, \ldots, x_d]^T, \tag{1}$$

where $\hat{y}$ is the model output, x is the vector of inputs, $\hat{f}$ is the estimated input-output functional relation. We assume that the true function $y = f(x)$ is not known.

2.1. Linear Regression

The simplest model that we consider is given by the linear multiple regression equation:

$$\hat{y} = w_0 + w_1 x_1 + w_2 x_2 + \ldots + w_d x_d , \tag{2}$$

where $w_0, w_1, \ldots, w_d$ are the regression parameters.

2.2. Multilayer Perceptron

A multilayer perceptron (MLP) [8] is shown in Fig. 1. This is a feed-forward network of artificial neurons with sigmoid activation functions. The number of layers in an MLP is usually two or three (layers of neurons, input layer is not counted).

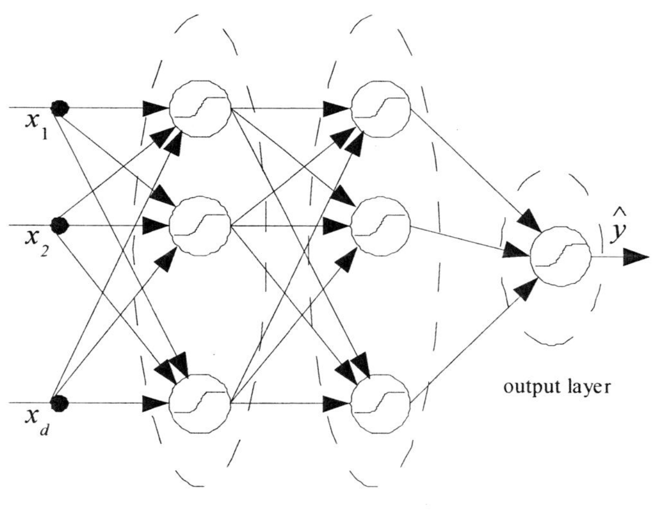

Figure 1. Multi-layer perceptron with 2 hidden layers.

The MLP is described by the following equation [18]:

$$\hat{y} = \Gamma^{[n_L]}(W^{[n_L]}\Gamma^{[n_L-1]}(W^{[n_L-1]} \ldots \Gamma^{[1]}(W^{[1]}x + w_0^{[1]}) + \ldots + w_0^{[n_L-1]}) + w_0^{[n_L]}), \tag{3}$$

where $W^{[l]}$ is the matrix $(n^{[l]} \times n^{[l-1]})$ of the synaptic weights of the l-th layer, $n^{[l]}$ is the number of neurons in the l-th layer, $l = 1, \ldots, n_L$, $n^{[0]} = d$, $n^{[n_L]} = n_y$, n_L is the number of layers, x is the input vector $(d \times 1)$, $w_0^{[l]}$ is the vector $(n^{[l]} \times 1)$ of bias parameters of the l-th layer, $\Gamma^{[l]}$ is a nonlinear operator implemented by a set of sigmoid activation functions of the l-th layer.

2.3. Radial Basis Function Network

A radial basis function network (RBFN) [1, 6] always contains two layers: the hidden layer of basis functions and the output layer (see Fig. 2).

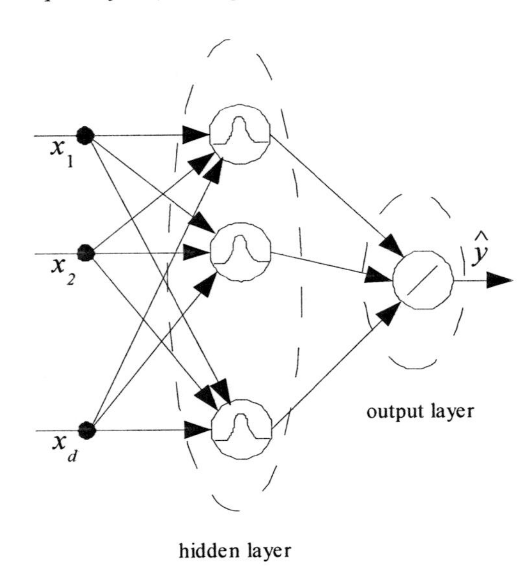

Figure 2. Radial basis function network.

The RBFN model is described by the equation

$$\hat{y} = \sum_{i=1}^{n_h} w_i \phi(\|x - v_i\|) + w_0, \tag{4}$$

where n_h is the number of neurons (basis functions) in the hidden layer, w_i is the weight of the i-th basis function, ϕ is the basis function (usually Gaussian), v_i is the vector $(n_d \times 1)$ of the center of the i-th basis function, w_0 is the bias parameter.

The most important advantage of the RBF networks consists in shorter training time as compared to the conventional multilayer networks trained by means of the error backpropagation technique. At the same time, the construction and training of an RBF network usually requires the use of at least two procedures. The first one is for the setting of the basis function parameters and the second one is for the training of the output layer weights.

It was shown in [9] that the RBFNs and neuro-fuzzy systems are functionally equivalent under minor restrictions, so similar training procedures can be used for both. The RBFNs are also considered to be more transparent than the MLP models because the radial basis functions have a clear interpretation in terms of clustering of the input space in contrast to the neurons of the MLPs, especially with more than one hidden layer.

2.4. Generalized Regression Neural Network

A generalized regression neural network (GRNN) [19] has a similar architecture to that of the RBFN but it has as many hidden layer units (basis functions) as the number of samples in the training data set. So each sample from the training data set is memorized in the GRNN as a prototype.

2.5. Neo-fuzzy Neuron

The neo-fuzzy neuron (NFN) was proposed as a simple neuro-fuzzy architecture with very fast learning and guarantee for the convergence to the global minimum of the error surface [25]. It is also very well suited for hardware implementations, including purely analog circuits [17]. This neuron has been successfully applied to the problems of time series prediction, signal filtering, and restoration [20].

The NFN model is described by the expressions

$$\hat{y} = \sum_{i=1}^{d} f_i(x_i), \quad f_i(x_i) = \sum_{h=1}^{m_i} \mu_{i,h}(x_i) w_{i,h}, \tag{5}$$

where $f_i(x_i)$ is the nonlinear synapse of the i-th input, m_i is the number of membership functions per input i, $\mu_{i,h}(x_i)$ is the h-th membership function (MF) at the i-th input, $w_{i,h}$ is the tunable synaptic weight of the respective MF. These weights are tuned automatically by an optimization procedure that is used to fit the model to the data. The architecture of an NFN is shown in Fig. 3.

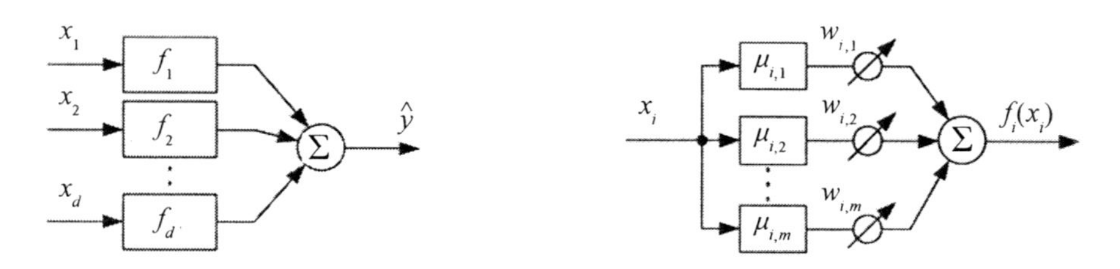

Figure 3. Neo-fuzzy neuron (*left*) and its nonlinear synapse (*right*).

The MFs in a neo-fuzzy neuron are fixed and equidistantly spaced. They are chosen such that at maximum two adjacent MFs in each synapse fire at a time and their sum is always unity:

$$\sum_{h=1}^{m_i} \mu_{i,h}(x_i) = 1, \quad i = 1, \ldots, d. \tag{6}$$

The functions of nonlinear synapses $f_i(x_i)$ are piecewise-linear since they are superpositions of triangular membership functions. An example of an approximation of a univariate nonlinear function by a nonlinear synapse is shown in Fig. 4 where the unknown approximated function is plotted as dotted line.

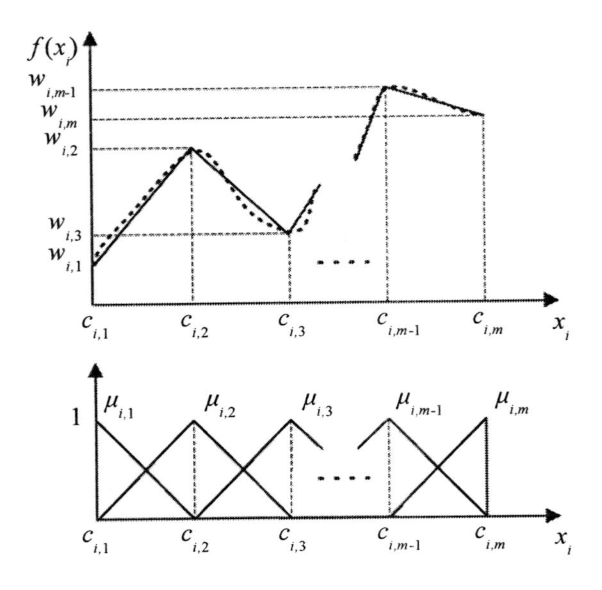

Figure 4. Approximation of a univariate nonlinear function by a nonlinear synapse.

Each nonlinear synapse is a single-input single-output fuzzy inference system. It contains fuzzy rules in the form:

$$\text{IF } x_i \text{ IS } X_{i,h} \text{ THEN } f_i = w_{i,h}, \quad i = 1,\ldots,d, \quad h = 1,\ldots,m_i, \tag{7}$$

where $X_{i,h}$ is the h-th linguistic label (fuzzy set) on the i-th input. The fuzzy sets are defined by the respective triangular MFs $\mu_{i,h}(x_i)$, and represent such linguistic values as “SMALL”, “MEDIUM”, “LARGE”, etc, depending on a particular fuzzy partition which can be chosen based on the domain knowledge and experience of the person that designs the model. In such a way, prior knowledge can be incorporated into the model and the neo-fuzzy model itself possesses inherent interpretability along with nonlinear approximation capabilities.

A disadvantage of the NFN model consists in the fact that it assumes that the nonlinearities of the inputs are separable. So the NFN model is *not* a universal approximator, in contrast to the neural nets [7] considered above or to the conventional fuzzy systems [16, 21]. But it possesses better approximation properties than the linear models, as it has more degrees of freedom allowing good piecewise-linear approximation for many processes and imposing very low requirements on the computational resources. The most important advantage of the NFN model is its extremely fast learning which can be performed in the online mode with very simple weight update rules [25, 2], or in just a single operation with the linear least-squares formula [3].

2.6. Neuro-fuzzy Kolmogorov's Network

The neuro-fuzzy Kolmogorov's network (NFKN) is a universal approximator based on the NFNs [14]. The NFKN architecture [4, 15] is comprised of two layers of neo-fuzzy neurons (NFNs) [25] and is described as follows:

$$\hat{f}(x_1,\ldots,x_d)=\sum_{l=1}^{n} f_l^{[2]}(o^{[1,l]}), \quad o^{[1,l]}=\sum_{i=1}^{d} f_i^{[1,l]}(x_i), \quad l=1,\ldots,n, \tag{8}$$

where n is the number of hidden layer neurons, $f_l^{[2]}(o^{[1,l]})$ is the l-th nonlinear synapse in the output layer, $o^{[1,l]}$ is the output of the l-th NFN in the hidden layer, $f_i^{[1,l]}(x_i)$ is the i-th nonlinear synapse of the l-th NFN in the hidden layer.

The NFKN architecture was named after the famous Kolmogorov's theorem on the representation of functions of multiple variables as a two-level superposition of univariate functions [13]. Indeed, the NFKN makes approximate representations of multivariate functions by a two-level superposition (8) of the nonlinear synapses (univariate approximators) of its hidden and output layers (see Fig. 5).

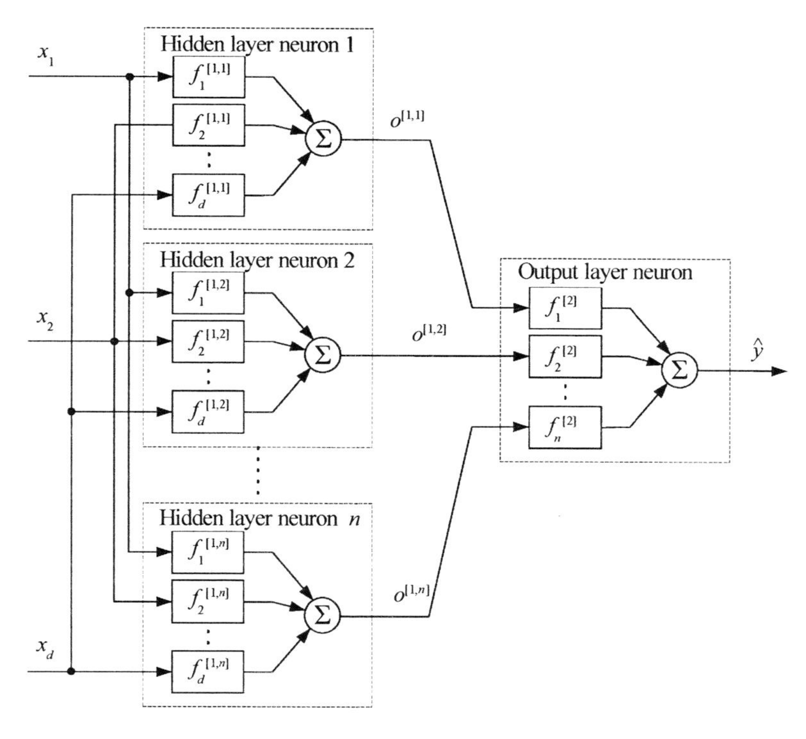

Figure 5. NFKN with d inputs and n hidden layer neurons.

The equations for the hidden and output layer synapses are

$$f_i^{[1,l]}(x_i)=\sum_{h=1}^{m_{1,i}}\mu_{i,h}^{[1]}(x_i)w_{i,h}^{[1,l]},\quad f_l^{[2]}(o^{[1,l]})=\sum_{j=1}^{m_{2,l}}\mu_{l,j}^{[2]}(o^{[1,l]})w_{l,j}^{[2]},$$
$$l=1,\ldots,n,\quad i=1,\ldots,d, \tag{9}$$

where $m_{1,i}$ and $m_{2,l}$ are the number of membership functions (MFs) per input i and l in the hidden and output layers respectively, $\mu_{i,h}^{[1]}(x_i)$ and $\mu_{l,j}^{[2]}(o^{[1,l]})$ are the MFs, $w_{i,h}^{[1,l]}$ and $w_{l,j}^{[2]}$ are the tunable weights. The MFs in the NFKN at each input in the hidden and output layers are shared between all neurons of the respective layer.

The outputs of the NFKN are computed via the following two-stage fuzzy inference procedure:

$$\hat{y}=\sum_{l=1}^{n}\sum_{j=1}^{m_{2,l}}\mu_{l,j}^{[2]}\left[\sum_{i=1}^{d}\sum_{h=1}^{m_{1,i}}\mu_{i,h}^{[1]}(x_i)w_{i,h}^{[1,l]}\right]w_{l,j}^{[2]}. \tag{10}$$

Since the nonlinear synapses represent local piecewise-liner approximations, the overall NFKN model is also piecewise-linear and does not involve any operations other than addition and multiplication in computing its input-output mapping. So a trained NFKN model can be easily implemented even on embedded processors with very limited computational resources.

The description (10) corresponds to the following two-level fuzzy rule base:

$$\text{IF } x_i \text{ IS } X_{i,h} \text{ THEN } o^{[1,1]}=w_{i,h}^{[1,1]}d \text{ AND}\ldots\text{AND } o^{[1,n]}=w_{i,h}^{[1,n]}d,$$
$$i=1,\ldots,d,\quad h=1,\ldots,m_{1,i}, \tag{11}$$

$$\text{IF } o^{[1,l]} \text{ IS } O_{l,j} \text{ THEN } \hat{y}=w_{l,j}^{[2]}n,\quad l=1,\ldots,n,\quad j=1,\ldots,m_{2,l}, \tag{12}$$

where $X_{i,h}$ and $O_{l,j}$ are the antecedent fuzzy sets in the first and second level rules, respectively. Thus, the NFKN is an interpretable model, and the domain knowledge can be used in designing the model in a similar way as for the NFNs.

3. Experiments

In this section, we describe the application of the models introduced in section 2 to the prediction of heart rate in healthy test subjects. The prediction is based on the measurements of physical activity, posture, and respiration. A comparison of the models w.r.t. their prediction quality is presented.

3.1. Participants

Participants were 18 healthy adults (6 men and 12 women), mean age: 39.27 years, standard deviation (SD): 11.67; mean body mass index: 22.78, SD: 3.22. These test subjects were recruited by means of postings and local newspaper advertisements. Participants on medication with direct effects on the autonomic nervous system were excluded. The study was approved by the local ethics committee for medical research and participants received a reimbursement of 200 CHF (approximately 160 USD).

3.2. Data Acquisition

Physiological recordings were performed by means of the LifeShirt System (VivoMetrics, Inc., Ventura, CA, USA). The LifeShirt is a non-invasive ambulatory monitoring system consisting of a data recorder, a garment with embedded sensors for continuous monitoring of ECG, respiration, 3-D accelerometry and other functions and a software package (VivoLogic®) for offline signal analysis. For detailed description of the LifeShirt system, see [22, 24].

The data recorder can be carried in a waist pack and the data is written to a flash memory card. Data were recorded at 10 Hz (accelerometry), 200 Hz (ECG), and 50 Hz (respiratory pattern), respectively. The following measures were automatically derived by means of the VivoLogic Software for the present analysis: heart rate ('HR', beats/min), motion ('AccM', summed absolute values of acceleration along 2 axes), posture ('AccP'), minute ventilation ('Vent', liters inhaled/min), tidal volume ('Vt', ml), breaths per minute ('Br/M').

3.3. Procedure

On arrival in the laboratory (between 8 and 9am), participants gave written informed consent. Subsequently, the sensors and the LifeShirt were attached, the data recording was started and the respiration signal was calibrated against a known fixed volume. Before leaving the laboratory, participants were instructed to pursue their regular everyday activities and to stop the recording after awakening the next morning. To avoid artifacts, they were advised not to pursue any sports.

3.4. Modeling

For modelling purposes, the data were downsampled to 30 sec sampling period (1/30 Hz). The posture channel was divided into 3 separate channels for supine, lateral, and upright body positions. With the other channels, it resulted in 7 input variables (one for 'AccM', three for 'AccP', and another three for the 'Vent', 'Vt', and 'Br/M' respectively). The dependent variable was the heart rate ('HR').

Mean length of a data set for one test subject was 2583.6 samples (21.53 hours), SD: 177.33 samples (1.48 hours). An example of a downsampled heart rate recording is shown in Fig. 6. A clear circadian pattern is visible: the heart rate is reduced and remains more stable during the night, while during the day it is on average higher due to the opposite phase of the circadian clock and has more variations because of physical activity, emotions, etc.

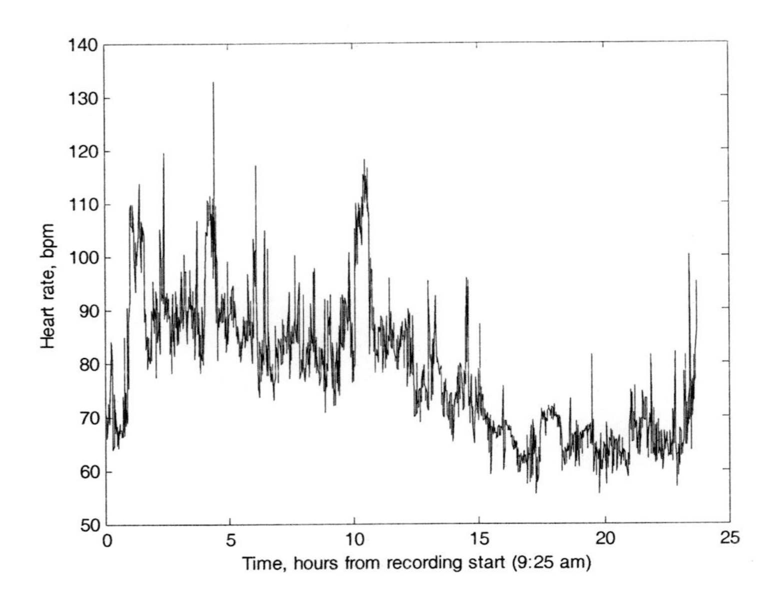

Figure 6. Heart rate recording for test subject N18.

The parameters of the compared models are listed in Table 1. The column 'Architecture' shows the number of inputs (first value), the number of outputs (last value), and the number of neurons in the respective intermediate layers (if applicable). The fewer parameters a model has, the better it is from the point of view of computational complexity.

Table 1. Models used for heart rate prediction (best values in bold, second best in italics)

Model	Activation function	Architecture	Number of parameters
Linear	Linear	**7-1**	**8**
MLP	Sigmoid	7-10-10-1	201
RBFN	Gaussian	7-25-1	201
GRNN	Gaussian	7-1292*-1	10336*
NFN	Triangular	**7-1**	*38*
NFKN	Triangular	*7-4-1*	104

*average value

For the linear model the parameters were the respective regression coefficients from (2). For the MLP, NFN, and NFKN models the only adjustable parameters were their synaptic weights, while for the RBFN and GRNN models these included also the respective prototype vectors v_i sized (7×1) for each basis function ϕ in (4). The radii of the basis functions were fixed and equal to 2.7 (RBFN) and 0.75 (GRNN) for all the basis functions. The NFN model had 8 MFs per 'AccM', 'Vent', 'Vt', and 'Br/M' inputs, and 2 MFs per each of the 3 'AccP' inputs. The NFKN model had 3 MFs per 'AccM', 'Vent', 'Vt', and 'Br/M' inputs, 2 MFs per each of the 3 'AccP' inputs in the hidden layer, and 8 MFs per each of the 4 synapses in its output layer. All model structures were chosen based on best performance.

The data sets for each subject were divided into two parts: one with samples with odd numbers and another one with samples with even numbers. For each of the compared models we performed two runs of testing: in the first run the samples with odd numbers were used for training, the samples with even numbers for checking, and in the second run vice versa. In each of the two runs the training and checking procedures were repeated 18 times with the respective training and checking data sets for all the 18 test subjects. For each subject, the data were standardized as follows: the means of all the variables in the training set were subtracted from both the training and checking sets, and then both the training and checking sets were divided by the respective standard deviations of the training set.

3.5. Learning Algorithms

The parameters of all the models during the training phase were estimated by minimizing the sum of squared errors on the training data set

$$E = \sum_{k=1}^{N} [y(k) - \hat{y}(k)]^2 = (Y - \hat{Y})^T (Y - \hat{Y}), \tag{13}$$

where N is the number of samples in the training set, $y(k)$ is the real output (target value) for the input vector $x(k) = [x_1(k), \ldots, x_d(k)]^T$ corresponding to the k-th sample from the training set, $\hat{y}(k)$ is the model output for the k-th sample, Y is the vector of target values, and $\hat{Y}$ is the vector of model outputs:

$$Y = [y(1), y(2), \ldots, y(N)]^T, \quad \hat{Y} = [\hat{y}(1), \hat{y}(2), \ldots, \hat{y}(N)]^T. \tag{14}$$

The models and their training procedures were implemented in Matlab 7.0. The parameters of the linear model and the NFN were found with the conventional linear least squares method.

The MLP, RBFN, and GRNN models were programmed with the Matlab Neural Network Toolbox. For training the MLP, the resilient propagation algorithm was used (function *trainrp*), which is considered to be one of the best training procedures for the MLPs, combining high speed and high precision. The RBFN and GRNN models were constructed with the *newrb* and *newgrnn* functions, respectively.

For the NFKN model, a separate Matlab toolbox was developed. The training of the NFKN model is based on the hybrid algorithm which involves linear least-squares optimization for the output layer and gradient-based learning of the hidden layer weights. The overall training algorithm is very fast and does not involve any nonlinear operations [15].

3.6. Results

The coefficient of determination R^2, indicating the proportion of "explained variance", was used to estimate the prediction quality. The R^2 index was computed as

$$R^2 = 1 - \frac{\sum_{k=1}^{N} e^2(k)}{\sum_{k=1}^{N} (y(k) - \bar{y})^2}, \quad \bar{y} = \frac{1}{N} \sum_{k=1}^{N} y(k), \tag{15}$$

where k is the number of a measurement in the sample Y, $e(k) = y(k) - \hat{y}(k)$ is the modelling error (residual) for the measurement k, and $\bar{y}$ is the mean value of the sample Y. The performance of the models in the first and second runs is summarized in Tables 2 and 3, respectively. R^2_{trn} stands for training, and R^2_{chk} for checking.

The fourth and the last columns in Tables 2 and 3 show the ratio of the R^2 indices w.r.t. their respective standard deviations $SD(R^2)$. Good performance of a model is indicated by high values of R^2 and low values of $SD(R^2)$ resulting in high values of the ratio $R^2 / SD(R^2)$.

The values of R^2_{chk} for individual test subjects for the runs 1 and 2 are shown in Fig. 7 and 8, respectively. The R^2_{chk} curves for the linear models in Figs. 7 and 8 are below all the respective curves for the nonlinear models, so the R^2_{chk} for the nonlinear models was always higher than that for the linear ones.

To test the significance of these results we used the nonparametric Wilcoxon matched pairs test on R^2_{chk} for all 18 test subjects. In both runs, the p-values for R^2_{chk} of each nonlinear model compared to the linear one are smaller than 0.0002, i.e. the results presented here are highly significant in the statistical sense.

Note that the NFN and the NFKN models had fewer parameters than the other nonlinear models. Nevertheless, the simplest NFN model with only 38 parameters outperformed the most complex GRNN models with more than 10000 parameters on average, and the NFKN model with 104 parameters performed like the RBFN and MLP models with 201 parameters each.

Table 2. Performance of the models in the 1st run (best values in bold, second best in italics)

Model	Mean R^2_{trn}	$SD(R^2_{trn})$	Mean $\frac{R^2_{trn}}{SD(R^2_{trn})}$	Mean R^2_{chk}	$SD(R^2_{chk})$	Mean $\frac{R^2_{chk}}{SD(R^2_{chk})}$
Linear	0.7622	0.0552	13.7986	0.7632	0.0566	13.4851
MLP	*0.8269*	**0.0411**	**20.1334**	**0.8145**	**0.0453**	**17.9920**
RBFN	0.8215	0.0440	18.6637	*0.8092*	0.0483	16.7421
GRNN	0.8131	0.0478	17.0092	0.7969	0.0527	15.1196
NFN	0.8020	0.0449	17.8543	0.7915	0.0496	15.9639
NFKN	**0.8320**	*0.0427*	*19.4870*	0.8042	*0.0461*	*17.4329*

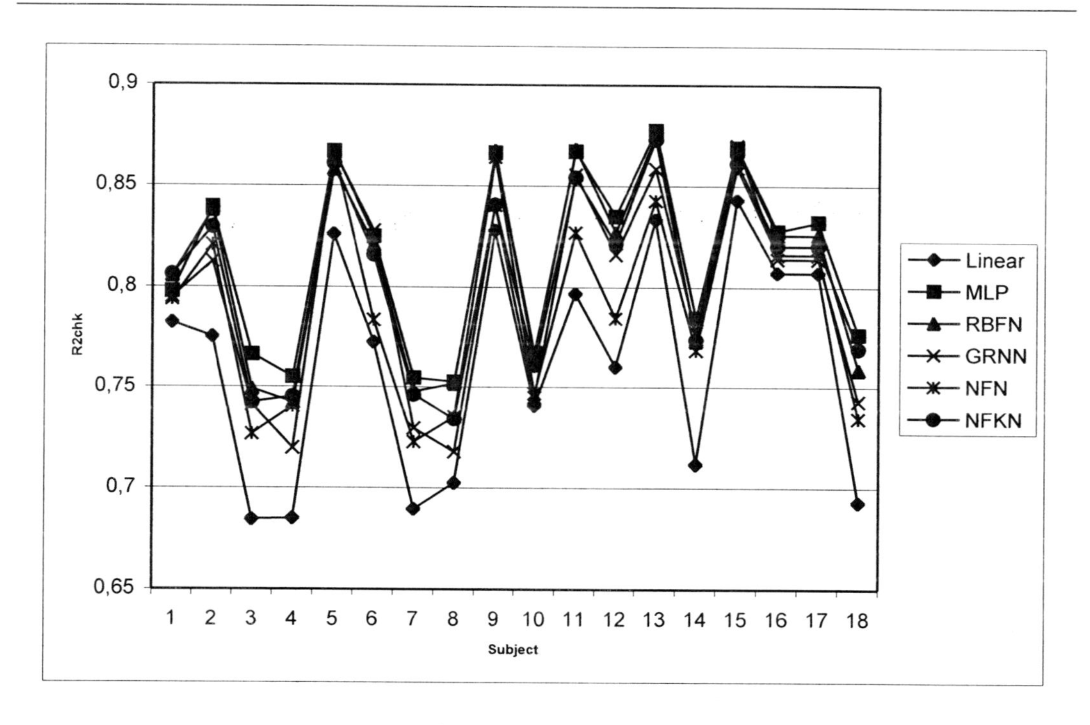

Figure 7. Values of R^2_{chk} for the 1st run for the compared models.

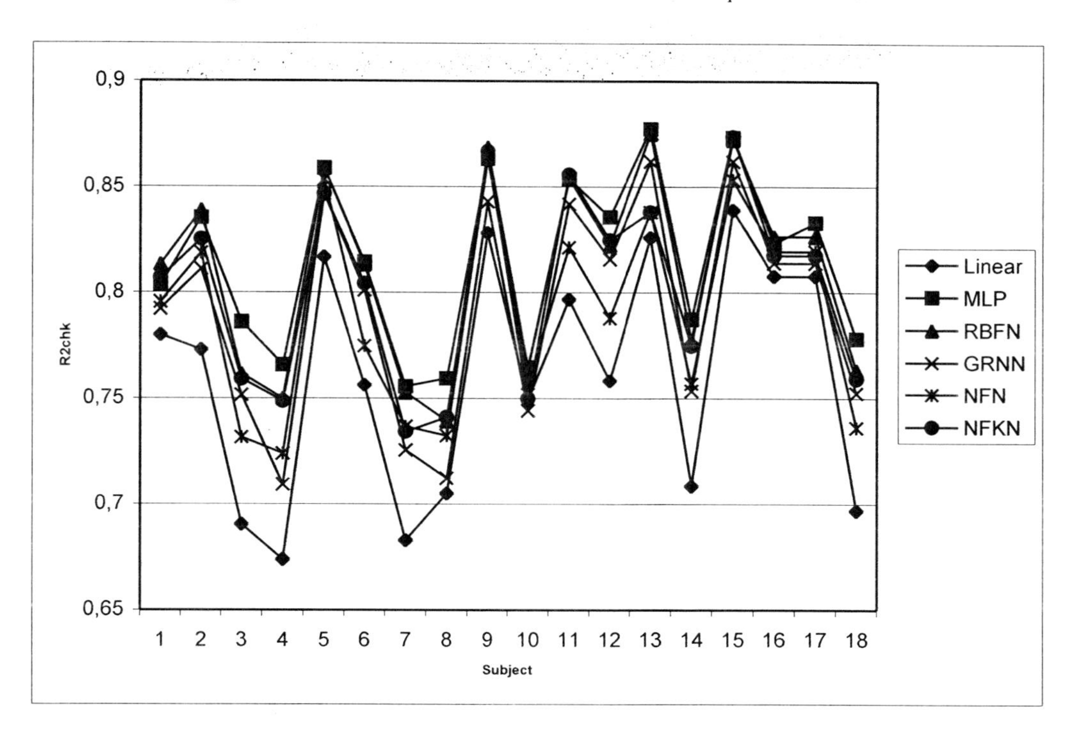

Figure 8. Values of R^2_{chk} for the 2nd run for the compared models.

Table 3. Performance of the models in the 2nd run (best values in bold, second best in italics)

Model	Mean R^2_{trn}	$SD(R^2_{trn})$	Mean $\frac{R^2_{trn}}{SD(R^2_{trn})}$	Mean R^2_{chk}	$SD(R^2_{chk})$	Mean $\frac{R^2_{chk}}{SD(R^2_{chk})}$
Linear	0.7645	0.0561	13.6238	0.7610	0.0558	13.6458
MLP	*0.8318*	**0.0424**	**19.6337**	**0.8150**	**0.0411**	**19.8062**
RBFN	0.8230	0.0470	17.5276	*0.8099*	0.0465	17.3986
GRNN	0.8161	0.0503	16.2402	0.7918	0.0511	15.4938
NFN	0.8030	0.0485	16.5533	0.7900	0.0476	16.6039
NFKN	**0.8332**	*0.0452*	*18.4324*	0.8024	*0.0455*	*17.6224*

4. CONCLUSION

The results presented in this chapter indicate that nonlinear models provide better accuracy in predicting human heart rate at the circadian scale than conventional linear models. These results are significant from the statistical point of view as confirmed by the nonparametric Wilcoxon test.

Neuro-fuzzy models, such as the NFN and the NFKN, are especially attractive because they provide the same level of accuracy as neural networks but have low computational complexity of both the input-output mapping and the training procedures, and are more transparent (interpretable). Another important advantage of the considered neuro-fuzzy models (the NFN and the NFKN) is their capability of working in high dimensions of input space because their computational complexity is increased *linearly* with the increase of the input dimension [4]. This is important for the development of complex models with many input variables.

Based on the presented results, it is expected that neural and neuro-fuzzy techniques described here will find application in analyzing human circadian rhythms. Linear techniques do not appear to provide reliable results for ambulatory monitoring where test subjects live their normal everyday life rather than stay in the controlled laboratory environment [12].

For the reliable detection of circadian parameters such as the phase of the internal clock based on indirect measurements of physiological parameters from, e.g., cardiovascular, respiratory, and temperature recordings, it will be necessary to solve a number of problems related to data preparation for the nonlinear models and estimating variance from between-individual differences in responsivity to behavioral variations such as physical activity. These problems include reliable calibration of the recording devices, signal filtering, and adequate preprocessing.

Acknowledgements

This work was supported by the 6th Framework Project EUCLOCK (No. 018741) and the Swiss National Science Foundation (grant 105311-105850).

REFERENCES

[1] Billings, S.A. & Hong, X. (1998). Dual-orthogonal radial basis function networks for nonlinear time series prediction. *Neural Networks,* **11**, 479-493.

[2] Bodyanskiy, Ye., Kokshenev, I., & Kolodyazhniy, V. (2003). An adaptive learning algorithm for a neo fuzzy neuron. *Proc. 3rd Int. Conf. of European Union Society for Fuzzy Logic and Technology (EUSFLAT'2003), Zittau, Germany*, 375-379.

[3] Bodyanskiy, Ye., Kokshenev, I., & Kolodyazhniy, V. (2004). *A learning fuzzy nonlinear filter. Proc. 9th Biennial Baltic Electronics Conference (BEC'2004). Tallinn Univ. of Technology, Tallinn, Estonia,* 141-144.

[4] Bodyanskiy, Ye., Kolodyazhniy, V., & Otto, P. Neuro-Fuzzy Kolmogorov's Network for Time Series Prediction and Pattern Classification. In: Furbach U. editor. *Lecture Notes in Computer Science Series: Lecture Notes in Artificial Intelligence*, Vol. 3698, Berlin Heidelberg: Springer-Verlag; 2005; 191-202.

[5] Borgelt, C., Klawonn, F., Kruse, R., & Nauck, D. Neuro-Fuzzy-Systeme. Von den Grundlagen künstlicher Neuronaler Netze zur Kopplung mit Fuzzy-Systemen. Reihe Computational Intelligence. Braunschweig: Vieweg; 2003.

[6] Chang, E.S., Chen, S., & Mulgrew, B. (1996). Gradient radial basis neural networks for nonlinear time series prediction. *IEEE Trans. on Neural Networks,* **7**, 190-194.

[7] Cybenko, G. (1989). Approximation by superpositions of a sigmoidal function. *Math. of Control, Signals, and Systems,* **2**, 303-314.

[8] Haykin, S.: Neural Networks: A Comprehensive Foundation. Upper Saddle River: Prentice Hall; 1999.

[9] Jang, J.-S. R., & Sun, C.-T. (1993). Functional equivalence between radial basis function networks and fuzzy inference systems. *IEEE Transactions on Fuzzy Systems,* **4**,156-159.

[10] Jang, J.-S. R., Sun, C.-T., & Mizutani, E. Neuro-Fuzzy and Soft Computing – A Computational Approach to Learning and Machine Intelligence. Upper Saddle River, NJ: Prentice Hall; 1997.

[11] Jin, Y. Advanced Fuzzy Systems Design and Applications. Heidelberg: Physica-Verlag; 2003.

[12] Klerman, E.B., Lee, Y., Czeisler, C.A., & Kronauer, R.E. (1999). Linear demasking techniques are unreliable for estimating the circadian phase of ambulatory temperature data. *J. of Biological Rhythms* **14**, 260-274.

[13] Kolmogorov, A.N. (1957). On the representation of continuous functions of many variables by superposition of continuous functions of one variable and addition. *Dokl. Akad. Nauk SSSR,* **114**, 953-956.

[14] Kolodyazhniy, V., Bodyanskiy, Ye., & Otto, P. Universal Approximator Employing Neo-Fuzzy Neurons. In: Reusch B. editor. *Computational Intelligence, Theory and Applications. Advances in Soft Computing,* Vol. 33, Berlin Heidelberg: Springer-Verlag; 2005; 631-640.

[15] Kolodyazhniy, V., & Otto, P.: Neuro-Fuzzy Modelling Based on Kolmogorov's Superposition: a New Tool for Prediction and Classification. *Lecture Notes in Informatics*, Vol. P-72, Bonn: Gesellschaft für Informatik; 2005; 273-280.

[16] Kosko, B. (1992). Fuzzy systems as universal approximators. *Proc. 1st IEEE International Conference on Fuzzy Systems, San Diego, CA,* 1153-1162.

[17] Miki, T., & Yamakawa, T. Analog implementation of neo-fuzzy neuron and its on-board learning. In: Mastorakis N.E. (Ed.), *Computational Intelligence and Applications*, Piraeus: WSES Press; 1999; 144-149.

[18] Narendra, K.S., & Parthasarathy, K. (1990). Identification and control of dynamical systems using neural networks. *IEEE Trans. on Neural Networks,* **1**, 4-27.

[19] Specht, D.F. (1991). A Generalized Regression Neural Network. *IEEE Trans. on Neural Networks* **2**, 568-576.

[20] Uchino, E., & Yamakawa, T. Soft computing based signal prediction, restoration, and filtering. In: Da Ruan editor, Intelligent Hybrid Systems: Fuzzy Logic, Neural Networks, and Genetic Algorithms. Boston: Kluwer Academic Publishers; 1997; 331-349.

[21] Wang, L.-X. (1992) Fuzzy systems are universal approximators. *Proc.* **1***st IEEE International Conference on Fuzzy Systems, San Diego, CA,* 1163-1170.

[22] Wilhelm, F.H., Pfaltz, M.C., & Grossman, P. (2006). Continuous electronic data capture of physiology, behavior and experience in real life: towards ecological momentary assessment of emotion. *Interacting with Computers* **18**, 171-186.

[23] Wilhelm, F.H., & Roth, W.T. (1998). Using minute ventilation for ambulatory estimation of additional heart rate. *Biological Psychology,* **49**, 137-150.

[24] Wilhelm, F.H., Roth, W.T., & Sackner, M. (2003). The LifeShirt: an advanced system for ambulatory measurement of respiratory and cardiac function. *Behavior Modification,* **27**, 671-691.

[25] Yamakawa, T., Uchino, E., Miki, T., & Kusanagi, H. (1992). A neo fuzzy neuron and its applications to system identification and prediction of the system behavior. *Proc.* **2***nd International Conference on Fuzzy Logic and Neural Networks "IIZUKA-***92***", Iizuka, Japan,* 477-483.

In: Pattern Recognition in Biology
Editor: Marsha S. Corrigan, pp. 205-220
ISBN: 978-1-60021-716-6

Chapter 7

NEUROPHYSIOLOGICAL CORRELATES OF PATTERN RECOGNITION IN THE PERIPHERAL VISUAL FIELD

Hiroaki Shoji and Hisaki Ozaki
Laboratory of Physiology, Faculty of Education,
Ibaraki University, Mito 310-8512, Japan

ABSTRACT

Although it is difficult to recognize visual patterns precisely detected in the peripheral visual field, obtained information might serve an important role for an efficient visual search. Retinal signal triggers a rotation of eyeballs toward a blurred visual image in the peripheral visual field. At that time, features of a detected pattern might influence the direction of eyeball rotation. In this chapter, we focus our interest on the recognition of geometric figures in the peripheral visual field. The results of behavioral studies are reviewed in the first part of this chapter. Factors affecting visual discrimination are discussed, e.g., different retinal locations, shape of the pattern, deterioration of visual acuity at the fovea, and developmental effects on pattern recognition in the peripheral visual field. In the latter part of this chapter, the neurophysiological aspects of geometric figure discrimination are discussed. Geometric figure presentation in the peripheral visual field provokes electrical brain responses. The later component of response seems to be related to the perception and recognition of visual objects in the peripheral visual field. Furthermore, neurophysiological dynamics of pattern recognition in the peripheral visual field are discussed.

1. INTRODUCTION

Visual acuity at the central fovea is an important visual function. However, peripheral vision also has an important role in human visual function. Previous behavioral studies have demonstrated that visual acuity declines exponentially with increasing eccentricity, and that visual acuity at an eccentricity of 10° is 10% to 20% of visual acuity at the central fovea (Weymouth, 1958). This suggests that precise recognition of pattern features with peripheral vision might be difficult. However, information from the periphery is used to realize an efficient visual search.

To capture precisely objects that are projected on the peripheral retina, we must rotate the eyeballs. As a result, visual fixations and saccade eye movements are repeated successively during visual search. Visual information is extracted during fixation, which has the duration of around 200-300 ms in normal adults (Henderson and Hollingworth, 1998). At that time, features of a detected pattern might provide useful information on the next direction of eyeball rotation. Therefore, although the images are blurred, information from the peripheral retina plays an important role in visual cognition (Ikeda et al., 1979).

In this chapter, we focused our interest on the perception and recognition of visual objects in the peripheral visual field. The performance of pattern recognition within the visual field is influenced by various physical factors of the presented pattern. First, we will examine the ability of pattern recognition of geometric figures with different angularity at different eccentricities of the visual field. Second, developmental changes on the pattern recognition in the periphery will be discussed. Furthermore, we will describe that sensory performance in the peripheral visual field deteriorates due to some factors, e.g., myopia or developmental disorders such as mental retardation. In the latter part of this chapter, we will discuss on electrophysiological effects related to pattern discrimination performance in the periphery. Finally, the neurophysiological aspects of pattern recognition in the peripheral visual field will be discussed.

2. Pattern Perception in the Peripheral Visual Field

The poor ability of pattern recognition in the peripheral visual field is rooted deeply in the anatomical structure of the retina, as well as in the visual cortices in humans. For instance, the density of cone cells depends on retinal location, i.e., a high density in the fovea and low density in the periphery (Curcio et al., 1987; Østerberg, 1935). Furthermore, ganglion cells in the central fovea receive information from a smaller number of cone cells than those in the periphery. Cells in the primary visual cortex also receive information from a more restricted region in the foveal retina than they do from the peripheral retina (Slotnick et al., 2001). A large number of cortical neurons might engage in processing information from the foveal retina. These physiological characteristics may explain fine spatial resolution at the central fovea and low spatial resolution in the periphery.

2.1. Edge Detection in the Peripheral Visual Field

In order to evaluate pattern recognition in the peripheral visual field, we examined accuracy of geometric figure discrimination in the peripheral visual field (Shoji and Ozaki, 2006). Three sets of stimuli were used, i.e., a combination of one circle and either of three squares (square distractors), three hexagons (hexagon distractors), or three octagons (octagon distractors). The surface area of those polygons was identical to that of the circle with a diameter of 1.5°.

A fixation mark (crosshatch) was continually presented at the center of the CRT screen, which was placed 50 cm away from the participants. The duration of presentation of the geometric figures was 300 ms; the stimuli were presented at quadrant locations (Fig. 1). Eccentricity, that is, the distance between the crosshatch and the quadrant locations, varied in random order between 2° and 16°, in 2° steps in visual angle. Participants were instructed to find the location of the target (the circle) while they kept on fixating the crosshatch. And the participants responded orally, naming the one of four possible numbers that had been presented on the screen 700 ms after the offset of the geometric figures.

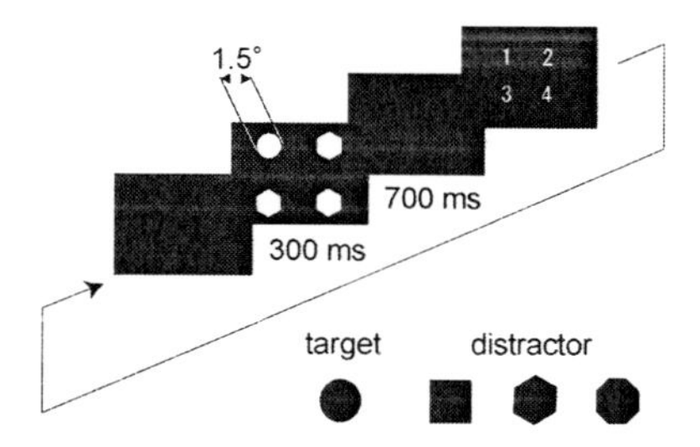

Figure 1. Time sequence of geometric figures displayed in the behavioral experiment (reprinted from Shoji and Ozaki, 2006). Three sets of stimuli were used, i.e., a combination of one circle and either of three squares (square distractors), three hexagons (hexagon distractors), or three octagons (octagon distractors). Stimuli were presented in each quadrant. Stimulus eccentricity varied randomly from 2° to 16°, at 2° intervals. Participants were instructed to detect the target (the circle) while they were fixating a crosshatch in the center of the CRT screen.

Fig. 2 shows the accuracy of circle (target) detection for 4 figures at each eccentricity. The performance of the target discrimination was influenced by both the eccentricity and the shape of the distractors. Although the target shown with the square distractors was accurately detected within 16° of visual angle, the accuracy of target detection for the hexagon and octagon distractors tended to decline with increasing eccentricity. When the angularity of the distractors became more obtuse, target detection performance tended to decline with increasing eccentricity.

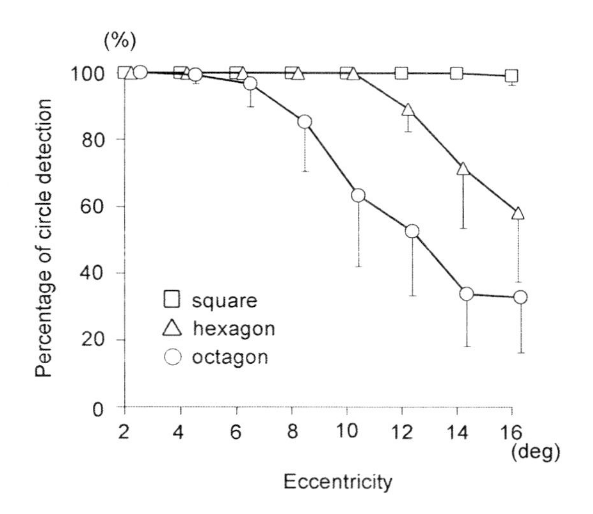

Figure 2. Accuracy of circle (target) detection from among 4 figures presented in each quadrant (reprinted from Shoji and Ozaki, 2006). Error bars indicate the standard deviation of the mean.

This behavioral task did not require participants to report the shape of the geometric figures, but participants were asked to detect a circle as the target among figural distractors. Therefore, sensitivity to the angular corner of the distractors would seem to have played an important role in finding out the target. When the angularity of the geometric figure is close to 180°, higher resolution will be needed for edge detection. Therefore, the required spatial resolution may depend on the angularity of the distractors. In addition, performance of orientation detection is higher for vertical and horizontal contours than for oblique ones, regardless of location in the visual field (Westheimer, 2003). The robust ceiling effect observed in target detection for square distractors might be due to the synergistic effect of angularity and contour.

2.2. Development of Pattern Perception in the Peripheral Visual Field

The anatomical elements of the visual system are basically present at birth, but they are still under development. Although early infants' vision at the central fovea is blurred, their poor visual acuity is remarkably improved with development during the first year of life. Although visual acuity in the retinal periphery is poorer than that in the retinal fovea at all ages, peripheral visual acuity is also dramatically improved in the early stage of life (Allen et al., 1996).

In order to utilize information efficiently from the peripheral retina in the visual search, developmental changes of visual system might be required through childhood. For instance, Cohen (1981) examined the number of eye fixation required to find out the target stimulus among distractors in the periphery (3° – 9°). The number of fixation in 5- and 8-year-old children was more frequent than that in adults, i.e., visual scanning in children was more inefficient than that in adults. These results suggest that the area utilizing useful information for visual search, which has been termed as 'conspicuity area' (Engel, 1971, 1974), 'functional visual field' (Ikeda and Takeuchi, 1975), 'useful visual field' (Shoji and Katada, 1998), or 'useful field of view' (Mackworth, 1965), might not be available in children compared to adults. Moreover, it might depend on the ability quickly extracting useful features from blurred image in the peripheral visual field.

In order to investigate this possibility, we examined pattern perception in the periphery of 55 participants (5.6- to 24.7-year-old), using one circle and three hexagons presentation, where participants were required to detect a circle quickly from among the three other distractors presented for 300 ms. Fig. 3 shows the individual performance of target (circle) detection at each eccentricity. The data in individuals without disabilities was plotted by plus signs. The plot data of people with mental retardation was also overlapped.

The accuracy of target detection tended to decline with increasing eccentricity. Pearson's correlation analysis between chronological age and the accuracy at each eccentricity was performed. Significant positive correlations were observed at eccentricities from 6° to 14° (6°: $r = .393$, $p<.004$; 8°: $r = .417$, $p<.003$; 10°: $r = .506$, $p<.001$; 12°: $r = .453$, $p<.002$; 14°: $r = .492$, $p<.001$).

These results indicate that pattern perception in the periphery might become more precise with increasing age throughout childhood. By around 12 years of age, pattern perception in the periphery appears to reach the adult level. Such perceptual ability would subserve efficient scanning of a scene.

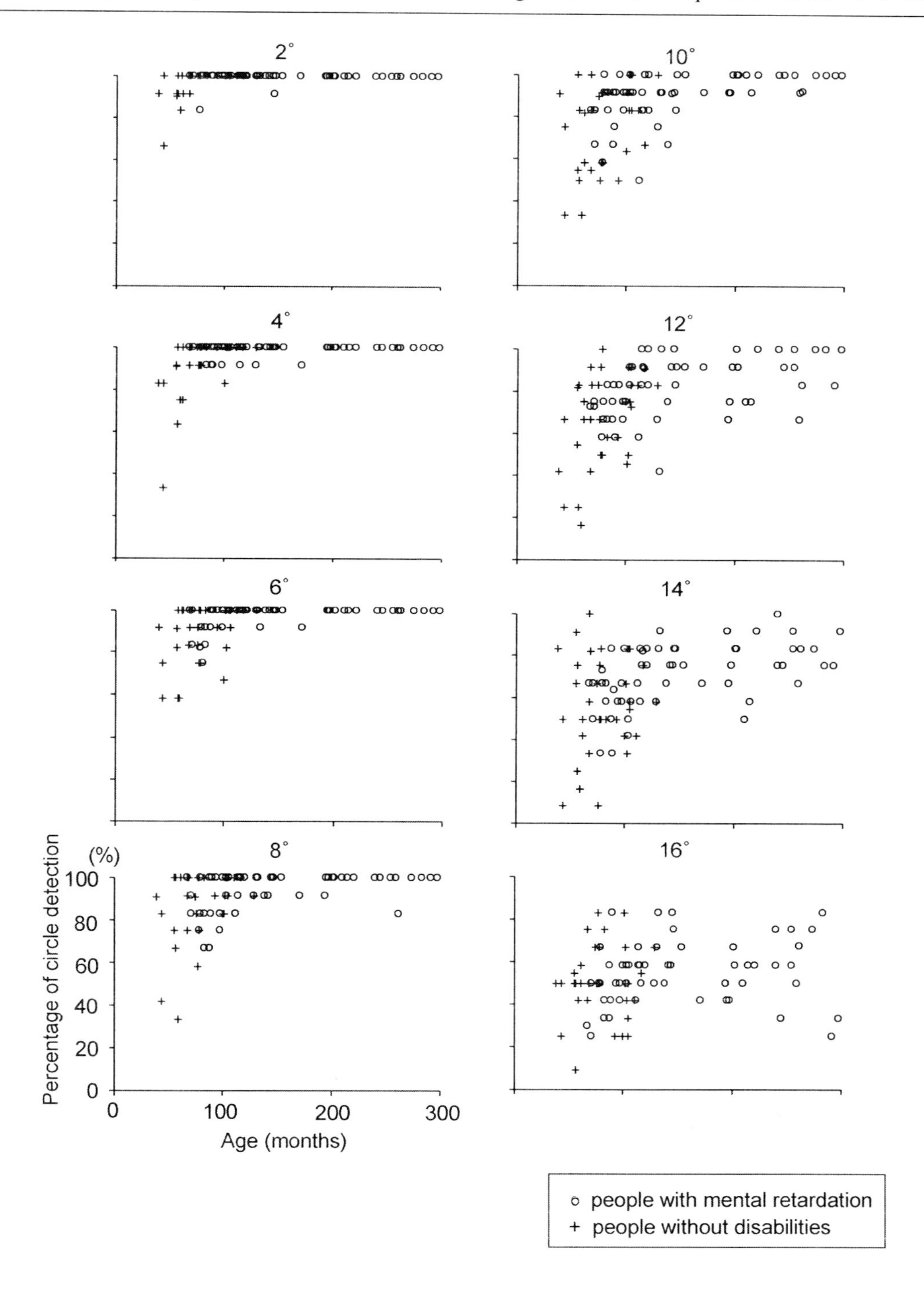

Figure 3. The plot data of percentage of target detection in individuals with mental retardation and without disabilities at each eccentricity. The plot data in individuals with mental retardation was indicated by plus signs; the plot data in individuals without disabilities was indicated by circles. Note that the plot data in people with mental retardation was arranged according to their mental age, but the plot data in people without disabilities was arranged according to their chronological age.

2.3. Pattern Perception in the Periphery in People with Mental Retardation

A different feature of visual scanning has been observed in people with mental retardation (Carlin et al., 2003; Dube et al., 1999). For instance, Boersma and Muir (1975) examined characteristics of visual scanning in people with mental retardation where participants were asked to find an inconsistent part in a picture, e.g., there is a part without snow in a snowy landscape. People with mental retardation tended to view whole parts of the picture including plain parts. As a result, their response time was longer than that in the normal control. Various cognitive factors seem to be concerned with such an inefficient visual search in people with mental retardation. One crucial factor can be their lowered pattern perception in the periphery.

In order to test this hypothesis, we examined their pattern perception in the peripheral visual field using the same paradigm of the previous section. People with mental retardation (n=33; chronological ages from 9.0 to 26.1-year-old; mental ages from 3.2 to 10.7-year-old), who were able to recognize geometric figures, participated in this experiment. In group comparisons analysis, Shoji and Katada (1998) demonstrated that the accuracy of target detection in the adolescents with mental retardation declined significantly at 10° and 12° compared with that in adolescents without disabilities.

Circles in Fig. 3 show plot data of the individual performance in people with mental retardation on target detection at each eccentricity. Note that the plot data in people with mental retardation was arranged according to their mental age, but the plot data in people without disabilities was arranged according to their chronological age.

Accuracy of target detection in people with mental retardation tended to decline with increasing eccentricity, similar to the data in people without disabilities. Pearson's correlation analysis between their mental ages and the accuracy at each eccentricity was performed. There were significant positive correlations at eccentricities from 2° to 12° (2°: $r = .533$, $p<.002$; 4°: $r = .542$, $p<.002$; 6°: $r = .415$, $p<.02$; 8°: $r = .378$, $p<.04$; 10°: $r = .380$, $p<.03$; 12°: $r = .434$, $p<.02$). These results suggest that the pattern recognition in the periphery in people with mental retardation might improve according to mental development.

In ongoing study, we are employing the same paradigm under free viewing condition in people with mental retardation. It is to demonstrate that the ability of pattern recognition in the periphery is related to efficient visual search. As a result, the poor performance of pattern recognition in the periphery brings about frequent eye movements for target capture. These results suggest that information from the periphery might play an important role in an efficient visual search.

2.4. Pattern Perception in the Periphery in People with Myopia

Visual acuity at an eccentricity of 10° might be as little as one-tenth, compared to acuity at the central fovea (Weymouth, 1958). Such poor spatial resolution in the periphery is related to the anatomical structure of the human eye, i.e., a high density of cone cells in the fovea, and low density in the periphery (Østerberg, 1935). In people with lower visual acuity at the central fovea, such as myopia, their peripheral spatial resolution might be similar to that in people with normal visual acuity.

We measured spatial resolution of conventional visual acuity by the improved Förster perimeter (Kawarai et al., 2000). Adults with myopia whose visual acuity at the central fovea was 0.1-0.3 participated in this measurement. [1] At the outside of perifovea, visual acuity of people with myopia was not significantly different from that in people with normal vision. On the other hand, when the participants with myopia were asked to detect a circle as the target among the hexagon distractors presented for 300 ms, their performance of the target discrimination significantly deteriorated in the peripheral visual field, compared with the performance of people with normal vision (Shoji et al., 2001; Shoji and Ozaki, 2004; Fig. 4). Although visual acuity is altered by the duration time of presentation of stimuli (Graham and Cook, 1937; Niwa and Tokoro, 1997), the perception of pattern shortly presented in the peripheral visual field might be influenced by visual acuity at the central fovea.

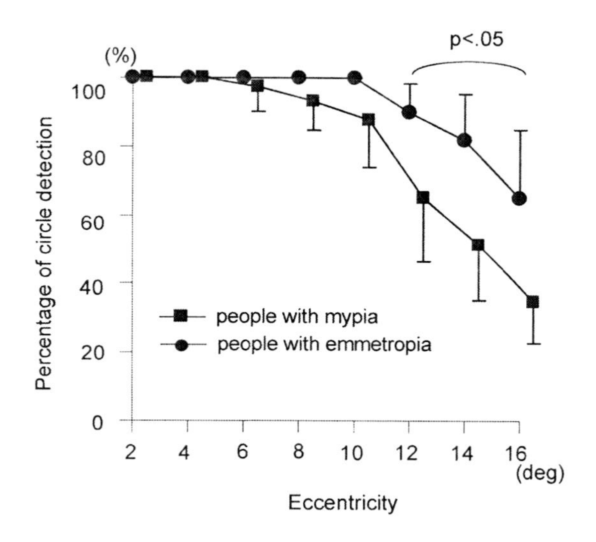

Figure 4. Percentage of target detection in people with myopia at each eccentricity. Error bars indicate the standard deviation of the mean. Accuracy in people with myopia was significantly lower than that in people with emmetropia at the outer eccentricities, i.e., 12°. These data have been redrawn from Shoji and Ozaki (2004).

3. Neurophysiological Changes due to Discrimination of Geometric Figures in the Periphery

The modern brain imaging techniques, e.g., functional magnetic resonance imaging (fMRI), positron emission tomography (PET), near infrared spectroscopy (NIRS), succeeded in visualizing blood flow and/or metabolism within the human brain. On the other hand, electroencephalography (EEG), evoked potential (EP), and event-related potential (ERP), which are the methods of recording the electrical signal produced by the synchronous firing

[1] In a preliminary study, we confirmed that participants with myopia could detect the circle and the hexagon used in the behavioral task where visual acuity is above 0.04 at the central fovea. Therefore, this means the participants with myopia will be able to discriminate the geometric figures presented at the central fovea.

of intracranial neuronal populations, have provided a lot of non-invasive information in the human brain (Ozaki and Lehmann, 2000). The electrophysiological tools are the method with various advantages compared to the modern brain imaging techniques; especially the high time resolution can offer information of the dynamic neural activities occurring within milliseconds or a few dozens of milliseconds. From the findings obtained by ERP measurement, we discuss on the neurophysiological aspects of pattern perception and/or recognition in the peripheral visual field.

3.1. Event-Related Potential (ERP) and Its Topographical Analysis

The ongoing EEG is recordable on the scalp through an amplifier. ERP is a tiny signal embedded in the ongoing EEG and it reflects neural activity related to the occurrences of specific cognitive events, i.e., sensory, motor, anticipation, cognition and so on. By averaging with stimulus onset across tens or hundreds of trials, we can extract the ERP signal; the background EEG irrelevant to events is gradually removed. In ERP waveforms, a series of positive and negative peaks with time-varying responses generated by the neural populations are observed.

EEG waveform depends on the reference channel; the waveform, even if it is identical to the EEG data, is drastically influenced by different reference electrodes, because EEG is recorded as potential differences between recording sites as function of time. There is no electric inactivity in any location served as reference electrode. However, if EEG data is transformed to the average reference, which is the mean of all electrodes at each time point, it is independent of the choice of the recording reference. The average reference can provide a reference-free EEG data.

Multichannel EEG recordings allow assessing the topographical distribution of electrical brain activity. In order to evaluate the spatial characteristics of electrical brain activity, Global Field Power (GFP) and the locations of the centers of gravity (centroids) of the positive and negative electrical field areas are used. GFP is computed as the spatial standard deviation of all amplitude values in the recording array, and it reflects electric strength of the activity at each time point (Lehmann and Skrandies, 1980). Evoked components can be identified in series of map as GFP peaks.

The spatial characteristics of the brain electric field are evaluated by the location of centroids that are calculated as the scalp position of the centers of gravity in the positive and negative areas of the scalp field (Lehmann, 1987). This computation is performed separately in the anterior-posterior and left-right direction (Fig. 5). If scalp topography shows changes by experimental conditions and patient groups, it suggests that different configurations of neural sources are activated. The difference of the field strength might imply the amount of synchronous activation of a neural population engaged in an event.

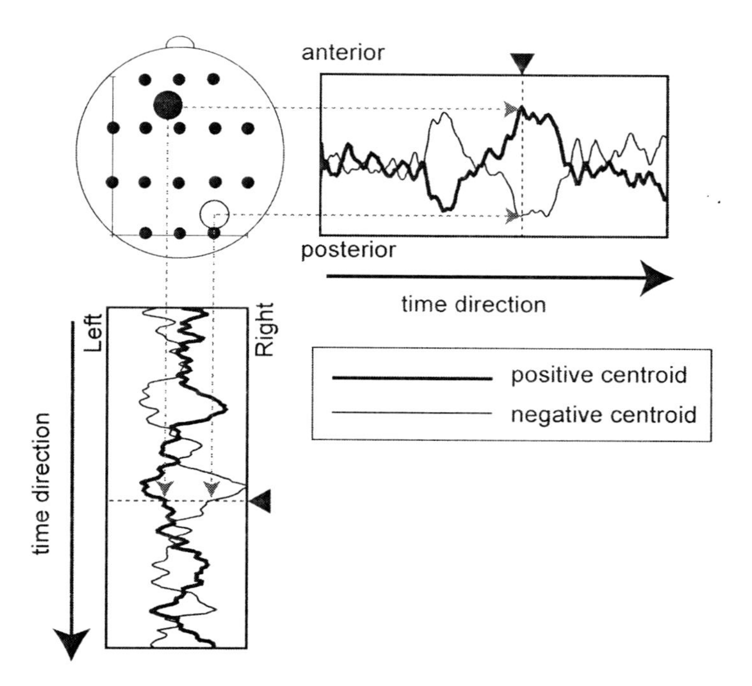

Figure 5. Example of topographic centroid analyses. The location of electrodes on the scalp was shown in the head schema. A closed circle and an open circle indicated the centroid location of the positive and negative areas at a recording point (shown by a triangle). The location of centroids is calculated as the scalp position of the centers of gravity in the positive and negative areas of the scalp field (Lehmann, 1987). The positive and negative centroids are divided in the anterior-posterior and left-right direction. The time course of locational change of the centroid is displayed separately in the anterior-posterior and left-right direction.

3.2. Topographic Change in ERP due to Presentation of Stimuli with Different Eccentricity

We examined neurophysiological aspects of geometric figure discrimination with different angularity at different eccentricities (Shoji and Ozaki, 2006). In order to examine perceptual processing with minimal interference by attention or other cognitive processes, a no-task procedure was used, in which participants were asked only to fixate the center of the screen. One of four figure sets, i.e., four squares, four hexagons, four octagons, and four circles, was presented randomly in the quadrant directions with an eccentricity of 4°, 8°, or 12° (Fig. 6). Stimuli were presented on the CRT screen for 300 ms with an inter-stimulus interval of 1200 ms. Electrical activities were recorded from 16 scalp locations (F3, Fz, F4, T3, C3, Cz, C4, T4, T5, P3, Pz, P4, T6, O1, Oz, O2). ERPs were obtained for each stimulus condition, and ERPs with the average reference were recomputed.

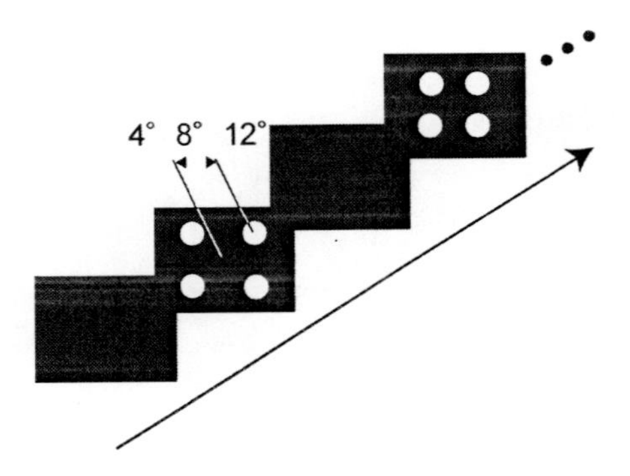

Figure 6. Time sequence of geometric figures displayed in the ERP experiment (reprinted from Shoji and Ozaki, 2006). Four squares, four hexagons, four octagons, or four circles were presented simultaneously in the quadrants at an eccentricity of 4°, 8°, and 12°. Participants were instructed to fixate a crosshatch in the center of the CRT during EEG recording.

Fig. 7 shows the grand mean ERP map series for various geometric figure presentations. In this figure, the mean value of each 20 ms epoch is displayed as a map. Map series from 70 to 310 ms are shown. GFP peak times were used to measure the latency of each ERP component (Fig. 7, left). In the individual data, a distinct first peak of GFP (P1 component) was found in the 60- to 100-ms range (the mean peak latency was 80 ms). The second clear peak of GFP (N1 component) appeared in the range for 110 - 170 ms (the mean peak latency was 140 ms). P1 component with posterior positivity and N1 component with posterior negativity were observed regardless of stimulus location and stimulus shape. The former posterior positivity was followed by posterior negativity, appearing as a polarity reversal with different topography in the ERP map series.

When the stimulus was presented at 4°, a P2 component with posterior positivity was observed at 200 ms after stimulation. Spatial characteristics of the brain electric field were shown by the positive and negative centroids (Fig. 7). The centroid locations remained anterior-positive/posterior-negative in the 110- to 160-ms range. Later, a polarity reversal occurred due to stimulus presentation in the inner visual field, i.e., at 4°. These polarity reversals explain the termination of the N1 component and the onset of P2 (see the outlined figures in the map series in Fig. 7). Such a crossing point with polarity reversal was also observed when stimuli were presented in the outer visual field, but with delayed time occurrence.

The spatial distribution of the ERP after 160 ms was noticeably affected by the stimulus location, regardless of the figural shape. Jedynak and Skrandies (1998) reported that the spatial distribution of electrical brain activity was related to retinal stimulus location. Brecelj et al. (1998) and Wang et al. (2001) examined the dipole location of early components of VEPs and VEFs, and found a retinotopic dipole location on the striate cortex. The retinotopic projection on the visual cortex in human beings has been reported in studies of people with brain damage (Holmes, 1945; Spalding, 1952). The projections from the fovea terminate in the occipital pole, while those from the peripheral field terminate, in the more anteriorly located calcarine sulcus. Thus, the topographic changes related to different retinal stimulus locations might be brought about by activation of different intracranial neuronal populations.

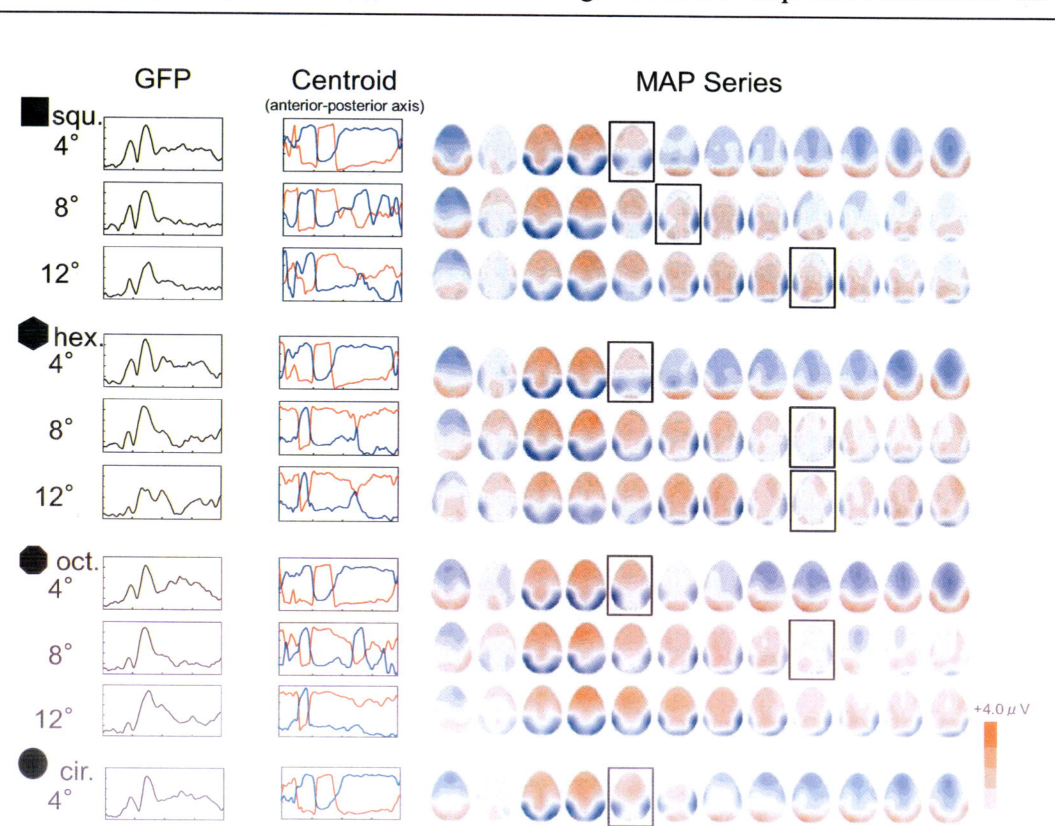

Figure 7. ERP map series and spatial characteristics for the stimulus conditions. Grand mean data of all participants. Global field power (GFP) and the centers of gravity (Centroid) in the anterior-posterior direction are displayed from 0 to 400 ms. The red and blue lines indicate the centroid location of the positive and negative electric fields. Mean values of each 20 ms ERP epoch from 70 to 310 ms are illustrated in the ERP map series (MAP Series). The outlined figures in the map series indicate the time of termination of the N1 component (anterior-positivity/posterior-negativity) and onset of the P2 component (anterior-negativity/posterior-positivity). These data have been redrawn from Shoji and Ozaki (2006).

3.3. Pattern Recognition in Peripheral Vision Evaluated by ERP Topography

ERP components from 200 ms after stimulation are influenced by the shape of the figure presented. When the stimulus was presented at 4°, an electric field with anterior-negativity/posterior-positivity (P2 component) was observed regardless of the figural shape. As eccentricity increased, this component became blurred. However, the P2 component was consistently driven when square figures were presented at outer eccentricity, i.e., 12°.

When polygons with a larger number of angles were presented, the crossing of centroid locations after 160 ms, which means the termination of the N1 component and the onset of P2, tended to be delayed compared to when the square was presented. The crossing point disappeared when the octagon and circle were presented at 12° in visual field. Although a clear GFP peak around 200 ms was observed, the spatial distribution of the ERP after 160 ms was sustained.

Studying contour perception, Brandeis and Lehmann (1989) reported that the perception of a subjective contour figure induced topographic changes in the anterior-posterior direction and an increase of field strength at latencies of 168 - 376 ms, even in passive looking. This is in line with our present data, in that a change in ERP topography occurred in a similar time range. These results suggest that cerebral processing around 200 ms might be related with the perception of contour and/or angularity of geometric figures.

Cells in the primary visual cortex receive information from a more restricted region in the foveal retina than they do from the peripheral retina (Slotnick et al., 2001). Therefore, a large number of cortical neurons might engage in processing information from the foveal retina, and contribute to obtain fine spatial resolution. In non-human primate studies, Dow et al. (1981) reported that vernier acuity thresholds correlated with the cortical magnification in V1. Findings by Dow et al. (1981) were recently supported by an fMRI study in humans (Duncan and Boynton, 2003). These findings suggest that the lower cortical magnification in the peripheral visual field might result in failure to recognize an edge with obtuse angles and/or oblique orientation contours. The varying cortical magnification of retinal input might be related to the edge-sensitive nature of the late P2 component.

3.4. The Lowered Pattern Recognition in the Periphery Evaluated by ERP Topography

As described in the previous sections, the performance of pattern recognition within the visual field is influenced by factors in the participants, such as mental retardation and myopia. The lowered pattern recognition in the periphery was observed in people with myopia and people with mental retardation. Then, does their lowered pattern recognition in the periphery cause some changes of neuroelectric signal? We compared their electrical brain responses elicited by repeated presentation of four hexagons in the quadrant directions with an eccentricity of 4°, 6°, 8°, 10°, or 12° (Shoji et al., 2001, 2002; See Fig. 6 on stimulus presentation procedure).

We would like to focus on the spatial distribution of the ERP after 160 ms, i.e., the termination of the N1 component (anterior-positivity/posterior-negativity) and onset of the P2 component (anterior-negativity/posterior-positivity). Fig. 8 shows the positive and the negative centroids in the anterior-posterior direction for people with and without mental retardation. Regardless of presence of mental retardation, the occurrence of the crossing point after 160 ms, which means the termination of the N1 component and the onset of P2, tended to be delayed when the more peripheral part of the visual field was stimulated. However, the occurrence of the crossing point was delayed more for people with mental retardation, regardless of eccentricities. Especially, the P2 component disappeared in people with mental retardation when geometric figures were presented at outer eccentricity, i.e., 12°.

Topographical changes of later components occurring around 200 ms are related to the pattern recognition in the peripheral visual field. Cerebral processing concerned with the N1 and P2 components was also observed in people with mental retardation, but with delayed time occurrence. Such a delayed processing – presumably due to a deficit in neuronal network connectivity in people with mental retardation – might have caused their poor pattern recognition in the peripheral visual field.

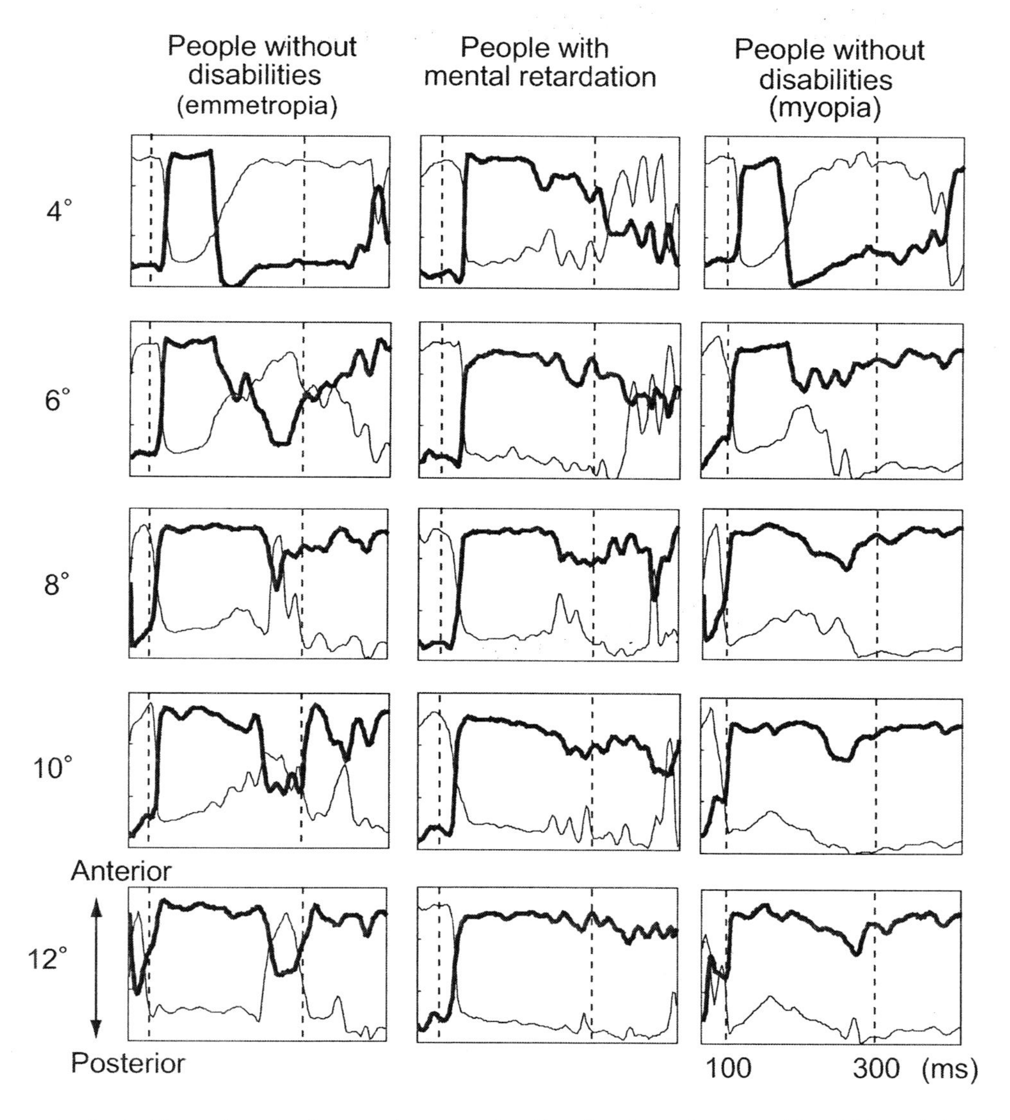

Figure 8. Centroid locations in the anterior-posterior direction in each group. The thick and thin lines indicate the centroid location of the positive and negative electric fields. A part of these data has been redrawn from Shoji et al. (2002).

On the other hand, in people with myopia, the poor pattern recognition in the periphery is also related to the spatial characteristics of electrical brain activity after 160 ms. Fig. 8 shows the positive and negative centroids in the anterior-posterior direction for people with myopia. The occurrence of the crossing point after 160 ms was clearly observed at 4° in the visual field (Fig. 8, right), corresponding to that in people with normal vision. However, the crossing point disappeared when geometric figures were presented at eccentricities above 6°; an electric field with anterior-positivity/posterior-negativity (N1 component) was sustained.

Kawabata and Adachi-Usami (1997) examined retinal function in myopic eyes by multifocal electroretinography (ERG). The ERG is the method of recording the eye's electrical response produced by a flash of black and white light stimuli in various visual fields. Thus, multifocal ERG provides information about the function of the rods and cones in the retina within the visual field. Kawabata and Adachi-Usami (1997) demonstrated that the ERG amplitudes for more peripheral retinal stimulation reduced in people with myopia, compared to the emmetropia group. Therefore, the changes of electrical brain activity corresponding to the poor pattern recognition in people with myopia might be caused by loss of cone photoreceptor function in the peripheral retina.

4. Conclusion

Topographical changes of later components occurring around 200 ms depend upon the retinal stimulus location and the angularity of geometric figures. Since a no-task procedure was used in the neurophysiological study, the findings cannot be explained simply by the poor allocation of spatial attention to the periphery. The findings suggest that recognition of an edge with obtuse angles or of oblique orientation contours becomes difficult due to lower cortical magnification in the peripheral visual field. Thus, cerebral processing concerned with the termination of the N1 component and the onset of P2 might play an important role in the perception and recognition of visual objects in the periphery.

Furthermore, the performance of pattern recognition in the periphery is influenced by various factors, e.g., the physical characteristic of stimuli and participant's background. The neurophysiological changes corresponding to the poor performance in people with mental retardation and people with myopia reflected the topographical changes of the N1 and P2 component. This implies that underlying neural generators may be different or differently activated.

However, functional changes of a sensory system are possible even in the matured brain of human adults. Previous studies on human visual perception have demonstrated that the ability of sensory discrimination in the peripheral visual field improved by short-term training in adults (Crist et al., 1997). Furthermore, Shoji and Skrandies (2006) also revealed that perceptual learning in the peripheral visual field causes changes of electrical brain activity. Such neurophysiological changes may reflect functional human brain plasticity. The future direction of these studies is to clarify neurophysiological changes induced by training for people with poor pattern recognition in the peripheral visual field.

References

Allen, D., Tyler, C. W., & Norcia, A. M. (1996). Development of grating acuity and contrast sensitivity in the central and peripheral visual field of the human infant. *Vision Research*, **36**, 1945-1953.

Boersma F. J., & Muir W. (1975). *Eye movements and information processing in mentally retarded children*. Rotterdam: Rotterdam University Press.

Brandeis, D., & Lehmann, D. (1989). Segments of event-related potential map series reveal landscape changes with visual attention and subjective contours. *Electroencephalography and Clinical Neurophysiology,* **73**, 507-519.

Brecelj, J., Kakigi, R., Koyama, S., & Hoshiyama, M. (1998). Visual evoked magnetic responses to central and peripheral stimulation: simultaneous VEP recordings. *Brain Topography,* **10**, 227-237.

Carlin, M. T., Soraci, S. A., Strawbridge, C. P., Dennis, N., Loiselle, R., & Chechile, N. A. (2003). Detection of changes in naturalistic scenes: comparisons of individuals with and without mental retardation. *American Journal of Mental Retardation,* **108**, 181-193.

Cohen, K. M. (1981). The development of strategies of visual search. In D. F. Fisher, R. A. Monty, & J. W. Senders (Eds.), *Eye movement: Cognition and visual perception* (pp. 271-288). Hillsdale, NJ: Lawrence Erlbaum Associates.

Crist, R. E., Kapadia, M. K., Westheimer, G., & Gilbert, C. D. (1997). Perceptual learning of spatial localization: specificity for orientation, position, and context. *Journal of Neurophysiology,* **78**, 2889-2894.

Curcio, C. A., Sloan, K. R. J., Packer, O., Hendrickson, A. E., & Kalina, R. E. (1987). Distribution of cones in human and monkey retina: individual variability and radial asymmetry. *Science,* **236**, 579-582.

Dow, B. M., Snyder, A. Z., Vautin, R. G., & Bauer, R. (1981). Magnification factor and receptive field size in foveal striate cortex of the monkey. *Experimental Brain Research,* **44**, 213-228.

Dube, W. V., Lombard, K. M., Farren, K. M., Flusser, D. S., Balsamo, L. M., & Fowler, T. R. (1999). Eye tracking assessment of stimulus overselectivity in individuals with mental retardation. *Experimental Analysis of Human Behavior Bulletin,* **17**, 8-14.

Duncan, R. O., & Boynton, G. M. (2003). Cortical magnification within human primary visual cortex correlates with acuity thresholds. *Neuron,* **38**, 659-671.

Engel, F. L. (1971). Visual conspicuity, directed attention and retinal locus. *Vision Research,* **11**, 563-576.

Engel, F. L. (1974). Visual conspicuity and selective background interference in eccentric vision. *Vision Research,* **14**, 459-471.

Graham, C. H., & Cook, C. (1937). Visual acuity as a function of intensity and exposure time. *American Journal of Psychology,* **49**, 654-661.

Henderson, J. M., & Hollingworth, A. (1998). Eye movements during scene viewing: an overview. In G. Underwood (Ed.), *Eye guidance in reading and scene perception* (pp. 295-312). Amsterdam: Elsevier.

Holmes, G. (1945). The organization of the visual cortex in man. *Proceedings of the Royal Society of London. Series B: Biological Sciences,* **132**, 348-361.

Ikeda, M., & Takeuchi, T. (1975). Influence of foveal load on the functional visual field. *Perception and Psychophysics,* **18**, 255-260.

Ikeda, M., Uchikawa, K., & Saida, S. (1979). Static and dynamic functional visual fields. *Journal of Modern Optics,* **26**, 1103-11113.

Jedynak, A., & Skrandies, W. (1998). Functional perimetry combined with topographical VEP analysis. *International Journal of Neuroscience,* **93**, 117-132.

Kawabata, H., & Adachi-Usami, E. (1997). Multifocal electroretinogram in myopia. *Investigative Ophthalmology and Visual Science,* **38**, 2844-2851.

Kawarai, Y., Koyama, Y., Sekita, Y., & Shoji, H. (2000). Relationships between spatial distribution of visual acuity and useful visual field for the discrimination of Hiragana characters. *Bulletin of the Faculty of Education, Ibaraki University (Educational Sciences),* **49**, 113-122. (in Japanese)

Lehmann, D. (1987). Principles of spatial analysis. In A. Gevins, & A. Remond (Eds.), *Handbook of Electroencephalography and Clinical Neurophysiology,* Vol. 1: *Methods of Analysis of Brain Electrical and Magnetic Signals* (pp. 309-354). Amsterdam: Elsevier.

Lehmann, D., & Skrandies, W. (1980). Reference-free identification of components of checkerboard-evoked multichannel potential fields. *Electroencephalography and Clinical Neurophysiology,* **48**, 609-621.

Mackworth, N. H. (1965). Visual noise causes vision tunnel. *Psychonomic Science,* **3**, 67-68.

Niwa, K., & Tokoro, T. (1997). Mesurement of temporal summation of visual acuity with use of modified tachistoscope. *Japanese Journal of Ophthalmology,* **41**, 403-408.

Østerberg, G. (1935). Topography of the layer of rods and cones in the human retina. *Acta Ophthalmologica. Supplement,* **6**, 1-103.

Ozaki, H., & Lehmann, D. (2000). EEG reconsidered: from neuroelectric signals to human conscious experience. *Japanese Journal of Clinical Neurophysiology,* **28**, 15-17.

Shoji, H., & Katada, A. (1998). Discrimination of geometric figures: useful visual field of adolescents with mental retardation. *Japanese Journal of Special Education,* **36(6)**, 23-29. (in Japanese)

Shoji, H., Katada, A., & Ozaki, H. (2001). Useful visual field and ERP topography in the condition of low visual acuity. Abstract, *Clinical Neurophysiology,* **112**, S94.

Shoji, H., & Ozaki, H. (2004). Useful visual field and event-related potentials in different visual acuity. *Clinical Electroencephalography,* **10**, 636-642. (in Japanese)

Shoji, H., & Ozaki, H. (2006). Topographic change in ERP due to discrimination of geometric figures in the peripheral visual field. *International Journal of Psychophysiology,* **62**, 115-121.

Shoji, H., Shinoda, H., & Ozaki, H. (2002). The useful visual field and ERP topography in persons with intellectual disabilities. In K. Hirata,Y. Koga,K. Nagata, & K. Yamazaki (Eds.), *Recent Advantage in Human Brain Mapping* (pp. 723-728). Amsterdam: Elsevier.

Shoji, H., & Skrandies, W. (2006). ERP topography and human perceptual learning in the peripheral visual field. *International Journal of Psychophysiology,* **61**, 179-187.

Slotnick, S. D., Klein, S. A., Carney, T., & Sutter, E. E. (2001). Electrophysiological estimate of human cortical magnification. *Clinical Neurophysiology,* **112**, 1349-1356.

Spalding, J. M. (1952). Wounds of the visual pathway. Part II. The striate cortex. *Journal of Neurochemistry,* **15**, 169-183.

Wang, L., Barber, C., Kakigi, R., Kaneoke, Y., Okusa, T., & Wen, Y. (2001). A first comparison of the human multifocal visual evoked magnetic field and visual evoked potential. *Neuroscience Letters,* **315**, 13-16.

Westheimer, G. (2003). The distribution of preferred orientations in the peripheral visual field. *Vision Research,* **43**, 53-57.

Weymouth, F. W. (1958). Visual sensory units and the minimal angle of resolution. *American Journal of Ophthalmology,* **46**, 102-113.

In: Pattern Recognition in Biology
Editor: Marsha S. Corrigan, pp. 221-240
ISBN: 978-1-60021-716-6

Chapter 8

TOLL BRIDGE BETWEEN THE OUTSIDE WORLD AND SELF: A ROAD TO SAFETY OR A PASS TO ATHEROSCLEROSIS?

***L. M. Pinchuk*[1]*, G. V. Pinchuk*[2]*, G. T. Pharr*[1] *and S. R. Lee*[1]**
[1]Department of Basic Science, College of Veterinary Medicine,
Mississippi State University
[2]Department of Sciences and Mathematics,
Mississippi University for Women

ABSTRACT

For many decades, one of the persisting paradigms of immunology stated that cells of the adaptive immune system recognize specific antigens with the help of clonally distributed antigen receptors, while cells of the innate immune system simply recognize "something foreign." In the late 1990's, however, this paradigm started to change. It became apparent that various cells, including cells of the immune system, express a family of germline-encoded receptors homologous to Drosophila protein Toll (Toll-like receptors, TLRs). These receptors recognize particular pathogen-associated molecular patterns (PAMPs) as well as the so-called "danger-associated molecular patterns" (DAMPs). The interaction between PAMPs and DAMPs on the one hand and TLRs on the other leads to induction of the innate and the specific (adaptive) immune responses. In this review, we discuss data on the tissue distribution of different TLRs, and on the ligands with which they interact. We analyze events that follow the interaction of these ligands with TLRs expressed on immune cells (B and T lymphocytes, macrophages, dendritic cells) and on cells outside of the immune system. We briefly discuss how abnormal TLR signaling can lead to disease. Finally, we review some recent studies pointing to the role of TLRs in such widespread and dangerous human disease as atherosclerosis.

INTRODUCTION

Immunology has been traditionally viewed as a science that studies mechanisms of defense against infectious agents and, more broadly, from "nonself." In this regard, specific recognition of "nonself" in its huge diversity has been, traditionally, a major focus of immunology. For several decades, immunology has been largely concerned with entities called antigens, the latter being self or "nonself" ("foreign"). It studied antibodies – molecules that are clonally distributed among B lymphocytes and, through their clone-specific variable regions, recognize specific antigens – that is, particular combinations of amino acids or sugar or nucleic acid residues. They, in complex with adjunct molecules, serve as antigen receptors of B lymphocytes. Another major interest of immunologists has been molecules called T cell antigen receptors (TCR) – molecules that are clonally distributed among T lymphocytes and recognize a peculiar kind of antigens - specific peptides "presented" to the T lymphocytes by cells of the host with the help of the polymorphic major histocompatibility complex (MHC) molecules.

One of the most accepted paradigms of immunology stated that normally, a vertebrate organism does not contain functional B or T lymphocytes that respond to "self" antigens. T and B lymphocytes with antigen receptors specific to "self" antigens are continuously produced, because the construction of antigen receptors is antigen-independent, and, essentially, stochastic. Antigen receptors are the result of somatic rearrangement of germline-encoded segments of variable region genes. However, T and B lymphocytes whose antigen receptors are capable of binding self antigens are, according to this paradigm, efficiently eliminated by processes of central and peripheral negative selection in lymphoid organs. On the other hand, the organism normally has myriads of B and T lymphocyte clones that can, and do, respond to all these molecular entities (antigens) when these entities are "foreign". Clones that produce foreign antigen-specific receptors are not eliminated by negative selection. On the contrary, when they recognize their specific antigens, they are activated, expand by mitosis, undergo differentiation and become efficient in eliminating foreign antigens from the internal environment of the host [1].

In spite of the tremendous progress in our understanding of how exactly antibodies and TCR recognize their specific molecular entities, certain important questions remained unanswered. By mid-1980's, evidence began to accumulate in favor of the view that T lymphocytes are not activated in response to recognition of specific foreign peptide-self MHC complexes unless cells of the host that "present" these molecular entities are "prepared" in a special way, so that they express molecules known as co-stimulators. Moreover, were it not for co-stimulators, the event of recognition of foreign antigens by clonally distributed antigen receptors would inevitably cause an irreversible "silencing" of lymphocytes, known as clonal anergy [2].

Two theories emerged, trying to explain this phenomenon. One of them, commonly known as "danger theory" or "danger model," offered by Polly Matzinger in the late 1980's and most explicitly formulated in 1994, stated that the goal of the immune system is not as much to recognize "nonself" as to respond to danger. According to Matzinger, the immune system is triggered to work (particularly, T lymphocytes are activated) not when a "nonself" invades the "self," but when some endogenous mediators, produced by "endangered" (stressed, damaged, or dying) self cells which cause the upregulation of co-stimulatory molecules on cells that "present" molecular moieties to the immune system [3]. Hydrophobic portions of molecules, for example, might represent such endogenous "danger signals" [4]. The alternative theory, attempting to explain the necessity of co-stimulator expression, was offered by Charles Janeway Jr. in 1989. According to Janeway, putative receptors exist on host cells: receptors that he proposed to call "pattern recognition receptors" (PRRs). These specialized receptors bind an array of molecular moieties common to pathogenic microorganisms ("pathogen-associated molecular patterns," or PAMPs). The binding of PAMPs to PRRs initiates signaling that eventually leads to co-stimulator expression and immune cell activation [5].

Apparently, identification of real, "material" PRRs and PAMPs was in order. Indeed, in 1997, Ruslan Medzhitov and co-workers presented evidence suggesting that in vertebrates, Toll-like receptors (TLRs) may be PRRs and their ligands may be PAMPs [6]. However, as we now know, these same receptors also bind a certain range of endogenous molecular patterns and can be thus activated following tissue damage or trauma. This puts TLRs into the "intersection" between Matzinger's "danger theory" and Janeway's "PAMP theory," claiming that TLRs recognize not only PAMPs, but also DAMPs – "danger-associated molecular patterns" [7].

In this review, we attempt to consolidate and summarize some recent data on TLRs, and to show that knowledge about these molecules helps us to understand not only how the immune system works, but also how atherosclerosis - one of the gravest human diseases, whose pathogenesis has traditionally been thought to unwind outside of the immune system, - emerges and develops.

What Are Toll-Like Receptors and What Do They Recognize

TLRs are a large family of molecules expressed by a wide variety of mammalian as well as non-mammalian cells. They received their name because they display a degree of amino acid homology to a Drosophila protein called Toll (German for "jazzy," "groovy," or "cool"). Toll was discovered as a transmembrane protein that plays an important role in pattern formation during Drosophila embryonic development. It became apparent that it also controls the induction of potent anti-microbial (especially anti-fungal) factors in adult Drosophila flies [8-10]. TLRs became an object of much interest to immunologists since it was discovered that they have an interleukin-1 (IL-1) receptor cytoplasmic domain as a part of their cytoplasmic portion. This IL-1 domain is instrumental in signaling through TLRs, making these receptors a link between pathogen invasion and tuning of the host defense systems [11, 12].

The TLR family includes at least ten receptors (TLR 1 to TLR 10) in humans and about 15 receptors in mice [13]. The range of mammalian cells that express TLRs includes cells of the immune system, such as macrophages, dendritic cells, natural killer (NK) cells, B lymphocytes and regulatory T lymphocytes (reviewed in [14]); such epithelial cells as cells of the endometrium, of the intestinal vascular endothelium, and alveolar epithelium; such connective tissue cells as fibroblasts and mast cells; skeletal muscle cells, smooth muscle cells, microglia and brain neurons [reviewed in 15]. This extremely broad pattern of expression of TLRs suggests that their function is not limited to triggering or tuning of the immune system or defense mechanisms. Interestingly, while some TLRs are usually found on the external cell membrane, other TLRs (especially TLR3, 7, 8, and 9) are often expressed intracellularly [reviewed in 15], suggesting that they participate in intracellular trafficking of their ligands.

Different TLRs have different ligands, both exogenous (which are, predominantly, microbial products), and endogenous. Thus, TLR4 recognizes lipopolysaccharide (LPS), which is a major endotoxin of the outer membrane of Gram-negative bacteria [16-18]. TLR4 also recognizes lipoteichoic acid present on the outer membrane of Gram-negative bacteria, bacterial heat-shock proteins and flavolipins, viral envelope proteins, and some plant molecules. In addition, recent studies showed that this receptor recognizes a number of molecules expressed by the host, including heat-shock proteins, fibronectin, and hyaluronan (reviewed in [19]). Another representative of the TLR family, TLR2, interacts with the major component of cell walls of Gram-positive bacteria, peptidoglycan [20, 21]. This receptor also recognizes a number of bacterial lipoproteins and glycolipids, lipoarabinomannan present in the cell wall of the genus *Mycobacterium*, LPS found in bacteria of the genera *Porphyromonas* and *Leptospira*, zymozan of yeast, GPI anchor of *Trypanosoma cruzi*, outer membrane protein A of *Klebsiella*, porin of *Neisseria*, and at least three heat-shock proteins of the host (HSP60, HSP70, and HSGp69) [reviewed in 22, 23]. TLR1 recognizes bacterial triacyl lipopeptides [24]. TLR9 recognizes bacterial unmethylated CpG DNA dinucleotides [25] and *Plasmodium malariae* pigment hemozoin [26]. TLR3 recognizes viral dsRNA [27], and TLR7 and TLR8 recognize viral ssRNA [28-30]. TLR11 recognizes a protozoan molecule known as profilin-like protein [31], and some yet unidentified components of uropathogenic bacteria [32]. TLR5 recognizes bacterial flagellin [33]. TLR6 participates in the recognition of bacterial diacylated lipopeptide and peptidoglycan [34]. Molecular patterns recognized by TLR10 are still not identified [19].

Certain combinations of TLRs are known to recognize particular ligands. For example, dimers formed by TLR2 and TLR1 recognize bacterial triacetylated lipopeptides and a Mycobacteria-derived lipoarabinomannan [35], as well as a soluble factor produced by *Neisseria meningitidis* [36]. Dimers formed by TLR 2 and TLR 6 interact with diacetylated lipopeptides of *Mycoplasma* [37], as well as with lipoteichoic acid expressed on the surface of bacteria that belong to the genera *Streptococcus* and *Staphylococcus* [38]. TLR2 acts together with its co-receptor, CD36 [39].

Structurally, TLRs are type 1 integral membrane proteins that have diverse leucine-rich repeat motifs in their extracellular portions. These motifs (or the TLR "ectodomain") participate in ligand molecular pattern binding (reviewed in [22, 23]). On the other hand, the cytoplasmic portion of each known TLR is a sequence of amino acids highly homologous to the cytoplasmic portion of the interleukin-1 receptor (IL-1R). This part of TLR is known as the Toll/IL-1 receptor domain, or TIR [11, 12, 15, 22, 23]. No high-affinity binding has been observed between the TLR ectodomain and the various TLR ligands [11, 15] and it has been postulated that ligands that interact with TLRs do so together with "co-ligand" molecules like MD1, MD2 and RP105 [22, 23]. This brief and incomplete description of TLRs and their ligands merely hints at the notion that TLRs are a large and sophisticated family of molecules interacting with a certain large, and yet not unlimited and perhaps not disorderly, array of molecular patterns.

How Do Toll-Like Receptors Transduce Signals

Signaling through TLR is quite complex. It is generally accepted that it involves a number of adaptor molecules, of which the best known are the myeloid differentiation factor 88

(MyD88), IL-1 receptor-associated protein kinases (IRAKs), the transforming growth factor-beta activated kinase 1 (TAK1), TAK1-binding proteins 1 and 2 (TAB1 and TAB2), and the tumor necrosis factor receptor-associated factor 6 (TRAF6) (reviewed in [11, 19, 22, 23]). For example, when TLR4 expressed on the surface of a macrophage interacts with its ligand (e.g., bacterial LPS), it may recruit MyD88 [12]. This adaptor molecule binds to the TIR domain of the TLR. The product of this binding, a "hybrid" between MyD88 and TIR called "TIR-domain containing adaptor protein" (TIRAP), recruits another molecule, a serine –threonine protein kinase termed IL-1 receptor-associated kinase 4 (IRAK-4). The latter then becomes phosphorylated, and binds another kinase, IRAK-1, which associates with the tumor necrosis factor receptor-activated factor 6 (TRAF6) (44) (and probably with a series of related molecules from the TRAF family, see [22, 40]). The activated TRAF6 together with the bound IRAK-1, in turn, dissociates from IRAK-4, and then binds and phosphorylates TAK1, TAB1 and TAB2. After this event, IRAK-1 dissociates from the TRAF6. Then, the TRAF6-TAK1-TAB1-TAB2 complex, together with the two ubiquitin-conjugating enzymes called Ubc13 and Uev1A, activates two downstream signaling pathways. Of these, one involves kinases called IKK (for "IkappaB kinase kinase"), and eventually causes activation and nuclear translocation of the transcription factor NF-κB. The other pathway leads to phosphorylation and activation of a group of the so-called mitogen-activated protein kinases (MAP), particularly the p38 mitogen-activated protein kinase (MAPK). The net result of the MyD88-dependent signaling pathway is the NF-κB-dependent activation of production of pro-inflammatory cytokines and chemokines [19, 22, 40].

However, because LPS activates Jnk, MAPK, and NF-κB in macrophages from MyD88 knockout mice [41, 42], it was concluded that signaling through TLR4 (and other TLRs) can proceed via the MyD88-dependent or via a MyD88-independent pathway. The latter may involve other adaptor molecules, called TRIF (Toll-IL-1 receptor domain-containing adaptor inducing interferon-β), and TRIF-related adaptor molecule (TRAM). The result of TRAM binding to TIR is the activation of TRIF. The overall result of the activation of this MyD88-independent signaling pathway is recruitment of caspases and inducing the production of type 1 interferons [43-45].

Because of the activation of interferon and other pro-inflammatory cytokine production, both of the above-mentioned pathways lead to up-regulation of costimulatory molecules, making the activated macrophage capable of activating T cells. Thus, the recognition of a bacterial endotoxin by TLRs expressed by the macrophage produces the well-known "adjuvant" effect and serves as an early trigger of the adaptive immunity [14].

TLR signaling is subject to both positive and negative regulation. A number of molecules called interferon regulatory factors (IRF) modulate both MyD88-dependent and –independent TLR signaling [46-50]. ATP and cAMP also seem to be negative regulators of TLR4 signaling. These inhibitors of TLR signaling in macrophages act by inactivating TRAF6 or by inhibiting such intermediate activators of TLR signaling as TBK-1 (TRAF family member-associated NF-κB-activator binding kinase-1) and RIP-1 (receptor-interacting protein-1) [51]. Yet another potent negative regulator of TLR signaling in macrophages is C5a, a cleavage fragment of the complement protein C5. This molecule in murine macrophages inhibits the TLR4-induced synthesis of a number of cytokines, notably IL-12, by acting through the extracellular signal-regulated kinase (ERK) and the phosphoinositide 3-kinase (PI3K). Thus, TLR signaling results not merely in the production of such major molecules of the innate immunity as TNF-α and type I interferons, but is also regulated by such major participants of innate immunity as complement and type I interferons (reviewed in [40, 52]).

How Do Toll-Like Receptors Function in Host Defense

Within recent years, astonishingly strong evidence has been obtained indicating that signaling through TLR plays a crucial role in host defense. TLR signaling is important for both non-specifics (innate) as well as in specific (adaptive) immunity. As a short but impressive illustration of the role TLRs play in the innate immunity, we will mention recent report by Montminy et al. [53] demonstrating the importance of TLR4 signaling in the fight with one of the deadliest infections known. These investigators showed that *Yersinia pestis*, the bacterium that causes bubonic and pneumonic plague, cannot be purged from the infected host in a timely fashion because it produces a variant form of lipopolysaccharide (LPS) unable to interact with TLR4. When *Y. pestis* culture was genetically modified by introducing plasmids that contain PLpxL (an enzyme that changes their LPS into a form that can interact with TLR4), the bacteria lost their pathogenic properties.

Dendritic cells (DC), which have received enormous attention within the recent decade because of their unique properties of key sentinels alerting both innate and adaptive immune responses, utilize TLR signaling for their various functions (reviewed in [14]). The DC network consists of several subsets that differ in location, surface markers, and cytokine profiles [54]. While the so-called myeloid or "conventional" DC express a wider variety of TLRs, the so-called plasmacytoid DC (PDC) of the human origin express only TLR7 and TLR9 [55]. Both of these receptors are localized in the endosomal membrane. Interaction of TLR7 and 9 with their natural ligands (viral ssRNA and bacterial unmethylated CpG-rich DNA, respectively) leads to signaling through a MyD88-dependent pathway which, in the case of TLR9, involves an enzyme DNA-dependent protein kinase (DNA-PK). The net result of this signaling in PDC is production of massive amounts of IFN-α and IFN-β, as well as up-regulation of costimulatory molecules [14, 55-59]. Thus, it is TLR signaling that makes a PDC into an efficient stimulator of innate antiviral immunity as well as into a professional antigen-presenting cell.

Similarly, monocytes seem to be "made" into actively functional immune cells after signaling through TLR. Human peripheral blood monocytes express TLR and produce IL-6 in response to the TLR signaling induced by bacterial fimbriae [60]. In an interesting study by Farina et al. it was found that some (although not all) human peripheral blood monocytes produced TNF-α, and a few produced IL-10 in response to TLR signaling. Strikingly, one of the outcomes of TLR signaling in monocytes was up-regulation of CD83 (a marker of mature DC), and of SLAM ("signaling lymphocytic activation molecule" - a marker of activated monocytes capable of antigen presentation) [61]. In our recent work, using a proteomics approach, we have identified TLR1 and TLR6 in the membrane fraction of bovine monocytes [62]. We also showed that bovine viral diarrhea virus infection significantly upregulates type I interferon production and the expression of TLR3 and TLR7 in these cells of the innate immune system [63, 64].

Signaling through TLR on cells of the innate immune system affects the adaptive immunity. Thus, TLR signaling in DC affects T cell functions. It appears that conventional or $CD11c^+$ DC respond to LPS (a TL4 agonist) and to flagellin (a TLR5 agonist) by producing

IL-12, a well-known stimulator of the T helper 1 (T_h1) T cell differentiation pathway. Conversely, TLR signaling on these cells leads to the activation of IL-10 production followed by stimulation of the T helper 2 (T_h2) pathway. PDC strongly stimulate T_h1 cells and inhibit T_h2 cells, thus contributing to the "polarizing" of T cell responses in immunity [56-58]. Interestingly, DC are able to compartmentalize microbial particles and apoptotic self cells into distinct phagosomes, and the efficiency of antigen processing in these phagosomes is dependent on the presence of TLR ligands in the phagocytosed "cargo" [58]. TLRs also induce actin remodeling in DC upon their interaction with their ligands, which enhances antigen capturing and accelerates DC maturation [59]. Experiments with CD45-deficient mice showed that CD45 is a negative regulator of TLR signaling in DC [65]. Most intriguingly, the absence of signaling through TLR in DC of mice kept in germ-free conditions leads to enhanced apoptosis of autoreactive T cells, possibly due to stimulation of regulatory T cells (see below) [56].

TLRs play a very important role in the function of B lymphocytes. Naïve murine B cells express TLR4 and a member of the TLR family called RP105 or CD180. The latter, which seems to be unique for B cells, initiates signaling that leads to B cell activation in the presence of a known B cell marker transmembrane protein, CD19 [66]. Human B cells express TLR9 after cross-linking of their antigen receptors. Interaction between TLR9, induced by this cross-linking, with its ligand, CpG-rich bacterial DNA, enhances B cell proliferation, antibody secretion and antibody class switching [67]. Signaling through TLR4 in B cell progenitors accelerates their maturation, while signaling through TLR2 arrests or retards it [68]. TLR10, the "orphan" receptor without its known ligand, is expressed on human B cells (and also on PDC) and is associated with functional MyD88, suggesting its role in cell activation [69].

In T lymphocytes, TLRs also serve as powerful modifiers of their function and differentiation. MyD88-deficient mice showed a profound deficiency in T_h1 but not in T_h2 function [70], suggesting that molecular patterns recognized by TLRs (at least in those cases when this is followed by the activation of the MyD88-dependent pathway) skew the differentiation of naïve T helper cells (T_h0) towards the T_h1 subset. On the other hand, T_h1-dependent, macrophage-mediated immunity against *L. monocytogenes* seems to be only partially MyD88- and TLR2-dependent [71]. Interestingly, several functional TLRs (TLR4, -5, -7, and -8) are expressed by a special subpopulation of T cells that co-express CD4, CD25, and low levels of CD45 ($CD4^+CD25^+CD45^{low}$), referred to as "regulatory T cells" or Treg [56, 72, 73]. These cells are known to be suppressors of various immune responses. Recent studies have shown that their immunosuppressive function is TLR4-dependent [72]. However, in certain circumstances (body injury) they can be immunostimulatory [73], and TLR8 signaling reverses their immunosuppressive activity [74].

How Can Toll-Like Receptor Involvement Lead to Diseases

In recent years, a number of studies appeared where abnormal signaling through mutated TLRs or their adaptor molecules or their signal-transducing molecules was linked to human disease. Infectious diseases were first to receive attention in this regard. For example, Agnese et al. have demonstrated that a single point mutation that results in the aspartic acid to glycine substitution in the position 299 of the TLR4 molecule (the D299G polymorphism) leads to a decreased response to bacterial lipopolysaccharide and thus increases the risk of Gram-negative infections [75]. The D299G polymorphism has been strongly implicated into the development of septic shock [76] and in the so-called systemic inflammatory syndrome [77]. Similarly, point mutations resulting in a polymorphism of TLR2 (the R753Q polymorphism) are considered a risk factor for staphylococcal infections [78] and tuberculosis [79]. Polymorphisms in TLR signal-transducing molecules such as IRAK-4 [80, 81] and Csp-12 [82] are thought to predispose patients, especially children, to severe recurrent bacterial infections and sepsis.

Most interestingly, bacterial LPS, even when it signals through "normal" wild-type TLR, can induce a state of tolerance or unresponsiveness in human monocytes, probably contributing to severe complications of Gram-negative bacterial infections [83, 84]. This induced unresponsiveness, which may result in a decreased activity of the innate as well as of the adaptive immune system (because monocytes are precursors of both macrophages and dendritic cells, thus playing an extremely important role in antigen presentation), cannot be explained by mutations in TLR or in their signal-transducing molecules. Moreover, there is no down-modulation of surface TLR4 expression in human monocytes tolerized to LPS. If anything, there is an up-regulation of one of the adaptor signal-transducing molecules, IRAK-M [84], or abnormalities in the association of TLR with MyD88 [83].

Impairment of TLR signaling can also profoundly deregulate the innate and the adaptive immune system by causing apoptosis of interdigitating and follicular dendritic cells [85] and lymphocytes (both B and T) [86]. Although it has been long known that various blood cells undergo apoptosis in septic patients, recently it has been established that LPS can directly cause apoptosis through TLR4 and signaling through a distinct pathway [87, 88].

Mutations in an adaptor molecule called NEMO (for "NF-κB essential modulator"), which participates in TLR signaling, cause severe bacterial infections, anhydrosis, sparse hair [89], and immunodeficiency resembling the classical hyper-IgM syndrome (the disease known to be caused by a mutation in the CD40 ligand). The condition caused by these NEMO mutations, which can be point mutations or gene rearrangements [89, 90] is inherited, like the classical hyper-IgM syndrome, in an X-linked recessive fashion. Also like the classical hyper-IgM syndrome, it is accompanied by serious abnormalities in T cell immunity like the inability of the patients' T cells to direct B cell class switch and affinity maturation [91, 92]. A similar condition, manifested as recurrent infections, T cell dysfunction, anhydrosis and lack of hair, albeit inherited in an autosomal dominant fashion, is caused by two point mutations in the signal transduction factor $I_{\kappa}B_{\alpha}$ [93]. In patients with this genetic defect, immunodeficiency is more severe and results in a complete loss of T cell memory [93].

TLRs have been recently strongly implicated in the development of autoimmune diseases. Traditionally, immune reactions against self antigens used to be viewed as an abnormal activity of the adaptive immune system. Recently, however, a notion started to emerge that views autoantigens as "autoadjuvants" – triggers of the innate immune system with subsequent activation of the adaptive immunity to self [94]. For example, in patients with systemic lupus erythematosus (SLE), plasmacytoid dendritic cells capture immune complexes consisting of DNA or RNA released by necrotic or apoptotic cells, and anti-nucleic acid antibodies. These immune complexes end up in endosomes and bind TLR7 or TLR9, which triggers signaling transduction cascades and results in massive type I interferon production by these cells [95, 96]. These same cells actively present the captured immune complexes to those T lymphocytes with the appropriate specific antigen receptors that have escaped negative selection at early stages of their development. High concentrations of IFN-α are able to promote symptoms of SLE because under these conditions, the survival of self-reactive B and T lymphocytes is stimulated, and the autoantibody production and self-specific cytolytic T cell activity are enhanced. With more self cells destroyed, more self antigen becomes available and the "vicious circle" is established [94].

This model of TLR involvement in SLE is strongly supported by data from human studies where single-nucleotide polymorphisms (SNP) associated with type I interferon signaling pathway was eradicated. These studies showed that patients with SNP in IRF5 – a regulatory factor that specifically inhibits TLR7-induced type I interferon production – are at high risk for SLE [97, 98]. Also, Ro, a ribonucleoprotein well known to be a target self antigen in SLE patients, has properties of a TLR7 and TLR8 agonist [99]. Inbred mouse strains that are commonly used as animal models of SLE – lpr/lpr, Ali5, 564lgi, 56R, and Yaa – a genetic knockout of TLR3, TLR7, TLR9, or a knockout of MyD88 causes either exacerbation or reduction of symptoms of renal disease (reviewed in [94]). Mice transgenic for heavy- and light chain variable regions of the so-called 564 antibody on the C57BL/6 genetic background (564Igi mice) spontaneously produce RNA-specific self antibodies resembling those found in SLE patients, but when the 564Igi mice are made TLR7-deficient; they cease to produce these antibodies [100].

In other diseases that have been long viewed as autoimmune, signaling through TLRs also seems to be an important link to pathogenesis. Experimental allergic encephalomyelitis, which is considered an animal model for multiple sclerosis in humans, develops in rodents after injections of neuroglial self antigen, particularly myelin oligodendrocyte glycoprotein. Studies have shown that in mice deficient for MyD88- or TLR9 develop only a very mild EAE compared to genetically sufficient (control) mice [101]. A mutant mice strain K/BxN spontaneously develops a condition resembling EAE, and the transfer of serum from these mice to intact mice is sufficient to induce symptoms of EAE. However, mice that are genetically deficient for TLR4 develop only mild and completely reversible arthritis when injected with sera from the diseased mice [102]. In experiments of Rasschaert et al. exposure of primary Wistar rat pancreatic β-cells to viral double-stranded viral RNA and interferons induces apoptosis of these cells and upregulates the expression of TLR3 mRNA, pointing to the role of TLR3 signaling in diabetes mellitus [103].

Recent reports also indicate that signaling through TLR4 may have a role in the development of late onset Alzheimer's disease [104] and that signaling through TLR2 can contribute to the development of renal ischemia/reperfusion injury [105]. Sequence variants in TLR gene cluster TLR6-TLR1-TLR10 are associated with an increased risk for prostate cancer [106].

TOLL-LIKE RECEPTORS AND ATHEROSCLEROSIS

Atherosclerosis is a pathologic process that affects approximately 1.7% of the entire population of the USA, or approximately 4.6 million people, every year. It is a form of arteriosclerosis (hardening of arteries) in which thickening and hardening of the vessel is caused by the accumulation of lipid-laden macrophages within the arterial wall. This process is, in its essence, inflammatory [107, 108], and it is considered a leading contributor to serious and potentially lethal coronary artery diseases (heart attacks, myocardial infarctions) or cerebrovascular diseases (stroke) [109].

According to the currently accepted scenario [109], atherosclerosis begins early in life. Its onset is believed to be related to micro-injuries of the arterial endothelium by infection, oxidative stress or serum protein unbalance (especially elevated levels of serum fibrinogen and C-reactive protein). Later in life, these injuries can be aggravated by smoking, diabetes, increased blood pressure and high/low density lipoprotein unbalance. The injured endothelial cells make smaller than normal amounts of antithrombotic and vasodilating cytokines. Instead, they release pro-inflammatory cytokines including TNF-α, interferons, IL-1, toxic oxygen radicals, and heat shock proteins. Also, growth factors stimulating proliferation of smooth muscle cells are released, and adhesion molecules are up-regulated. This is accompanied by adhesion of lipid-laden macrophages to the inflamed endothelium and results in the formation and progression of the development of typical atherosclerotic plaques. The latter in their progression undergo stages of the fatty streak, fibrous plaque, and, finally, the so-called "complicated lesion."

The role of infection as a very important factor in the pathogenesis of atherosclerosis has been extensively studied for many decades. As early as the late 1800s, Hektoen pointed to the link between atherosclerosis and tuberculosis [110], and a century later Benditt et al. showed that a herpes virus that causes Marek's disease can directly induce atherosclerotic plaques in chickens [111]. In humans, infection with the human cytomegalovirus (HCMV) was shown to lead to hyperproduction of chemokines such as RANTES and MCP-1, which, in turn, induced migration of arterial smooth muscle cells to the "neointimal space," thus causing progression in the development of atherosclerotic plaques [112]. More recently, as the list of infectious agents implicated into atherogenesis continued to expand, a notion was formed that not any particular microorganism but the whole "infectious burden" is an etiological factor for atherosclerosis [113].

Obviously, TLR signaling in endothelial cells is important for the development of atherosclerosis. Some pathogenic microorganisms that have long been associated with high risk of atherosclerosis (for example, *Chlamydia pneumoniae*) can directly bind to TLR2 expressed on endothelial cells, and stimulate the development of atherosclerotic plaques in a TLR2-dependent fashion [114]. Also, *Porphyromonas gingivalis*, a bacterium that has been long known as a causative agent of periodontitis and is considered a serious risk factor for atherosclerosis, has been shown to produce a protein called GroEL, a homologue of the heat-shock protein HSP60. This bacterial protein directly binds endothelial cell-expressed TLR2 and TL4 and activates TLR-dependent signal transduction pathways [115]. On the other hand, fimriae and LPS of *P. gingivalis* also bind to TLR [116]. Preliminary results indicate that *P. gingivalis* or its heat-shock protein can even serve as a vaccine against atherosclerosis in animal model studies [117].

Studies have shown that slightly chemically modified low density proteins (which have long been associated with the development of atherosclerotic plaques and artery occlusion) can directly bind some TLR, especially TLR4 and TLR2, expressed on arterial endothelial cells [118]. While TLR4-deficient mice show only a slight reduction in the development of atherosclerotic plaques, the mice that lack MyD88 by genetic modification are characterized by a dramatic delay in the development of atherosclerosis [119]. It has been suggested that during atherogenesis, MyD88 is recruited into signaling by TLR2. Interestingly, endothelial cells increase their TLR2 expression when the flow of blood in arteries changes from laminar to "disturbed" [120], consistent with the known role of the altered hemodynamics in atherosclerosis and further linking TLRs with atherogenesis. Furthermore, two distinct subsets of human endothelial cells have been identified, one expressing and the other not expressing TLR2. The former includes those cells that are known to be associated with atherosclerotic plaque formation (coronary artery endothelial cells), while the latter includes those that are not (human microvascular endothelial cells, umbilical vein endothelial cells and some other) [121].

It seems very plausible that endogenous TLR ligands play an important role in atherosclerosis. In an elegant study by Mullick et al., when low density lipoprotein receptor (LDR)-deficient *ldr-/ldr-* mice were crossed with mice deficient for TLR2, the resulting LDR- and TLR2-deficient hybrid mice were resistant to atherosclerosis, implicating TLR2 in the pathogenesis of this disease. However, when *ldr-/ldr-* mice were sublethally irradiated and transplanted with bone marrow from TLR2-deficient mice, they developed severe atherosclerosis, suggesting that TLR2 expressed on non-bone marrow-derived cells and presumably interacting with some endogenous ligand(s) plays a pivotal role in the development of this disease [122].

Conclusion

In this review, we have highlighted only a few aspects of TLR function and of the involvement of these molecules in pathogenesis of diseases. However, even this brief survey might give readers an idea about complexity of the TLR family and its ligands. It is remarkable that the expanding knowledge about these molecules provides increasingly precise mechanistic explanations of the work of the host defense system and its possible impairments leading to disease. New discoveries regarding TLRs and their ligands helped to reconcile the two seemingly opposing views on the initial steps by which the host's defense system becomes alert – that is, the "danger theory" and the "pathogen-associated molecular patterns" theory. These findings give investigators plenty of material in which to formulate new hypothesis concerning the evolution of host defense mechanisms [123]. Of course, other mechanisms, which are entirely TLR-independent, are now being revealed as well and seem to be of a great interest also [124]. Future studies will lead to an even clearer understanding of these mechanisms and to better prevention and treatment of human diseases [125].

References

[1] Pinchuk, G. V. (2001). *Shaum's Outline of Theory and Problems of Immunology*. New York et al.: McGraw-Hill.

[2] Schwartz, R. H. (1990). A cell culture model for T lymphocyte clonal anergy. *Science,* **248**, 1349-1356.

[3] Matzinger, P. (1994). Tolerance, danger, and the extended family. *Annu. Rev. Immunol.*, **12**, 991-1045.

[4] Seong, S., and Matzinger, P.(2004). Hydrophobicity: an ancient danger-associated molecular pattern that initiates innate immune responses. *Nat. Rev. Immunol.*, **4,** 469-478.

[5] Janeway, C. A. Jr. (1989). Approaching the asymptote? Evolution and revolution in immunology. *Cold Spring Harb. Symp. Quant. Biol.,* **1**, 1-13.

[6] Medzhitov, R., Preston-Hurlburt, P., and Janeway, C. A. Jr. (1997). A human homologue of the Drosophila Toll protein signals activation of adaptive immunity. *Nature,* **388**, 394-397.

[7] Bianchi, M. E. (2007). DAMPs, PAMPs and alarmins: all we need to know about danger. *J. Leukoc. Biol.,* **81**, 1-5.

[8] Stein, D., Roth, S., Vogelsang, E., and Nusslein-Volhard, C. (1991). The polarity of the dorsoventral axis in the Drosophila embryo is defined by an extracellular signal. *Cell,* **65**, 725-735.

[9] Lemaitre, B., Nicolas, E., Michaut, L., Reichhart, J. M., and Hoffmann, J. A. (1996). The dorsoventral regulatory gene cassette spatzle/Toll/cactus controls the potent antifungal response in Drosophila adults. *Cell,* **86**, 973-983.

[10] Gay, N. J., and Keith, F. J. (1991). Drosophila Toll and IL-1 receptor. *Nature,* **351**, 355-356.

[11] Akira, S. (2003). Toll-like receptor signaling. *J. Biol. Chem.,* **278**, 38105-38108.

[12] Medzhitov, R., Preston-Hurlburt, P., Kopp, Stadlen, E. A., Chen, C., Ghosh, S., and Janeway, C. A. Jr. (1998). MyD88 is an adapter protein in the hToll/IL-1 receptor family signaling pathways. *Mol. Cell.,* **2**, 253-258.

[13] Creagh, E. M., and O'Neill, L. A. (2006). TLRs, NLRs and RLRs: a trinity of pathogen sensors that co-operate in innate immunity. *Trends Immunol.*, **27**, 352-357.

[14] Germain, R. N. (2004). An innately interesting decade of research in immunology. *Nature Med.,* **10**, 1307-1320.

[15] Hopkins, P. A., and Srisandan, S. (2005). Mammalian Toll-like receptors: to immunity and beyond. *Clin. Exp. Immunol.*, **140**, 395-407.

[16] Poltorak, A., He, X., Smirnova, I., Liu, M. Y., Van Huffel, C., Du, X., Birdwell, D., Alejos, E., Silva, M., Galanos, C., Freudenberg, M., Ricchiardi-Castagnioli, P., Layton, B., and Beutler, B. (1998). Defective LPS signaling in C3H/HeJ and C57BL/10ScCr mice: mutations in Tlr4 gene. *Science,* **282**, 2085-2088.

[17] Hoshino, K., Takeuchi, O., Kawai, T., Sanjo, H., Ogawa, T., Takeda, Y., Takeda, K., and Akira, S. (1999). Toll-like receptor 4 (TLR4)-deficient mice are hyporesponsive to lipopolysaccharide: evidence for TLR4 as the Lps gene product. *J. Immunol.*, **162**, 3749-3752.

[18] Arbour, N. C., Lorenz , E., Schutte, B. C., Zabner, J., Kline, J. N., Jones, M., Frees, K., Watt, J. L., and Schwartz, D. A. (2000). TLR4 mutations are associated with endotoxin hyporesponsiveness in humans. *Nature Genet.*, **25**, 187-191.

[19] Pandey, S., and Agrwal, D. K. (2006). Immunobiology of Toll-like receptors: emerging trends. *Immunol. Cell Biol.,* **84**, 333-341.

[20] Takeuchi, O., Hoshino, K., Kawai, T., Sanjo, H., Takada, H., Ogawa, T., Takeda, K., and Akira, S. (1999). Differential roles of TLR2 and TLR4 in recognition of gram-negative and gram-positive bacterial cell wall components. *Immunity,* **11**, 443-451.

[21] Iwaki, D., Mitsuazawa, H., and Murakami, S. (2002). The extracellular toll-like receptor 2 domain directly binds peptidoglican derived from Staphylococcus aureus. *J. Biol. Chem.,* **277**, 24315-24320.

[22] Akira, S., Uematsu, S., and Takeuchi, O. (2006). Pathogen recognition and innate immunity. *Cell,* **124**, 783-801.

[23] Kawai, T., and Akira, S. (2006). TLR signaling. *Cell Death and Differentiation*, **13**, 816-825.

[24] Takeuchi, O., Sato, S., Horiuchi, T., Hoshino, K., Takeda, K., Dong, Z., Modlin, R.L., and Akira, S. (2002). Role of Toll-like receptor 1 in mediating immune response to microbial lipoproteins. *J. Immunol.*, **169**, 10-14.

[25] Bauer, S., Kirsching, C. J., Hacker, H., Redecke, V., Hausmann, S., Akira, S., Wagner, G., and Lipford, G. B. (2001). Human TLR9 confers responsiveness to bacterial DNA via species specific CpG motif recognition. *Proc. Natl. Acad. Sci. USA*, **98**, 9237-9242.

[26] Coban, C., Ishii, K. J., Kawai, T., Hemmi, H., Sato, S., Uematsu, S., Yamamoto, M., Takeuchi, O., Itagaki, S., Kumar, N., Horii, T., and Akira, S. (2005). Toll-like receptor 9 mediates innate immune activation by malaria pigment hemozoin. *J. Exp. Med.*, **201**, 19-25.

[27] Alexopoulou, L., Holt, A. C., Medzhitov, R., and Flavell R. A. (2001). Recognition of double-stranded RNA and activation of NF-kappaB by Toll-like receptor 3. *Nature*, **413**, 732-738.

[28] Heil, F., Hemmi, H., Hochrein, H., Ampenberger, F., Kirsching, C., Akira, S., Lipford, G., Wagner H., and Bauer, S. (2004). Species-specific recognition of single-stranded RNA via toll-like receptor 7 and 8. *Science*, **303**, 1526-1529.

[29] Lund, J. M., Alexopoulou, L., Sato, A., Karow, M., Adams, N. C., Gale, N. W., Iwasaki, A., and Flawell, R. A. (2004). Recognition of single-stranded RNA virus by toll-like receptor 7. *Proc. Natl. Acad. Sci. USA*, **101**, 5598-5603.

[30] Hemmi, H., Kaisho, T., Takeuchi, O., Sato, S., Sanjo, H., Hoshino, K., Horiuchi, T., Tomizawa, H., Takeda, K., and Arira, S. (2002). Small anti-viral compounds activate immune cells via the TLR7 MyD88-dependent signaling pathway. *Nat. Immunol.*, **3**, 196-200.

[31] Yarovinsky, F., Zhang, D., Andersen, J. F.., Bannenberg, G. L., Serhan, C. N., Hayden, M. S., Hieny, S., Sutterwala, F. S., Flavell, R. A., Ghosh, S., and Sher, A. (2005). TLR11 activation of dendritic cells by a protozoan profilin-like protein. *Science* **308**, 1626-1629.

[32] Zhang, D., Zhang, G., Hayden, M. S., Greenblatt, M. B., Bussey, C., Flavell, R. A., and Ghosh, S. (2004). A toll-like receptor that prevents infection by uropathogenic bacteria. *Science*, **303**, 1522-1536.

[33] Hiyashi, F., Smith, K. D., Ozinsky, A., Hawn, T. R., Yi, E. C., Goodlett, D. R., Eng, J. K., Akira, S., Underhill, D. M., and Aderem, A. (2001). The innate response to bacterial flagellin is mediated by toll-like receptor 5. *Nature*, **410**, 1099-10103.

[34] Nakao, Y., Funami, K., Kikkawa, S., Taniguchi, M., Nishiguchi, N., Fukumori, Y., Seya, T., and Matsumoto, M. (2005). Surface-expressed TLR6 participates in the recognition of diacylated lipopeptide and peptidoglycan in human cells. *J. Immunol.*, **174**, 1566-1573.

[35] Tapping, R. I., and Tobias, P. S. (2003). Mycobacterial lipoarabinomannan mediates physical interactions between TLR1 and TLR2 to induce signaling. *J. Endotoxin Res.*, **9**, 264-268.

[36] Wyllie, D. H., Kiss-Toth, E., Visintin, A., Smith, S. C., Boussouf, S., Segal, D. M., Duff, G. W., and Dower, S. K. (2000). Evidence for an accessory protein function for Toll-like receptor 1 in anti-bacterial responses. *J. Immunol.*, **165**, 7125-7132.

[37] Into, T., Kimura, K., Yasuda, M., Kataoka, H., Inoue, N., Hasele, A., Takeda, A., Akira, S., and Shibata, K. (2004). Stimulation of human Toll-like receptor (TLR) 2 and TLR6 with membrane lipoproteins of Mycoplasma fermentans induces apoptotic cell death after NF-κB activation. *Cell Microbiol.*, **6**, 187-199.

[38] Henneke, P., Morath, S., Uematsu, S., Weichert, S., Pfizenmaier, M., Takeuchi, O., Muller, A., Poyart, C., Akira, S., Berner, R., Teti, G., Hartung, T., Trieu-Cuot, P., Casper, D. L., and Golenbock, D. T. (2005). Role of lipoteichoic acid in the phagocyte response to group B streptococcus. *J. Immunol.*, **174**, 6449-6455.

[39] Hoebe, K., Geprge, P., Rutschmann, S., Du, X., Mudd, S., Crozat, K., Sovath, S., Shamel, L., Hartung, T., Zahringer, U., and Beutler, B. (2005). CD36 is a sensor of diacylglycerides. *Nature,* **43**:, 523-527.

[40] Miggin, S. M., and O'Neill, L. A. J. (2006). New insights into regulation of TLR signaling. *J. Leukoc. Biol.*, **80**, 220-226.

[41] Takeuchi, O., Kaufmann, A., Grote, K., Kawai, T., Hoshino, K., Morr, M., Muhlradt, P. F., and Akira, S. (2000). Preferentially the R-stereoisomer of the mycoplasmal lipopeptide macrophage-activating lipopeptide-2 activates immune cells through a toll-like receptor-2- and MyD88-independent signaling pathway. *J. Immunol.*, **164**, 554-557.

[42] Nilson, N., Nonstad, U., and Khan, N. (2004). Lipopolysaccharide and double-stranded RNA upregulate toll-like receptor 2 independently of myeloid differentiation factor 88. *J. Biol. Chem.*, **249**, 39727-39735.

[43] Horng, T, Barton, G. M., Flavell, R. A., and Medzhitov, R. (2002). The adaptor molecule TIRAP provides signaling specificity for Toll-like receptors. *Nature,* **420**, 329-333.

[44] Hoebe, K., Du, X., Georgel, P., Janssen, E., Tabeta, K., Kim, S. O., Goode, J., Lin, P., Mann, N., Mudd, S., Crozat, K., Sovath, S., Han, J., and Beutler, B. (2003). Identification of Lps2 as a key transducer of MyD88-independent TIR signaling. *Nature,* **424**, 743-748.

[45] Sato, S., Sugiuama, M., Yamamoto, M., Watanabe, Y., Kawai, T., Takeda, K., and Akira, S. (2003). Toll/IL-1 receptor domain-containing adaptor inducing IFN-beta (TRIF) associates with TNF receptor-associated factor 6 and TANK-binding kinase 1 and activates two distinct transcription factors NF-kappa B and IFN-regulatory factor-3 in the Toll-like receptor signaling. *J. Immunol.*, **171**, 4304-4310.

[46] Kawai, T., Sato, S., Ishii, K. J., Coban, C., Hemmi, H., Yamamoto, M., Terai, K., Matsuda, M., Inoue, J., Uematsu, S., Takeuchi, O., and Akira, S. (2004). Interferon-alpha induction through Toll-like receptors involves a direct interaction of IRF7 with MyD88 and TRAF6. *Nat. Immunol.,* **5**, 1061-1068.

[47] Takaoka, A., Yanai, H., Kondo, S., Duncan, G., Negishi, H., Mizutani, T., Kano, S.I., Honda, K., Ohba, Y., Mak, T. W., and Taniguchi, T. (2005). Integral role of IRF-5 in the gene induction programme activated by Toll-like receptors. *Nature,* **434**, 243-249.

[48] Honda, Y., Yanai, H., Negishi, H., Asagiri, M., Sato, M., Mizutani, T., Shimada, N., Ohba, Y., Takaoka, A., Yoshida, N., and Taniguchi, T. (2005). IRF-7 is the master regulator of type-I interferon-dependent immune responses. *Nature,* **434**, 772-777.

[49] Shoenemeyer, A., Barnes, B. J., Mancl, M. E., Latz, E., Goutagny, N., Pitha, P. M., Fitzgerald, K. A., and Golenbock, D. T. (2005). The interferon regulatory factor IRF5 is a central mediator of TLR7 signaling. *J. Biol. Chem.*, **280**, 17005-17012.

[50] Kawai, T., Takeuchi, O., Fujita, T., Inoue, J., Muhlradt, P. F., Sato, S., Hoshino, K., and Akira, S. (2001). Lipopolysaccharide stimulates the MyD88-independent pathway and results in activation of IFN-regulatory factor 3 and the expression of a subset of lipopolysaccharide-inducible genes. *J. Immunol.*, **167**, 5887-5894.

[51] Kawai, T., Takahashi, K., Sato, S., Coban, C., Kumar, H., Kato, A., Ishii, K. J., Takeuchi, O., and Akira, S. (2005). IPS-1; an adaptor triggering RIG-I- and Mda5-mediated type I interferon induction. *Nat. Immunol.*, **6**, 981-988.

[52] Kagan, J. C., and Medzhitov, R. (2006). Phosphoinositide-mediated adaptor recruitment controls Toll-like receptor signaling. *Cell,* **125**, 943-955.

[53] Montminy, S. W., Khan, N., McGrath, S., Walkowicz, M. J., Sharp, F., Conlon, J. E., Fusake, K., Kusumoto, S., Sweet, C., Miyake, K., Akira, S., Cotter, R. J., Goguen, J. D., and Lien, E. (2006). Virulence factors of *Yersinia pestis* are overcome by a strong lipopolysaccharide response. *Nat. Immunol.*, **7**, 1066-1073.

[54] Liu, Y. J., Kanzler, H., Soumelis, V., and Gilliet, M. (2001). Dendritic cell lineage, plasticity and cross regulation. *Nat. Immunol.*, **2**, 585-589.

[55] Collonna, M., Trinchieri, G., and Liu, Y. J. (2004). Plasmacytoid dendritic cells in immunity. *Nat. Immunol.,* **5**, 1219-1226.

[56] Xu, D., Liu, H., and Komai-Koma, M. (2004). Direct and indirect role of Toll-like receptors in T cell mediated immunity. *Cell. Molec. Immunol.*, **1**, 239-246.

[57] Kawai, T., and Akira, S. (2005). Innate immune recognition of viral infection. *Nat. Immunol.,* **7**, 131-137.

[58] Blander, J. M., and Medzhitov, R. (2006). Toll-dependent selection of microbial antigens for presentation by dendritic cells. *Nature,* **440**, 808-812.

[59] West, M. A., Wallin, R. P. A., Matthews, S. P., Svensson, H. G., Zaru, R., Ljunggren, H.-G., Prescott, A. R., and Watts, C. (2004). Enhanced dendritic cell antigen capture via Toll-like receptor-induced actin remodeling. *Science,* **305**,1153-1157.

[60] Ogawa, T., Asai, Y., Hashimoto, M., and Uchida, H. (2002). Bacterial fimbriae activate human peripheral blood monocytes utilizing TLR2, CD14 and CD11a/CD18 as cellular receptors. *Eur. J. Immunol.*, **32**, 2543-2550.

[61] Farina, C., Thiel, D., Semlinger, B., Hohlfeld, R., and Meinl, E. (2004). Distinct responses of monocytes to Toll-like receptor ligands and inflammatory cytokines. *Int. Immunol.*, **16**, 799-809.

[62] Lee, S.-R., Pharr, G. T., Cooksey, A., McCarthy, F. M., Boyd, B. L., and Pinchuk, L. M. (2006). Differential detergent fractionation for non-electrophoretic bovine peripheral blood monocyte proteomics reveals proteins involved in professional antigen presentation. *Dev. Comp. Immunol.*, **30**, 1070-1083.

[63] Pinchuk, L. M., Lee, S.-R., Pharr, G. T., and Pinchuk, G. V. BVDV infection affects antigen uptake and cytokine expression in bovine monocytes. In: Proceedings of the 87th Annual Meeting of the Conference of Research Workers in Animal Diseases, Chicago; December 3-5 2006; p. 116 (abstract P79).

[64] Lee, S.-R., Pharr, G. T., Boyd, B. L., and Pinchuk, L. M. Bovine viral diarrhea viruses modulate Toll-like receptor, type I IFN, and pro-inflammatory cytokine gene expression (submitted).

[65] Piercy, J., Petrova, S., Tchillian, E. Z., and Beverley, P. C. (2006). CD45 negatively regulates tumour necrosis factor and interleukin-6 production in dendritic cells. *Immunology,* **118**, 250-256.

[66] Yazawa, N., Fujimoto, M., Sato, S., Miyake, K., Asano, N., Nagai, Y., Takeda, K., Akira, S., Tedder, T. F., and Tamaki, K. (2003). CD19 regulates innate immunity by the toll-like receptor RP105 signaling in B lymphocytes. *Blood,* **102**, 1374-1380.

[67] Leadbetter, E. A., Rifkin, I. R., Hohlbaum, A. M., Beaudette, B. C., and Rothstein, M. J. (2002). Chromatin-IgG complexes activate B cells by dual engagement of IgM. *Nature,* **416**, 603-607.

[68] Hayashi, E. A., Akira, S., and Nobrega, A. (2005). Role of TLR in B cell development: signaling through TLR4 promotes B cell maturation and is inhibited by TLR2. *J. Immunol.*, **174**, 6639-6647.

[69] Hasan, U., Chaffois, C., Gaillard, C., Sauliner, V., Merck, E., Tancredi, S., Guiet, C., Briere, F., Vlach, J., Lebeque, S., Trinchieri, G., and Bates, E. M. (2005). Human TLR10 is a functional receptor, expressed by B cells and plasmacytoid dendritic cells, which activates gene transcription through MyD88. *J. Immunol.*, **174**, 2942-2950.

[70] Schnarre, M., Barton, G. M., Holt, A. C., Takeda, K., Akira, S., and Medzhitov, R. (2001). Toll-like receptors control activation of adaptive immune responses. *Nat. Immunol.,* **2**, 947-950.

[71] Edelston, B. T., Unanue, E. R. (2002). MyD88-dependent but Toll-like receptor 2-independent innate immunity to Listeria: no role for either in macrophage listericidal activity. *J. Immunol.*, **169**, 3869-3875.

[72] Caramalho, I., Lopez-Carvalho, T., Ostler, D., Zelenay, S., Haury, M., and Demengeot, J. (2003). Regulatory T cells selectively express Toll-like receptors and are activated by lipopolysaccharide. *J. Exp. Med.*, **197**, 403-411.

[73] Murphy, T. J., Choileain, N. N., Zang, Y., Mannick, J. A., Lederer, J. A. (2005). $CD4^{+}CD25^{+}$ regulatory T cells control innate immune reactivity after injury. *J. Immunol.,* **174**, 2957-2963.

[74] Peng, G., Guo, Z., Kiniwa, Y., Voo, K. S., Peng, W., Fu, T., Wang, D., Li, Y., Wang, Y., and Wang, R.-F. (2005). Toll-like receptor 8-mediated reversal of CD4+ regulatory T cell function. *Science,* **309**, 1380-1384.

[75] Agnese, D. M., Calvano, J. E., Hahm, S. J., Coyle, S. M., Corbett, S. A., Calvano, S. E., and Lowry, S. F. (2002). Human toll-like receptor 4 mutations but not CD14 polymorphisms are associated with an increased risk of gram-negative infections. *J. Infect. Dis.,* **186**, 1522-1524.

[76] Lorenz, E., Mira, J., Frees, K., and Schwartz, D. (2002). Relevance of mutations in the TLR4 receptor in patients with Gram negative septic shock. *Arch. Int. Med.,* **162**, 1028-1032.

[77] Child, N. J., Yang, I. A., Pulletz, M. C., de Courcy-Godler, K., Andrews, A. L., Pappachan, V. J., and Holloway, J. W. (2003). Polymorphisms in Toll-like receptor 4 and the systemic inflammatory response syndrome. *Biochem. Soc. Trans.,* **31**, 652-653.

[78] Lorenz, E., Mira, J., Cornish, K., Arbour, N., and Schwartz, D. A. (2000). A novel polymorphism in the toll-like receptor 2 gene and its potential association with staphylococcal infection. *Infect. Immun.*, **68**, 6398-6401.

[79] Ogus, A. C, Yoldas, B., Ozdemir, T., Uguz, A., Olcen, S., Keser, I., Coskun, M., Cilli, A., and Yegin, O. (2004). The Arg753Gin polymorphism of the human Toll-like receptor 2 gene in tuberculosis disease. *Eur. Respir. J.,* **23**, 219-223.

[80] Picard, C., Puel, A., Bonnet, M., Ku, C.-L., Bustamante, J., Yang, K., Soudais, C., et al. (2003). Pyogenic bacterial infections in humans with IRAK-4 deficiency. *Science,* **299**, 2076-2079.

[81] Medvedev, A. E., Lentschat, A., Kuhns, D. B., Blanco, J. C. G., Salkowski, C., Zhang, S., Arditi, M., Gallin, J. I., and Vogel, S. N. (2003). Dstinct mutations in IRAK-4 confer hyporesponsiveness to lipopolysaccharide and interleukin-1 in a patient with recurrent bacterial infections. *J. Exp. Med.,* **198**, 521-531.

[82] Saleh, M., Vaillancourt, J. P., Graham, R. K., Huyck, M., Srinivasula, S. M., Alnermi, E. S., Steinberg, M. H., Nolan, V., Baldwin, C. T., Hotchkiss, R. S., Buchman, T. G., Zehnbauer, B. A., Hayden, M. R., Farrer, L. A., Roy, S., and Nicholson, D. W. (2004). Differential modulation of endotoxin responsiveness by human caspase-12 polymorphisms. *Nature,* **469**, 75-79.

[83] Medvedev, A. E., Lentschat, A., Wahl, L. M., Golenbock, D. T., and Vogel, S. N. (2002). Dysregulation of LPS-induced Toll-like receptor 4-MyD88 complex formation and IL-1 receptor-associated kinase 1 activation in endotoxin-tolerant cells. *J. Immunol.,* **169**, 5209-5216.

[84] Escoll, P., Del Fresno, C., Garcia, L., Valles, G., Lendinez, M. J., Arnalich, F., and Lopez-Collazo, E. (2003). Rapid up-regulation of IRAK-M expression following a second endotoxin challenge in human monocytes and in monocytes isolated from septic patients. *Biochem. Biophys. Res. Commun.*, **311**, 465-472.

[85] Tinsley, K. W., Grayson, M. H., Swanson, P. E., Drewry, A. M., Chang, K. C., Karl, I. E., and Hotchkiss, R. S. (2003). Sepsis induces apoptosis and profound depletion of splenic interdigitating and follicular dendritic cells. *J. Immunol.*, **171**, 909-914.

[86] Hotchkiss, R. S., Tinsley, K. W., Swanson, P. E., Schmieg, R. E. Jr., Hui, J. J., Chang, K. C., Osborne, D. F., Freeman, B. D., Cobb, J. P., Buchman, T. G., and Karl, I. E. (2001). Sepsis-induced apoptosis causes progressive profound depletion of B and $CD4^+$ T lymphocytes in humans. *J. Immunol.*, **166**, 6952-6963.

[87] Bannerman, D. D., Tupper, J. C., Erwert, R. D., Winn, R. K., and Harlan, J. M. (2002). Divergence of bacterial lipopolysaccharide pro-apoptotic signaling downstream of IRAK-1. *J. Biol. Chem.*, **277**, 8048-8053.

[88] Bannerman, D. D., Erwert, R. D., Winn, R. K., and Harlan, J. M. (2002). TIRAP mediates endotoxin-induced NF-κB activation and apoptosis in endothelial cells. *Biochem. Biophys. Res. Commun.,* **295**, 157-162.

[89] Smai, A., Courtois, G., Vabres, P., Yamaoka, S., Heuertz, S., Munnich, A., Israel, A., Heiis, N .S., and Klauck, S. M. (2000). Genomic rearrangement in NEMO impairs NF-κB activation and is a cause of incontinentia pigmenti. *Nature,* **405**, 466-472.

[90] Puel, A., Picard, C., Ku, C. L., Smahi, A., and Casanova, J. L. (2004). Inherited disorders of NF-κB-mediated immunity in man. *Curr. Opin. Immunol.*, **16**, 34-41.

[91] Jain, A., Ma, C. A., Liu, S., Brown, M., Cohen, J., and Strober, W. (2001). Specific missence mutations in NEMO result in hyper-IgM syndrome with hypohydrotic ectodermal dysplasia. *Nat. Immunol.*, **2**, 223-228.

[92] Ku, C.-L., Yang, K., Bustamante, J., Puel, A., von Bemuth, H., Santos, O. F., Lawrence, T., Chang, H.-H., Al-Mousa, H., Picard, C., and Casanova, J.-L. (2005). Inherited disorders of human Toll-like receptor signaling: immunological implications. *Immunol. Rev.*, **203**, 10-26.

[93] Courtois, G., Smahi, A., Reichenbach, J., Dofflinger, R., Cancrini, C., Bonnett, M., Puel, A., Chable-Bessia, C., Yamaoka, S., Feinberg, J., Dupuis-Girod, S., Bodemer, C., Livadiotti, S., Novelli, F., Rossi, P., Fischer, A., Israel, A., Munnich, A., LeDeist, F., and Casanova, J. L. (2003). A hypermorphic $I_{\kappa}B_{\alpha}$ mutation is associated with autosomal dominant anhidrotic ectodermal dysplasia and T cell immunodeficiency. *J. Clin. Invest.*, **112**, 1105-1115.

[94] Marshak-Rothstein, A. (2006). Toll-like receptors in systemic autoimmune disease. *Nat. Rev. Immunol.,* **6**, 823-835.

[95] Kadowaki, N., Ho, S., Antonenko, S., Malefyt, R. W., Kastelein, R. A., Bazan, F., and Liu, Y. J. (2001). Subsets of human dendritic cell precursors express different Toll-like receptors and respond to different microbial antigens. *J. Exp. Med.*, **194**, 863-869.

[96] Lovgren, T., Eloranta, M. L., Bave, U., Alm, G. V., and Ronnblom, L. (2004). Induction of interferon-a production in plasmacytoid dendritic cells by immune complexes containing nucleic acid released by necrotic or late apoptotic cells and lupus IgG. *Arthritis Rheum.,* **50**, 1861-1872.

[97] Sigurdsson, S., Nordmark, G., Goring, H. H., Lindroos, K., Wiman, A. C., Sturfelt, G., Jonsen, A., et al. (2005). Polymorphisms in the tyrosine kinase 2 and interferon regulatory factor 5 genes are associated with systemic lupus erythematosus. *Am. J. Hum. Genet.,* **76**, 528-537.

[98] Graham, R. R., Kozyrev, S. V., Baechler, E. C., Reddy, M. V., Plenge, R. M., Bauer, J. W., Ortmann, W. A., Koeuth, T., Gonzalez Escribano, M. F., et al. (2006). A common haplotype of interferon regulatory factor 5 (IRF5) regulates splicing and expression and is associated with increased risk of systemic lupus erythematosus. *Nat. Genet.,* **38**, 550-555.

[99] Vollmer, J., Tluk, S., Schmitz, C., Hamm, S., Jurk, M., Forsbach, A., Akira, S., Kelly, K. M., Reeves, W. H., Bauer, S., and Krieg, A. M. (2006). Immune stimulation mediated by autoantigen binding sites within small nuclear RNAs involves Toll-like receptors 7 and 8. *J. Exp. Med.,* **202**, 1575-1585.

[100] Prinz, M., Garbe, F., Schmidt, H., Mildner, A., Gutcher, I., Wolter, K., Piesche, M., et al. (2006). Innate immunity mediated by TLR9 modulates pathogenicity in an animal model of multiple sclerosis. *J. Clin. Invest.*, **116**, 456-464.

[101] Choe, J.-Y., Crain, B., Wu, S. R., and Corr, M. (2003). Interleukin-1 receptor dependence of serum transferred arthritis can be circumvented by Toll-like receptor 4 signaling. *J. Exp. Med.,* **197**, 537-542.

[102] Rasschaert, J., Ladiere, L., Urbain, M., Dogusan, Z., Katabua, B., Sato, S., Akira, S., Gysemans, C., Mathieu, C., and Eizirik, D. L. (2005). Toll-like receptor 3 and STAT-1 contribute to double-stranded RNA+interferon-gamma-induced apoptosis in primary pancreatic beta-cells. *J. Biol. Chem.*, **280**, 33984-33991.

[103] Minoretti, P., Gazzaruso, C., Vito, C. D., Emanuele, E., Bianchi, M., Coen, E., Reino, M., and Geroldi, D. (2006). Effect of functional Toll-like receptor 4 Asp299Gly polymorphism on susceptibility to late-onset Alzheimer's disease. (*Neurosci. Lett.*, **391**, 147-149.

[104] Leemans, J. C., Stockman, G., Claessen, N., Rouschop, K. M., Teske, G. J., Kirschning, C. J., Akira, S., van der Poll, T., Weening, J. J., and Florquin, S. J. (2005).

Renal-associated TLR2 mediates ischemia/reperfusion injury in the kidney. *Clin. Invest.*, **115**, 2894-2903.

[105] Sun, J., Wiklund, F., Zheng, S. L., Chang, B., Balter, K., Li, L., Johansson, J. E., Li, G., Adami, H. O., Liu, W., Tolin, A., Turner, A. R., Meyers, D. A., Isaacs, W. B., Xu, J., and Gronberg, H. (2005). Sequence variants in Toll-like receptor gene cluster (TLR6-TLR1-TLR10) and prostate cancer risk. *J. Natl. Cancer Inst.*, **97**, 525-532.

[106] Libby, P. (2002). Inflammation in atherosclerosis. *Nature*, **420**:, 868-874.

[107] Hansson, G. K. (2005). Inflammation, atherosclerosis, and coronary artery disease. *New England J. Med.,* **352**, 1685-1695.

[108] McCance, K. L., and Huether, S. E. (2006). Pathophysiology: the Biologic Basis for Disease in Adults and Children. Elsevier-Mosby, St. Louis, MO, (5th ed.), p.1780 .

[109] Hektoen, L. (1896). The vascular changes of tuberculosis meningitis. *J. Exp. Med.,* **1**, 112-136.

[110] Benditt, E. P., Barrett, T., and McDougal, J. K. (1983). Viruses in the ethiology of atherosclerosis. *Proc. Natl. Acad. Sci. USA,* **80**, 6386-6391.

[111] Streblow, D. N., Orloff, S. L., and Nelson, J. A. (2001). Do pathogens accelerate atherosclerosis? *J. Nutrition,* **131**, 2798S-2804S.

[112] Espinola-Klein, C., Rupprecht, H. J., Blankenberg, S., Bickel, C., Kopp, H., Rippin, G., Viktor, A., Hafner, G., Schlumberger, W., and Meyer, J. (2002). Impact of infectious burden on extent and long-term prognosis of atherosclerosis. *Circulation,* **105**, 15-25.

[113] Sasu, S., LaVerda, N., Qureshi, N., Golenbock, D. T., and Beasley, D. (2001). Chlamydia pneumoniae and chlamydial heat-shock protein 60 stimulate proliferation of human vascular smooth muscle cells via Toll-like receptor 4 and p44/42 mitogen-activated protein kinase activation. *Circ. Res.*, **89**, 244-253.

[114] Argueta, J. G. M., Shiota, S., Yamaguchi, N., Masuhiro, Y., and Hanazawa, S. (2006). Induction of Porphyromonas gingivalis GroEL signaling via binding to Toll-like receptors 2 and 4. *Oral Microbiol. Immunol.*, 21, 245-251.

[115] Hajishengaliis, G., Tapping, R. I., Harokopakis, E., Nishiyama, S.-I., Ratti, P., Schifferle, R. E., Lyle, E. A., Triantafilou, M., Triantaphilou, K., and Yoshimura, F. (2006). Differential interactions of fimbriae and lipopolysaccharide from Porphyromonas gingivalis with the Toll-like receptor 2-centered pattern recognition apparatus. *Cell. Microbiol.*, **8**, 1557-1570.

[116] Miyamoto, T., Yumoto, H., Takahashi, Y., Davey, M., Gibson III, F. C., and Genco, C. A. (2006). Pathogen-accelerated atherosclerosis occurs early after exposure and can be prevented via immunization. *Infect. Immun.*, **74**, 1376-1380.

[117] Miller, Y. I., Viriyakosal, S., Binder, C. J, Feramisco, J. R., Kirkland, T. N., and Witztum, J. L. (2003). Minimally modified LDL binds to CD14, induces macrophage spreading via TLR4/MD-2, and inhibits phagocytosis of apoptotoc cells. *J. Biol. Chem.,* **278**, 1561-1568.

[118] Michelen, K. S., Doherty, T. M., Shah, P. K., and Arditt, M. (2004). TLR signaling: an emerging bridge from innate immunity to atherosclerosis. *J. Immunol.,* **173**, 5901-5907.

[119] Mullaly, S. C., and Kubes, P. (2004). Toll gates and traffic arteries. *Circ. Res.,* **95**, 657-663.

[120] Doherty, T. M., Shah, P. K., and Arditi, M. (2005). Lipopolysaccharide, Toll-like receptors, and the immune contribution to atherosclerosis. *Atherosclerosis Thromb. Vasc. Biol.*, **25**, 38-44.

[121] Eldfeldt, K., Swedenborg, J., Hansson, G., and Yan, Z.-Q. (2002). Expression of Toll-like receptors in human atherosclerotic lesions: a possible pathway for plaque activation. *Circulation,* **105**, 1158-1174.

[122] Mullick, A. E., Tobias, P. S., and Curtiss, L. K. (2005). Modulation of atherosclerosis in mice by Toll-like receptor 2. *J. Clin. Invest.,* **115**, 3149-3156.

[123] Matzinger, P. (2007). Friendly and dangerous signals: is the tissue in control? *Nat. Immunol.,* **8**, 11-13.

[124] Wagner, H., and Bauer, S. (2006). All is not Toll: new pathways in DNA recognition. *J. Exp. Med.,* **203**, 265-268.

[125] Jiang, Z. H., and Koganty, R. R. (2003). Synthetic vaccines: the role of adjuvants in immune targeting. *Curr. Med. Chem.*, **10**, 1423-1439.

INDEX

A

B

C

D

E

F

G

H

I

J

K

L

M

N

O

P

Q

R

S

T

U

V

W

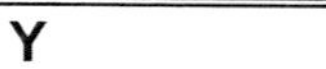